T0338517

Mathematics: Theory & Applications

Editors

Richard V. Kadison
Isidore M. Singer

Kjeld Knudsen Jensen
Klaus Thomsen

Elements of KK-Theory

Birkhäuser   1991
Boston • Basel • Berlin

Kjeld Knudsen Jensen
Department of Mathematics
University of Pennsylvania
Philadelphia, PA 19104
U.S.A.

Klaus Thomsen
Matematisk Institut
Ny Munkegade
DK-8000 Aarhus C
Denmark

Library of Congress Cataloging-in-Publication Data

Jensen, Kjeld Knudsen, 1961-
    Elements of KK-theory / Kjeld Knudsen Jensen and Klaus Thomsen.
        p.   cm. -- (Mathematics)
    Includes bibliographical references and index.
    ISBN 0-8176-3496-7 (alk. paper)
    1. KK-theory.   I.   Thomsen, Klaus, 1957-   . II. Title.
    III. Series: Mathematics (Boston, Mass.)
    QA612.33.J46   1991   91-17642
    512'.55--dc20   CIP

Printed on acid-free paper.

© Birkhäuser Boston 1991

ISBN 0-8176-3496-7
ISBN 3-7643-3496-7

Typset by ARK Publications, Inc., Newton, Massachusetts
Printed and bound by Edwards Brothers, Inc., Ann Arbor, Michigan.
Printed in the U.S.A.

9 8 7 6 5 4 3 2 1

CONTENTS

PREFACE

The KK-theory of Kasparov is now approximately twelve years old; its power, utility and importance have been amply demonstrated. Nonetheless, it remains a forbiddingly difficult topic with which to work and learn. There are many reasons for this. For one thing, KK-theory spans several traditionally disparate mathematical regimes. For another, the literature is scattered and difficult to penetrate. Many of the major papers require the reader to supply the details of the arguments based on only a rough outline of proofs. Finally, the subject itself has come to consist of a number of difficult segments, each of which demands prolonged and intensive study.

Our goal in writing this book is to deal with some of these difficulties and make it possible for the reader to "get started" with the theory. We have not attempted to produce a comprehensive treatise on all aspects of KK-theory; the subject seems too vital to submit to such a treatment at this point. What seemed more important to us was a timely presentation of the very basic elements of the theory, the functoriality of the KK-groups, and the Kasparov product. Our program consists of presenting both Kasparov's original approach where the product is derived as a complicated tensor product-like construct, and the more algebraic version due to Joachim Cuntz, in which the Kasparov product is viewed as (a generalization of) composition of *-homomorphisms, as well as a detailed description of how to pass from one approach to the other. In addition, we have included a chapter indicating the way in which the Kasparov version relates to the theory of C^*-extensions.

We have kept the prerequisites to a minimum. We refer to just three monographs:

R. Douglas, *Banach Algebra Techniques in Operator Theory*, Academic Press, New York, London, 1972.

R.V. Kadison, J. Ringrose, *Fundamentals in the Theory of Operator Algebras*, Vol. II, Academic Press, New York, 1986.

G.K. Pedersen, C^*-*algebras and their Automorphism Groups*, Academic Press, London, New York, San Francisco, 1979.

Aside from the basic theory of C^*-algebras up to the existence of quasi-central approximate identities, we must use tensor products of C^*-algebras and completely positive maps. For the first, we refer to Pedersen's monograph and, for the second, to Kadison and Ringrose. At a single, but very crucial point, we need the theory of Fredholm operators; we refer to the book of Douglas for this. The only algebraic prerequisite that is not completely elementary is a working knowledge of tensor products of modules over non-commutative rings.

Our list of references contains only material that has been used directly in the preparation of the text and some papers that are not contained in the references to Blackadar's book.

A full-scale introduction to KK-theory would have to describe the very close connection between KK-theory and K-theory as well as describing all the powerful tools available for calculating the KK-groups. It would also contain a number of examples and concrete calculations with the KK-groups and the Kasparov product. For this other material, the reader must turn to the literature, starting, for example, with our list of references.

Finally, we take the opportunity to thank R.V. Kadison for the generous help and encouragement we received from him during the preparation of the present volume. We also want to thank Ann Kostant for the excellent way she has prepared the manuscript for printing.

CHAPTER 1

Hilbert C^*-Modules

1.1. Hilbert C^*-Modules and Multiplier Algebras

Let B be a C^*-algebra with norm $\| \cdot \|$.

Definition 1.1.1. A *pre-Hilbert B-module* is a complex vector space E which is also a right B-module equipped with a map $< \cdot, \cdot >: E \times E \to B$ which is linear in the second variable and satisfies the following relations for all $b \in B$, $x, y \in E$:

(i) $< x, yb > = < x, y > b$,

(ii) $< x, y >^* = < y, x >$

(iii) $< x, x > \geq 0$

(iv) $x \neq 0$ implies $< x, x > \neq 0$.

Note that it is implicitly assumed in this definition that the abelian group structure of E coming from the B-module structure is the same as the vector space addition $+$. It is easy to see that scalar multiplication and the right B-module structure of a pre-Hilbert B-module are compatible in the sense that $(\lambda e)b = \lambda(eb) = e(\lambda b)$, $\lambda \in \mathbb{C}$, $e \in E$, $b \in B$.

Lemma 1.1.2. *Let E be a pre-Hilbert B-module, and set $\|e\| = \| < e, e > \|^{\frac{1}{2}}$, $e \in E$. Then E is a normed vector space, and the following inequalities hold* :

(1.1.1) $$\|eb\| \leq \|e\| \, \|b\|, \quad e \in E, \, b \in B,$$

(1.1.2) $$\|< e, f >\| \leq \|e\| \, \|f\|, \quad e, f \in E.$$

In fact, these inequalities hold even if condition (iv) *in the definition of a pre-Hilbert B-module is dropped.*

Proof. We prove the last inequality first. Let ϕ be a state on B such that $\phi(<e,f><e,f>^*) = \|<e,f>\|^2$. Since we can assume that $\|<e,f>\| \neq 0$, we can consider $a = <e,f>^* \|<e,f>\|^{-1}$. Using the usual Cauchy-Schwarz inequality we see that

$$\begin{aligned}
\|<e,f>\|^2 &= (\phi(<e,f>a))^2 \\
&= (\phi(<e,fa>))^2 \\
&\leq \phi(<e,e>) \; \phi(<fa,fa>) \\
&= \phi(<e,e>)\phi(a^* <f,f> a) \\
&\leq \|<e,e>\| \; \|a^* <f,f> a\| \\
&\leq \|<e,e>\| \; \|<f,f>\| \\
&= \|e\|^2\|f\|^2.
\end{aligned}$$

Thus (1.1.2) follows without using (iv) in the definition of a pre-Hilbert B-module.

To prove (1.1.1) note that

$$<eb,eb> = b^* <e,e> b \leq \|e\|^2\|b\|^2.$$

This immediately yields (1.1.1), still without employing (iv). Finally, the triangle inequality for the norm on E follows from (1.1.2) in the usual way:

$$\begin{aligned}
\|<e+f,e+f>\| &= \|<e,e> + <f,f> + <e,f> + <f,e>\| \\
&\leq \|e\|^2 + \|f\|^2 + 2 \|e\| \, \|f\| \\
&= (\|e\| + \|f\|)^2.
\end{aligned} \qquad \square$$

Definition 1.1.3. A *Hilbert B-module* is a pre-Hilbert B-module E which is complete in the norm

$$\|e\| = \|<e,e>\|^{\frac{1}{2}}, \; e \in E.$$

Observe that the inequalities obtained in Lemma 1.1.2 ensure that the B-valued "inner product" $<\cdot,\cdot>$ of a pre-Hilbert B-module E as well as the right B-module structure extend by continuity to make the completion of E into a Hilbert B-module. In this way the completion of a pre-Hilbert B-module is a Hilbert B-module.

Let E be a Hilbert B-module. Then the closed span of the set $\{<x,y>: x,y \in E\}$ is a two-sided ideal in B. We denote it by $<E,E>$.

Lemma 1.1.4. *Let E be a Hilbert B-module and let $\{u_i\}$ be a net in B with the property that $u_i^* = u_i$, $\|u_i\| \leq 1$ for all i, and $\lim xu_i = x$ for all $x \in\, <E, E>$. Then $\lim\ eu_i = e$ for all $e \in E$.*

Proof.

$$\| < eu_i - e, eu_i - e > \| =$$

$$\| u_i < e, e > u_i + < e, e > - < e, e > u_i - u_i < e, e > \| \to 0 \qquad \square$$

Remark 1.1.5. If E is a Hilbert B-module and A is a C^*-algebra containing $< E, E >$ as an ideal, then there is a way to make E into a Hilbert A-module without changing the "inner product". Namely, let $\{u_i\}$ be an approximate unit for $< E, E >$. Then the identity

$$< eu_j a - eu_i a, eu_j a - eu_i a > =$$

$$a^* u_j < e, e > u_j a + a^* u_i < e, e > u_i a$$

$$- a^* u_j < e, e > u_i a - a^* u_i < e, e > u_j a, \quad a \in A,\ e \in E,$$

shows that $\{eu_i a\}$ converges in E. We can define $ea = \lim eu_i a$, and it is straightforward to check that this makes E into a Hilbert A-module. Note that Lemma 1.1.4 shows that this is in fact the only possible way to make E into a Hilbert A-module with the same "inner product".

Example 1.1.6.

(i) B is a Hilbert B-module in itself. The "inner product" is given by $< a, b > = a^* b$. Note that every ideal in B is also a Hilbert B-module with this "inner product".

(ii)   Let E be the space $\oplus_1^\infty B$ of sequences in B that are eventually 0. Define $< \cdot, \cdot >$ on E by

$$< (a_1, a_2, a_3, \ldots), (b_1, b_2, b_3, \ldots) > = \sum_n a_n^* b_n.$$

Then E is a pre-Hilbert B-module and the Hilbert B-module which we obtain by completing it will be denoted H_B. Note that when $B = \mathbb{C}, H_B$ is just the Hilbert space $l^2(\mathbb{N})$. Accordingly we set $H_{\mathbb{C}} = \mathcal{H}$. For general B, H_B can be identified with the space of sequences (b_i) in B that are square summable in the sense that $\sum_{n=1}^\infty b_n^* b_n$ converges in B. In general, this is *not* the sequences in B with square summable norms.

(iii)   For any subset J of B the closed linear span of JB is a Hilbert B-module with the "inner product" inherited from B.

To a pair E_1, E_2 of Hilbert B-modules we associate the space $\mathcal{L}_B(E_1, E_2)$ consisting of the functions $T : E_1 \to E_2$ for which there is another map $T^* : E_2 \to E_1$, called the adjoint of T, such that $< Tx, y > = < x, T^*y >$ for all $x \in E_1, y \in E_2$. The existence of an adjoint implies that T is actually both linear and a B-module map, a fact which follows easily from the definition of a (pre-) Hilbert B-module . Furthermore, since the set $\{< Tx, y > = < x, T^*y >: \|x\| \leq 1\}$ is bounded in B for all $y \in E_2$, the Banach-Steinhaus theorem (the principle of uniform boundedness) implies that each $T \in \mathcal{L}_B(E_1, E_2)$ is bounded. Thus $\mathcal{L}_B(E_1, E_2)$ is a linear subspace of the Banach space of bounded linear mappings from E_1 to E_2 and inherits the operator norm $\|T\| = \sup\{\|Tx\| : \|x\| \leq 1\}$. The inequality

$$\|< Tx, y >\| = \| < x, T^*y > \|$$
$$\leq \|x\| \, \|T^*y\| \leq \|T^*\|, \quad \|x\| \leq 1, \, \|y\| \leq 1,$$

implies that $\|T\| \leq \|T^*\|$. But it is clear that $T^* \in \mathcal{L}_B(E_2, E_1)$ and that $T^{**} = T$, so by symmetry we get $\|T\| = \|T^*\|$. In particular, it is easy to conclude from this that $\mathcal{L}_B(E_1, E_2)$ is closed in the operator norm.

We write $\mathcal{L}_B(E)$ for $\mathcal{L}_B(E, E)$.

Lemma 1.1.7. $\mathcal{L}_B(E)$ *is a C^*-algebra.*

Proof.   It is straightforward to check that $\mathcal{L}_B(E)$ is a *-algebra, and it is well-known that the inequality $\|ST\| \leq \|S\| \, \|T\|$ holds, so it suffices to check the C^*-identity. As we already know that the involution is isometric, it suffices to show that $\|T\|^2 \leq \|T^*T\|$. This follows from

$$\|< Tx, Tx >\| = \|< x, T^*Tx >\| \leq \|T^*Tx\| \leq \|T^*T\|, \quad \|x\| \leq 1. \qquad \square$$

Remark 1.1.8.   There is an important family of representations of $\mathcal{L}_B(E)$. Let ϕ be a state of B. Then $N_\phi = \{x \in E : \phi(< x, x >) = 0\}$ is a Hilbert submodule of E and E/N_ϕ is equipped with the inner product $(x + N_\phi, y + N_\phi)_\phi = \phi(< x, y >)$, $x, y \in E$. The completion of E/N_ϕ in the corresponding norm is a Hilbert space $\mathcal{H}_\phi$. Since $< m(x), m(x) > \leq \|m\|^2 < x, x >$ for $m \in \mathcal{L}_B(E)$, $x \in E$, there is a representation $\pi_\phi : \mathcal{L}_B(E) \to \mathcal{B}(\mathcal{H}_\phi)$. $\pi_\phi(m)$, $m \in \mathcal{L}_B(E)$, is given by $\pi_\phi(m)(x + N_\phi) = m(x) + N_\phi$, $x \in E$, on E/N_ϕ and extended to $\mathcal{H}_\phi$ by continuity. Observe that for any faithful family S of states on B, the representation $\sum_{\phi \in S}^{\oplus} \pi_\phi$ is faithful.

Now return to the case of two Hilbert B-modules E_1, E_2. Every pair of elements, $x \in E_2$, $y \in E_1$, gives rise to a map $\Theta_{x,y} : E_1 \to E_2$ given by $\Theta_{x,y}(z) = x < y, z >$, $z \in E_1$. It is straightforward to check that $\Theta_{x,y} \in \mathcal{L}_B(E_1, E_2)$ and that $\Theta_{x,y}^* = \Theta_{y,x}$. We let $\mathcal{K}_B(E_1, E_2)$ denote the closed linear span of $\{\Theta_{x,y} : x \in E_2, y \in E_1\}$, and write $\mathcal{K}_B(E)$ for $\mathcal{K}_B(E, E)$. By checking on the linear generators, $\Theta_{x,y}$, it is easy to see that $\mathcal{L}_B(E_2)\mathcal{K}_B(E_1, E_2) \subseteq \mathcal{K}_B(E_1, E_2)$ and that $\mathcal{K}_B(E_1, E_2)\mathcal{L}_B(E_1) \subseteq \mathcal{K}_B(E_1, E_2)$. If E_3 is a third Hilbert B-module, we also have that $\mathcal{L}_B(E_2, E_3)\mathcal{K}_B(E_1, E_2) \subseteq \mathcal{K}_B(E_1, E_3)$ and $\mathcal{K}_B(E_1, E_2)\mathcal{L}_B(E_3, E_1) \subseteq \mathcal{K}_B(E_3, E_2)$. In particular, we have the following

Lemma 1.1.9. $\mathcal{K}_B(E)$ *is a closed two-sided ideal in* $\mathcal{L}_B(E)$.

Lemma 1.1.10. $\mathcal{K}_B(E_1, E_2) = \{m \in \mathcal{L}_B(E_1, E_2) : mm^* \in \mathcal{K}_B(E_2)\}$.

Proof. If $x, x_1 \in E_2$, $y, y_1 \in E_1$, then $\Theta_{x,y} \circ \Theta_{x_1,y_1}^* = \Theta_{x<y,y_1>,x_1} \in \mathcal{K}_B(E_2)$. This gives one inclusion. For the other, let $\{v_n\}$ be an approximate unit for $\mathcal{K}_B(E_2)$. The arguments from the proof of Lemma 1.1.7 give that $\|mm^*\| = \|m\|^2$ for all $m \in \mathcal{L}_B(E_1, E_2)$. Thus we get the equality

$$\|v_n m - m\|^2 = \|v_n mm^* v_n - v_n mm^* - mm^* v_n + mm^*\|,$$

for all $m \in \mathcal{L}_B(E_1, E_2)$. If $mm^* \in \mathcal{K}_B(E_2)$, we see that $\lim v_n m = m$. Since $v_n m \in \mathcal{K}_B(E_1, E_2)$ for all n, we see that $m \in \mathcal{K}_B(E_1, E_2)$ □

Note that $\mathcal{K}_{\mathbb{C}}(\mathcal{H})$ is just the usually compact operators, so we set $\mathcal{K}_{\mathbb{C}}(\mathcal{H}) = \mathcal{K}$. The preceding constructions apply in particular when B is considered as a Hilbert B-module (cf. example 1.1.6 (i)). It is easy to see that $\mathcal{K}_B(B) \simeq B$ under a *-isomorphism sending $\Theta_{x,y}$ to xy^*. This isomorphism and the inclusion $\mathcal{K}_B(B) \subseteq \mathcal{L}_B(B)$ give us an embedding of B in $\mathcal{L}_B(B)$. In the following we suppress this embedding and consider instead B as an ideal in $\mathcal{L}_B(B)$. This amounts to the identification of an element $b \in B$ with the element in $\mathcal{L}_B(B)$ defined as left multiplication by b. Under this identification $m(a) = ma$, $m \in \mathcal{L}_B(B), a \in B$, where ma is the product (= composition of operators) in $\mathcal{L}_B(B)$.

$\mathcal{L}_B(B)$ is isomorphic to the multiplier algebra of B, as introduced for example in [24]. Here we *define* the *multiplier algebra of B to be* $\mathcal{L}_B(B)$

as introduced above. In the following we use the more standard notation $\mathcal{M}(B)$ for the multiplier algebra; i.e. $\mathcal{M}(B) = \mathcal{L}_B(B)$.

Definition 1.1.11.   Let E be a Hilbert B-module. The semi-norms $\|\cdot\|_x$, $x \in E$, on $\mathcal{L}_B(E)$ given by $\|T\|_x = \|Tx\| + \|T^*x\|$, $T \in \mathcal{L}_B(E)$, define a locally convex topology on $\mathcal{L}_B(E)$ which we call the *strict topology*.

It is an easy exercise in functional analysis to use the completeness of E and the inequality (1.1.2) to show that $\mathcal{L}_B(E)$ is complete in the strict topology.

Definition 1.1.12.   Let E be a Hilbert A-module and F a Hilbert B-module. A map $\mu : \mathcal{L}_A(E) \to \mathcal{L}_B(F)$ is called *strictly continuous* when it is continuous with respect to the strict topologies of $\mathcal{L}_A(E)$ and $\mathcal{L}_B(F)$.

In the following proposition and in the rest of this book we use the overline to denote "closed linear span".

Proposition 1.1.13. *Let A be a C^*-algebra, E a Hilbert B-module and $\phi : A \to \mathcal{L}_B(E)$ a *-homomorphism. Then the following two conditions are equivalent :*

(i) *there is a projection $p \in \mathcal{L}_B(E)$ such that $\overline{\phi(A)(E)} = p(E)$,*

(ii) *there is a strictly continuous *-homomorphism $\underline{\phi} : \mathcal{M}(A) \to \mathcal{L}_B(E)$ extending ϕ.*

If these conditions are satisfied, $\underline{\phi}$ is the only strictly continuous extension of ϕ.

Proof.   Let $\{u_i\}$ be an approximate unit for A. Then $\{mu_i\}$ converges to m in the strict topology of $\mathcal{M}(A)$ for all $m \in \mathcal{M}(A)$. So if $\psi : \mathcal{M}(A) \to \mathcal{L}_B(E)$ is any strictly continuous extension of ϕ, we have $\psi(m) = \lim \psi(mu_i) = \lim \phi(mu_i)$. This proves the uniqueness of any strictly continuous extension of ϕ.

(i) $\Rightarrow$ (ii). Let $m \in \mathcal{M}(A), e \in E$. We assert that $\{\phi(mu_i)(e)\}$ and $\{\phi(u_im)(e)\}$ converge in E to the same limit. To see this note first that $p\phi(a) = \phi(a)p = \phi(a)$ for all $a \in A$. Thus $\phi(mu_i)(e) = \phi(mu_i)p(e)$ and $\phi(u_im)(e) = \phi(u_im)p(e)$ for all i. Now let $\epsilon > 0$ be given. Choose finite sets $\{a_k\} \subseteq A$ and $\{e_i\} \subseteq E$ such that $2\|m\| \, \|p(e) - \sum_k \phi(a_k)(e_k)\| < \epsilon$.

Note that

$$\|\phi(mu_i)(e) - \phi(mu_j)(e)\| \leq 2\|m\| \left\|p(e) - \sum_k \phi(a_k)(e_k)\right\|$$

$$+ \sum_k \|\phi(mu_i a_k)(e_k) - \phi(mu_j a_k)(e_k)\|.$$

Since $\phi(mu_i a_k) \to \phi(ma_k)$ in norm for all k, it follows that

$$\|\phi(mu_i)(e) - \phi(mu_j)(e)\|$$

is eventually less than ϵ. Thus $\{\phi(mu_i)(e)\}$ is Cauchy in E. In the same way we see that so is $\{\phi(u_i m)(e)\}$. Note that

$$\|\phi(mu_i)(e) - \phi(u_i m)(e)\| \leq 2\|m\| \left\|p(e) - \sum_k \phi(a_k)(e_k)\right\|$$

$$+ \sum_k \|\phi(mu_i a_k)(e_k) - \phi(u_i m a_k)(e_k)\|$$

for all i. Since $\phi(mu_i a_k) \to \phi(ma_k)$ and $\phi(u_i m a_k) \to \phi(ma_k)$ in norm for all k, we conclude that $\|\phi(mu_i)(e) - \phi(u_i m)(e)\| < \epsilon$ eventually. This proves the assertion.

For each $m \in \mathcal{M}(A)$ we can therefore define $\underline{\phi}(m) : E \to E$ by $\underline{\phi}(m)(e) = \lim \phi(u_i m)(e) = \lim \phi(mu_i)(e)$ for all $e \in E$. Since

$$< \underline{\phi}(m)(e), f > =$$
$$\lim < \phi(mu_i)(e), f > =$$
$$\lim < e, \phi(u_i m^*)(f) > = < e, \underline{\phi}(m^*)f >, \quad e, f \in F,$$

we conclude that $\underline{\phi}(m) \in \mathcal{L}_B(E)$. It is clear that $\underline{\phi} : \mathcal{M}(A) \to \mathcal{L}_B(E)$ is a linear map and the preceding calculation gives $\underline{\phi}(m^*) = \underline{\phi}(m)^*$, $m \in \mathcal{M}(A)$. Since

$$\underline{\phi}(mn)(e) = \lim_i \lim_j \phi(u_i m n u_j)(e)$$
$$= \lim_i \lim_j \phi(u_i m)\phi(n u_j)(e)$$
$$= \underline{\phi}(m)\underline{\phi}(n)(e)$$

for all $m, n \in \mathcal{M}(A)$, we conclude that $\underline{\phi}$ is a *-homomorphism. Since $\underline{\phi}(a)(e) = \lim \phi(au_i)(e) = \phi(a)(e), e \in E, a \in A$, we see that $\underline{\phi}$ is indeed an extension of ϕ.

To show that $\underline{\phi}$ is strictly continuous let $\{m_i\}$ be a net in $\mathcal{M}(A)$ converging to $m \in \mathcal{M}(A)$ in the strict topology. By the Banach-Steinhaus theorem, $\sup_i \|m_i - m\| \le M$ for some $M < \infty$. Let $e \in E$ and choose finite sets $\{a_k\} \subseteq A$ and $\{e_k\} \subseteq E$ such that $2M\|p(e) - \sum_k \phi(a_k)(e_k)\| < \epsilon$. Then

$$\|\underline{\phi}(m)(e) - \underline{\phi}(m_i)(e)\| \le M\|p(e) - \sum_k \phi(a_k)(e_k)\|$$

$$+ \sum_k \|\phi((m - m_i)a_k)(e_k)\|.$$

Since $m_i a_k \to m a_k$ for all k, we conclude that $\underline{\phi}(m_i)(e) \to \underline{\phi}(m)(e)$ in E. Since $\underline{\phi}(m)^* = \underline{\phi}(m^*)$, we conclude that $\underline{\phi}(m_i) \to \underline{\phi}(m)$ in the strict topology.

(ii) $\Rightarrow$ (i). Set $p = \underline{\phi}(1)$. Then clearly $\overline{\phi(A)(E)} \subseteq p(E)$. But if $e \in p(E)$ then $e = p(e) = \lim \underline{\phi}(u_i)(e) = \lim \phi(u_i)(e) \in \overline{\phi(A)(E)}$ since $u_i \to 1$ in the strict topology of $\mathcal{M}(A)$ and $\underline{\phi}$ is strictly continuous. $\square$

Lemma 1.1.14. *Let A be a C^*-algebra, E a Hilbert B-module and $\phi : A \to \mathcal{K}_B(E)$ a *-isomorphism. Then $\overline{\phi(A)(E)} = E$ and the strictly continuous extension $\underline{\phi} : \mathcal{M}(A) \to \mathcal{L}_B(E)$ of ϕ is a *-isomorphism.*

Proof. It follows from the definition of $\mathcal{K}_B(E)$ and (1.1.1) that we have $E < E, E > \subseteq \overline{\mathcal{K}_B(E)(E)}$. But $\overline{E < E, E >} = E$ by Lemma 1.1.4. Thus $\overline{\phi(A)(E)} = \overline{\mathcal{K}_B(E)(E)} = E$. To see that $\underline{\phi}$ is injective, assume that $m \in \mathcal{M}(A)$ and that $\underline{\phi}(m) = 0$. Then $\phi(mu_i) = \underline{\phi}(m)\phi(u_i) = 0$ for all u_i in an approximate unit for A. Thus $mu_i = 0$ for all i. Since $mu_i \to m$ in the strict toplogy, we conclude that $m = 0$.

To see that $\underline{\phi}$ is also surjective, let $m \in \mathcal{L}_B(E)$. Let us define a map $n : A \to A$ by $n(a) = \phi^{-1}(m\phi(a)), a \in A$. Since

$$n(a)^*b = \phi^{-1}(\phi(a)^*m^*\phi(b)) = a^*\phi^{-1}(m^*\phi(b)), \quad a, b \in A,$$

we see that n admits the adjoint $n^*(b) = \phi^{-1}(m^*\phi(b))$, $b \in A$. Thus $n \in \mathcal{L}_A(A) = \mathcal{M}(A)$. Since $\underline{\phi}(n)\phi(a) = \phi(n(a)) = m\phi(a)$ for all $a \in A$, we conclude that $\underline{\phi}(n) = m$. $\square$

Corollary 1.1.15. *Let A and B be C^*-algebras and $\phi : A \to \mathcal{M}(B)$ a *-homomorphism. Then the following two conditions are equivalent:*

(i) *there is a projection $p \in \mathcal{M}(B)$ such that $\overline{\phi(A)B} = pB$,*

(ii) *there is a strictly continuous *-homomorphism $\underline{\phi} : \mathcal{M}(A) \to \mathcal{M}(B)$ extending ϕ.*

When these conditions are satisfied $\underline{\phi}$ is the only strictly continuous extension of ϕ. Moreover, when ϕ is a $*$-isomorphism from A onto B, $\underline{\phi}$ is a $*$-isomorphism between $\mathcal{M}(A)$ and $\mathcal{M}(B)$.

Proof. Proposition 1.1.13 and Lemma 1.1.14. $\qquad\qquad$ $\square$

Corollary 1.1.16. *Let E be a Hilbert B-module. There is then a $*$-isomorphism $\psi : \mathcal{M}(\mathcal{K}_B(E)) \to \mathcal{L}_B(E)$ such that the diagram*

$$
\begin{array}{ccc}
 & \psi & \\
\mathcal{M}(\mathcal{K}_B(E)) & \to & \mathcal{L}_B(E) \\
\cup & & \cup \\
\mathcal{K}_B(E) & = & \mathcal{K}_B(E)
\end{array}
$$

commutes.

Proof. Apply Lemma 1.1.14 to the identity map on $\mathcal{K}_B(E)$. $\qquad$ $\square$

Lemma 1.1.17. *Let E be a Hilbert B-module. A net $\{m_i\} \subseteq \mathcal{L}_B(E)$ converges to $m \in \mathcal{L}_B(E)$ in the strict topology if and only if $m_i k \to mk$ and $m_i^* k \to m^* k$ in norm for all $k \in \mathcal{K}_B(E)$.*

Proof. If $m_i \to m$ in the strict topology, we know from the Banach-Steinhaus theorem that $\|m_i - m\|$ is uniformly bounded. So to show that $m_i k \to mk$ and $m_i^* k \to m^* k$ for all $k \in \mathcal{K}_B(E)$, it suffices to check for $k = \Theta_{x,y}$ for some $x, y \in E$. But in this case the thing is clear because $m_i x \to mx$ and $m_i^* x \to m^* x$ in E.

Conversely, assume that $m_i k \to mk$ and $m_i^* k \to m^* k$. Let ψ be the $*$-isomorphism of Corollary 1.1.16. Since $\|m_i - m\| = \|\psi^{-1}(m_i - m)\|$, the Banach-Steinhaus theorem again shows that $\|m_i - m\|$ is uniformly bounded. Since $\overline{\mathcal{K}_B(E)(E)} = E$ by Lemma 1.1.14, we see that $m_i e \to me$ and $m_i^* e \to m^* e$ for all $e \in E$. $\qquad\qquad$ $\square$

Definition 1.1.18. Two Hilbert B-modules E, F are *isomorphic* when there is a linear bijection $\psi : E \to F$ such that

$$(1.1.3) \qquad\qquad <\psi(e), \psi(f)> \; = \; <e, f>, \quad e, f \in E.$$

We write $E \approx F$ in this case and call ψ an *isomorphism*.

It follows from (1.1.3) that an isomorphism $\psi : E \to F$ is a B-module map. An isomorphism is easily seen to give an equivalence relation on the set of Hilbert B-modules.

Lemma 1.1.19.   *Let $\psi : E \to F$ be an isomorphism of Hilbert B-modules. Then the map $m \to \psi \circ m \circ \psi^{-1}$ is a *-isomorphism, $\mathcal{L}_B(E) \simeq \mathcal{L}_B(F)$, which maps $\mathcal{K}_B(E)$ onto $\mathcal{K}_B(F)$.*

Proof. Straightforward.   □

We digress momentarily to introduce a class of C^*-algebras that will be of main interest for us in the development of KK-theory.

Definition 1.1.20.   A C^*-algebra B is *σ-unital* if it contains a strictly positive element.

Recall that a positive element $h \in B$ is strictly positive if and only if $\phi(h) > 0$ for all states ϕ of B. In [24], 3.10.4-3.10.5, it is shown that B is σ-unital if and only if it contains a countable approximate unit; if h is a strictly positive element of norm ≤ 1, then $\{h^{\frac{1}{n}}\}$ will be a countable approximate unit. We need the following characterization of strictly positive elements.

Lemma 1.1.21.   *A positive element $h \in B$ is strictly positive if and only if hB is dense in B.*

Proof. If hB is not dense in B, the closure of Bh is a left ideal $\neq B$. By [24], 3.10.7, there is therefore a state of B which vanishes on the closure of Bh. In particular, this state vanishes on h, i.e. h is not strictly positive. Conversely, if h is not strictly positive there is a state of B which vanishes on h. But then it vanishes on Bh by the Cauchy-Schwarz inequality. Hence [24], 3.10.7, gives that neither hB nor Bh is dense in B.   □

Lemma 1.1.22.   *Let E be a Hilbert B-module, and T a positive element of $\mathcal{K}_B(E)$. Then T is strictly positive if and only if the range of T is dense in E.*

Proof. If T is strictly positive $T\mathcal{K}_B(E)$ is dense in $\mathcal{K}_B(E)$ by Lemma 1.1.21. As observed in the proof of Lemma 1.1.14, the span of $\mathcal{K}_B(E)E$ is dense in E. Thus TE must be dense in E if T is strictly positive. Conversely, assume that TE is dense in E. Then for every $x \in E$ there is a sequence $\{x_n\}$ in E such that $Tx_n \to x$. But then $\Theta_{x,y} = \lim \Theta_{Tx_n,y} = \lim T\Theta_{x_n,y}$ for all $y \in E$. Hence the closure of $T\mathcal{K}_B(E)$ equals $\mathcal{K}_B(E)$, i.e. T is strictly positive by 1.1.21.   □

Definition 1.1.23.   A Hilbert B-module E is called *countably generated* when there is a countable set $\{x_n\}$ in E such that span of the set $\{x_n b : n \in \mathbb{N}, b \in B\}$ is dense in E. A set $\{x_n\}$ in E with this property is called *a set of generators for E*.

Theorem 1.1.24. [Kasparov's stabilization theorem]. *If E is a countably generated Hilbert B-module, then $E \oplus H_B \approx H_B$.*

Proof. If B has no unit, we can consider E as a Hilbert $\widehat{B}$-module, where $\widehat{B}$ is the C^*-algebra obtained by adjoining a unit to B (cf. Remark 1.1.5.). If we can show that $E \oplus H_{\widehat{B}} \approx H_{\widehat{B}}$ as Hilbert $\widehat{B}$-modules, then the desired conclusion follows from the fact that H_B is isomorphic to $\overline{H_{\widehat{B}} B}$ and $E \oplus H_B$ to $\overline{(E \oplus H_{\widehat{B}})B}$. It suffices therefore to prove the theorem when B is unital.

Let $\{\eta_n\} \subseteq E$ be a countable set of generators for E, chosen such that for each $n \in \mathbb{N}$, $\eta_n = \eta_m$ for infinitely many other $m \in \mathbb{N}$. After normalizing each η_n we can assume that $\|\eta_n\| \leq 1$ for all n. Let ϵ_i be the element of H_B whose coordinates are all zero, except at the i-th place where there is 1, the unit in B. Define $T : H_B \to E \oplus H_B$ to be the element of $\mathcal{L}_B(H_B, E \oplus H_B)$ given by $T(\epsilon_i) = (2^{-i}\eta_i, 4^{-i}\epsilon_i), i \in \mathbb{N}$. Then $T = \sum_i 2^{-i}\Theta_{(\eta_i, 2^{-i}\epsilon_i), \epsilon_i} \in \mathcal{K}_B(H_B, E \oplus H_B)$. Fix $n \in \mathbb{N}$. For every other $m \in \mathbb{N}$ such that $\eta_n = \eta_m$ we have that $(\eta_n, 2^{-m}\epsilon_m) = T(2^m \epsilon_m) \in \operatorname{Ran} T$. Since there are infinitely many such m, we see that $(\eta_n, 0)$ is contained in the closure of Ran T. But then $(0, \epsilon_n) = 4^n(T(\epsilon_n) - 2^{-n}(\eta_n, 0))$ is also in this closure. Since T is a B-module map and $\{(\eta_n, 0), (0, \epsilon_n) : n \in \mathbb{N}\}$ generates a dense B-submodule of $E \oplus H_B$, we conclude that T has dense range. Note that

$$T^*T = \sum_{i,j} 2^{-(i+j)}\Theta_{\epsilon_i(<\eta_i,\eta_j>+<2^{-i}\epsilon_i, 2^{-j}\epsilon_j>), \epsilon_j}$$

$$= \sum_i 4^{-2i}\Theta_{\epsilon_i, \epsilon_i} + (\sum_i 2^{-i}\Theta_{(\eta_i, 0), \epsilon_i})^*(\sum_i 2^{-i}\Theta_{(\eta_i, 0), \epsilon_i})$$

$$\geq \sum_i 4^{-2i}\Theta_{\epsilon_i, \epsilon_i}.$$

The latter operator is obviously positive and has dense range in H_B so it is strictly positive by 1.1.22. Since T^*T dominates this element in $\mathcal{K}_B(H_B)$ it must also be strictly positive, i.e. have dense range by 1.1.22. Therefore $|T| = (T^*T)^{\frac{1}{2}}$ has also dense range. Define $V : H_B \to E \oplus H_B$ to be the element in $\mathcal{L}_B(H_B, E \oplus H_B)$ given by $V(|T|x) = Tx, x \in H_B$. Since

$$< V(|T|x), V(|T|y) > \; = \; < Tx, Ty >$$
$$= < x, T^*Ty > \; = \; < |T|x, |T|y >$$

for all $x, y \in H_B$ and RanV contains Ran$T$, we conclude that V defines the desired isomorphism. □

Corollary 1.1.25. *A Hilbert B-module E is countably generated if and only if $\mathcal{K}_B(E)$ is σ-unital.*

Proof. By Remark 1.1.5 we can assume that B is unital. Assume first that E is countably generated. Let $Q : E \oplus H_B \to E \oplus H_B$ be the projection onto E. Then $E \approx Q(E \oplus H_B)$. By Theorem 1.1.24, $E \oplus H_B \approx H_B$. So $E \approx P(H_B)$ for some $P \in \mathcal{L}_B(H_B)$ by Lemma 1.1.19. Thus $\mathcal{K}_B(E) \simeq \mathcal{K}_B(P(H_B))$ by the same lemma. The restriction of elements in $\mathcal{L}_B(H_B)$ to $P(H_B)$ maps $\Theta_{x,y}$ to $\Theta_{P(x),P(y)}, x, y \in H_B$, and it is easy to see that the restriction map gives an isomorphism $P\mathcal{K}_B(H_B)P \simeq \mathcal{K}_B(P(H_B))$ of C^*-algebras. Thus it suffices to show that $P\mathcal{K}_B(H_B)P$ is σ-unital. If $\{v_i\}$ is an approximate unit for $\mathcal{K}_B(H_B)$, then $\{Pv_iP\}$ is an approximate unit in $P\mathcal{K}_B(H_B)P$, so it suffices to show that $\mathcal{K}_B(H_B)$ is σ-unital. But this is easy; in the notation of the proof of Theorem 1.1.24, the element $\sum_i 2^{-i}\Theta_{\epsilon_i,\epsilon_i}$ is strictly positive since it has dense range.

Conversely, assume that $\mathcal{K}_B(E)$ is σ-unital and let $K \in \mathcal{K}_B(E)$ be a positive element with dense range. For each $n \in \mathbb{N}$ choose $x(n, i) \in E$, $i = 1, 2, 3, \ldots, M_n$, such that the distance from K to span $\{\Theta_{x(n,i),x(n,j)} : i, j = 1, 2, 3, \ldots, M_n\}$ is less than $\frac{1}{n}$. Then it is easy to see that $\{x(n, i) : n \in \mathbb{N}, i = 1, 2, 3, \ldots, M_n\}$ will be a set of generators for E. □

Theorem 1.1.26. *Let A be a σ-unital C^*-algebra and $\phi : A \to B$ a surjective *-homomorphism. Assume that $\mathcal{F} \subseteq \mathcal{M}(A)$ is a separable closed selfadjoint subspace of the multiplier algebra $\mathcal{M}(A)$ of A.*

Then the strictly continuous *-homomorphic extension $\underline{\phi} : \mathcal{M}(A) \to \mathcal{M}(B)$ of ϕ maps the C^*-algebra

$$\mathcal{A} = \{m \in \mathcal{M}(A) : mf - fm \in \ker \phi, \; f \in \mathcal{F}\}$$

onto the relative commutant,

$$\mathcal{M}(B) \cap \underline{\phi}(\mathcal{F})' = \{m \in \mathcal{M}(B) : mg = gm, g \in \underline{\phi}(\mathcal{F})\},$$

of $\underline{\phi}(\mathcal{F})$ in $\mathcal{M}(B)$.

Proof. The theorem is trivial if A and hence also B is unital. We give the proof when both A and B are non-unital. The remaining case, where B is unital while A is not, is left as an exercise, E 1.1.7.

Let z be a positive element of $\mathcal{M}(B) \cap \underline{\phi}(\mathcal{F})'$. It suffices to find $m \in A$ such that $\phi(m) = z$. Since $\mathcal{F}$ is separable we can find a sequence $f_1, f_2, f_3, \ldots$, with dense span in $\mathcal{F}$ and with $\lim \|f_n\| = 0$. Let $h \in A$ be a strictly positive element of A. Set $k = \phi(h)$ and $g_i = \underline{\phi}(f_i)$, $i \in \mathbb{N}$. If a non-zero positive linear functional ω on B vanishes on k, then $\omega \circ \phi$ is a non-zero positive linear functional which vanishes on h. This shows that k is strictly positive in B because h is strictly positive in A. Let D be the C^*-subalgebra of $\mathcal{M}(B)$ generated by k, z and the g_i's. Then D is separable and we can choose a countable approximate unit $\{u_n : n \in \mathbb{N}\}$ for $D \cap B$ which is quasi-central for D (cf. [24], 3.12.16.). Since $k \in D \cap B$ is strictly positive, $\{u_n\}$ is an approximate unit for B by Lemma 1.1.21. Set $z_n = z^{\frac{1}{2}} u_n z^{\frac{1}{2}}$. Then $\{z_n\}$ is an increasing sequence of positive elements in B such that $z_n \to z$ in the strict topology of $\mathcal{M}(B)$. In particular $z_n k \to zk$ in norm. Since $\{u_n\}$ is quasi-central for D we have furthermore that $\|z_n g_i - g_i z_n\| \to 0$ for all i. This convergence is uniform in i because $\|g_i\| \to 0$ and $\|z_n\| \le \|z\|$ for all n. After passing to a subsequence we can assume that $\|(z_{n+1} - z_n)k\| < 2^{-n}$ and $\sup_i \|[z_{n+1} - z_n, g_i]\| < 2^{-n}$ for all $n \in \mathbb{N}$.

We will construct by induction a sequence $\{y_n\} \subseteq A$ such that

(i) $0 \le y_n \le y_{n+1} \le \|z\| 1$, $n \in \mathbb{N}$,

(ii) $\phi(y_n) = z_n$, $n \in \mathbb{N}$,

(iii) $\|(y_{n+1} - y_n)h\| < 2^{-n}$, $n \in \mathbb{N}$,

(iv) $\|[y_{n+1} - y_n, f_i]\| < 2^{-n}$, $n \in \mathbb{N}, i \in \mathbb{N}$.

To start the induction note that $0 \le z_1 \le \|z\| 1$. Since ϕ is unital we conclude from [24], 1.5.10, that $z_1 = \phi(y_1)$ for some $y_1 \in A$ with $0 \le y_1 \le \|z\| 1$. Now assume that $y_1, y_2, \ldots, y_n$ has been constructed. To construct y_{n+1} note that $0 \le z_{n+1} - z_n \le \|z\| 1 - z_n$. Arguing as above we find $x \in A$ such that $0 \le x \le \|z\| 1 - y_n$. Set $a = y_n + x$. Then $0 \le y_n \le a \le \|z\| 1$

and $\phi(a) = z_{n+1}$. Let $\{v_\lambda\}$ be an approximate unit in ker ϕ which is quasi-central for $\mathcal{M}(A)$. Then

$$\lim_\lambda \left\| (a - y_n)^{\frac{1}{2}}(1 - v_\lambda)(a - y_n)^{\frac{1}{2}} h \right\| = \lim_\lambda \left\| (a - y_n)h(1 - v_\lambda) \right\|$$
$$= \| (z_{n+1} - z_n)k \|$$

by [24], 1.5.4. Similarly

$$\lim_\lambda \left\| [(a - y_n)^{\frac{1}{2}}(1 - v_\lambda)(a - y_n)^{\frac{1}{2}}, f_i] \right\| = \| [(z_{n+1} - z_n), g_i] \|$$

for all $i \in \mathbb{N}$. By choosing λ sufficiently large we obtain

$$\left\| (a - y_n)^{\frac{1}{2}}(1 - v_\lambda)(a - y_n)^{\frac{1}{2}} h \right\| < 2^{-n}$$

and

$$\left\| [(a - y_n)^{\frac{1}{2}}(1 - v_\lambda)(a - y_n)^{\frac{1}{2}}, f_i] \right\| < 2^{-n} \quad \text{for all } i \in \mathbb{N}.$$

Set $y_{n+1} = a - (a - y_n)^{\frac{1}{2}} v_\lambda (a - y_n)^{\frac{1}{2}}$. Then y_{n+1} has all the right properties and we can therefore construct the sequence $\{y_n\}$ by induction.

Since $\|y_n\| \leq \|z\|$ by (i) and $\{y_n h\}$ is convergent in A by (iii), we see from Lemma 1.1.21 that $\{y_n\}$ is Cauchy in the strict topology of $\mathcal{M}(A)$. Consequently there is a multiplier $m \in \mathcal{M}(A)$ such that $y_n a \to ma$ for all $a \in A$. By (ii) we get that $\phi(m)\phi(a) = \phi(ma) = \lim_n \phi(y_n a) = \lim_n z_n \phi(a) = z\phi(a)$ for all $a \in A$. It follows that $\phi(m) = z$. By (iv) $\{[y_n, f_i]\}$ converges in A to some element $b_i \in A$ for all $i \in \mathbb{N}$. On the other hand $\{[y_n, f_i]\}$ converges to $[m, f_i]$ in the strict topology. Thus $[m, f_i] = b_i \in A$ for all $i \in \mathbb{N}$. Since $\phi(b_i) = \lim_n \phi([y_n, f_i]) = \lim_n [\phi(y_n), \phi(f_i)] = [z, g_i] = 0$ for all $i \in \mathbb{N}$, we conclude that $m \in A$. $\qquad\square$

Remark 1.1.27. When A is abelian or $\mathcal{F} = \{0\}$ Theorem 1.1.26 asserts that $\phi : \mathcal{M}(A) \to \mathcal{M}(B)$ is surjective. In the abelian case, where $A = C_o(X)$ for some locally compact Hausdorff space X, we can identify B with $C_o(Y)$ for some closed subset Y of X such that ϕ becomes the map obtained by restricting functions to Y. In this case $\mathcal{M}(A)$ (resp. $\mathcal{M}(B)$) can be identified with the continuous bounded functions on X (resp. Y), (cf. E 1.1.6), and ϕ with the restriction of continuous bounded functions. The assertion of Theorem 1.1.26 is in this case that ϕ is surjective when X is σ-compact (this corresponds to $A = C_o(X)$ being σ-unital). That ϕ is surjective means precisely that every continuous bounded function on Y

can be extended to a continuous bounded function on X. Thus Theorem 1.1.26 is a non-commutative version of Tietze's extension theorem.

We conclude this section with the following useful proposition.

Proposition 1.1.28.   *Let A, C be arbitrary C^*-algebras, E_1, E_2 Hilbert A-modules and F_1, F_2 Hilbert C-modules. Let $\pi : A \to C$ be a $*$-homomorphism and $\phi_i : E_i \to F_i$ linear surjections such that*

(i) $< \phi_i(x), \phi_i(y) > = \pi(< x, y >)$, $x, y \in E_i$, $i = 1, 2$.

Then there is a unique linear map $\psi : \mathcal{L}_A(E_1, E_2) \to \mathcal{L}_C(F_1, F_2)$ satisfying

(ii) $\psi(m)(\phi_1(e)) = \phi_2(m(e))$, $m \in \mathcal{L}_A(E_1, E_2)$, $e \in E_1$.

If $E_1 = E_2, F_1 = F_2$ and $\phi_1 = \phi_2$, then $\psi : \mathcal{L}_A(E) \to \mathcal{L}_C(F)$ is a $*$-homomorphism mapping $\mathcal{K}_A(E)$ onto $\mathcal{K}_C(F)$.

Proof.   Let $m \in \mathcal{L}_A(E_1, E_2)$. Then we have $< m(e), m(e) > \le \|m\|^2 < e, e >$ for all $e \in E_1$. Next, two applications of (i) will then yield

$$< \phi_2(m(e)), \phi_2(m(e)) > = \pi(< m(e), m(e) >)$$
$$\le \|m\|^2 \pi(< e, e >) = \|m\|^2 < \phi_1(e), \phi_1(e) > .$$

Thus we can define a linear map $\psi(m) : F_1 \to F_2$ by $\psi(m)(\phi_1(e)) = \phi_2(m(e))$, $e \in E_1$. By symmetry we can define $\psi^*(m^*) : F_2 \to F_1$ by $\psi^*(m^*)(\phi_2(e)) = \phi_1(m^*(e))$, $e \in E_2$. Then

$$< \psi(m)(\phi_1(e)), \phi_2(f) > = < \phi_2(m(e)), \phi_2(f) >$$
$$= \pi(< m(e), f >)$$
$$= \pi(< e, m^*(f) >)$$
$$= < \phi_1(e), \phi_1(m^*(f)) >, \quad e \in E_1, f \in E_2,$$

showing that $\psi(m) \in \mathcal{L}_C(F_1, F_2)$ with adjoint $\psi(m)^* = \psi^*(m^*)$. It is clear that $\psi : \mathcal{L}_A(E_1, E_2) \to \mathcal{L}_C(F_1, F_2)$ is the map we are looking for. Uniqueness is obvious.

If $E_1 = E_2 = E$, $F_1 = F_2 = F$ and $\phi = \phi_1 = \phi_2$, then $\psi = \psi^*$ and now it follows straightforwardly from (ii) that ψ is a $*$-homomorphism in this case. To see that $\psi(\mathcal{K}_A(E)) \subseteq \mathcal{K}_C(F)$, we note that (i) implies that

$$< \phi(f), \phi(ea) > = \pi(< f, ea >)$$
$$= \pi(< f, e >)\pi(a)$$
$$= < \phi(f), \phi(e)\pi(a) >$$

for all e, $f \in E$, $a \in A$. Thus $\phi(ea) = \phi(e)\pi(a)$. It then follows that

$$\psi(\Theta_{x,y})(\phi(e)) = \phi(\Theta_{x,y}(e))$$
$$= \phi(x < y, e >)$$
$$= \phi(x)\pi(< y, e >)$$
$$= \phi(x) < \phi(y), \phi(e) >$$
$$= \Theta_{\phi(x),\phi(y)}(\phi(e))$$

for all $x, y \in E$, $e \in E$. Thus $\psi(\Theta_{x,y}) = \Theta_{\phi(x),\phi(y)}$. $\square$

It is clear that when $W : E \to F$ is an isomorphism of Hilbert A-modules, then the map $\mathcal{L}_A(E) \to \mathcal{L}_A(F)$ which Proposition 1.1.28 associates to W is the map sending $m \in \mathcal{L}_A(E)$ to $WmW^* \in \mathcal{L}_A(F)$.

1.1.29. *Notes and remarks.*

The notion of Hilbert B-modules goes back to Paschke, [23], who also proved the basic Lemma 1.1.2. The definition of $\mathcal{L}_B(E, F)$ and $\mathcal{K}_B(E, F)$ was given by Kasparov in [18]. In this paper he proved (among other things) Corollary 1.1.16 and Theorem 1.1.24. The proof of Kasparov's stabilization theorem given here is due to Mingo and Philips, [21], and they also proved Corollary 1.1.25. Another proof of the stabilization theorem can be found in [11]. Theorem 1.1.26 is a result of Olsen and Pedersen (cf. [22]).

Exercise 1.1

E 1.1.1

Show that a Hilbert $\mathbb{C}$-module E is a Hilbert space, that $\mathcal{L}_{\mathbb{C}}(E) = \mathcal{B}(E) =$ the bounded linear operators on E and that $\mathcal{K}_{\mathbb{C}}(E)$ is the ideal of compact operators. Show that on $\mathcal{B}(E)$, the strict topology is the same as the strong * topology.

E 1.1.2

(i) Show that there is a commutative diagram

$$\begin{array}{ccc} \mathcal{M}(B \oplus B) & \simeq & \mathcal{M}(B) \oplus \mathcal{M}(B) \\ \cup & & \cup \\ B \oplus B & = & B \oplus B. \end{array}$$

(ii) Show that there is a commutative diagram

$$
\begin{array}{ccc}
\mathcal{M}(M_2(B)) & \simeq & M_2(\mathcal{M}(B)) \\
\cup & & \cup \\
M_2(B) & = & M_2(B).
\end{array}
$$

E 1.1.3

Prove the following analog of Riesz's representation theorem : Let E be a Hilbert B-module and $T \in \mathcal{K}_B(E, B)$. Then there is an element $x \in E$ such that $Ty = <x, y>$, $y \in E$.

E 1.1.4

Show that $\mathcal{M}(B) = B$ if and only if B is unital.

E 1.1.5

Let E be a Hilbert B-module. Show that $\mathcal{K}_B(E)$ is dense in $\mathcal{L}_B(E)$ in the strict topology, that the product in $\mathcal{L}_B(E)$ is separately strictly continuous and jointly strictly continuous on norm bounded subsets.

E 1.1.6

Let X be a locally compact Hausdorff space and let $A = C_o(X)$ be the C^*-algebra of continuous functions on X which vanish at infinity. Prove that $\mathcal{M}(A)$ is *-isomorphic to the C^*-algebra of continuous bounded functions on X (with the supremum norm).

E 1.1.7

Prove Theorem 1.1.26 when B is unital and A is not.

E 1.1.8

Let $p \in \mathcal{M}(B)$ be a projection. Show that there is a commutative diagram

$$
\begin{array}{ccc}
\mathcal{M}(pBp) & \simeq & p\mathcal{M}(B)p \\
\cup & & \cup \\
pBp & = & pBp
\end{array}
$$

E 1.1.9

Let A and B be C^*-algebras and $j : A \to B$ an injective *-homomorphism mapping A onto a (closed two-sided) ideal in B.

(i) Prove that there is a unique *-homomorphism $\mu : \mathcal{M}(B) \to \mathcal{M}(A)$ such that $j(\mu(m)a) = mj(a)$, $a \in A$.

(ii) Describe $\ker \mu$.

(iii) Assume that $j(A)$ is an essential ideal in B, i.e. if $J \subseteq B$ is a non-zero ideal, then $j(A) \cap J \neq 0$. Prove that μ is then injective.

1.2. Constructions with Hilbert B-modules

1.2.1. *Direct sum.*

Let $E_1, E_2, \ldots, E_n$ be Hilbert B-modules. Then $\oplus_1^n E_i$ is a right B-module, and when we define a B-valued "inner product" by

$$< (a_1, a_2, \ldots, a_n), (b_1, b_2, \ldots, b_n) > = \sum_1^n < a_i, b_i >$$

we have turned $\oplus_1^n E_i$ into a Hilbert B-module. (That $\oplus_1^n E_i$ is complete follows from the inequality

$$\max_i \{ \| < a_i, a_i > \| \} \leq \left\| \sum_i < a_i, a_i > \right\| \leq \sum_i \| < a_i, a_i > \|.)$$

1.2.2. *Pushout.*

Now let E be a Hilbert B-module and $f : B \to A$ a surjective $*$-homomorphism between C^*-algebras. Define a Hilbert submodule N_f of E by $N_f = \{ x \in E : f(< x, x >) = 0 \}$. Set $E_f' = E/N_f$ and let $q : E \to E_f'$ denote the quotient map. Then E_f' is a right A-module when we define $q(x) f(b) = q(xb), x \in E, b \in B$, and when we define an A-valued "inner product" by $< q(x), q(y) > = f(< x, y >), x, y \in E, E_f'$ becomes a pre-Hilbert A-module. We let E_f denote the Hilbert A-module obtained from E_f' by completion.

1.2.3. *Internal tensor product.*

Let E be a Hilbert B-module, F a Hilbert A-module. Let $\phi : B \to \mathcal{L}_A(F)$ be a $*$-homomorphism. Then ϕ makes F into a left B-module : $bx = \phi(b)x, b \in B, x \in F$. Thus we can form the algebraic tensor product $E \otimes_B F$ which is a right A-module in the obvious way: $(x \otimes_B y)a = x \otimes_B ya$. We can define a map $< \cdot, \cdot > : E \otimes_B F \times E \otimes_B F \to A$ to be the unique map which is linear in the first variable and conjugate linear in the second, and satisfies

$$< x_1 \otimes_B x_2, y_1 \otimes_B y_2 > = < x_2, \phi(< x_1, y_1 >)y_2 >,$$

$x_1, y_1 \in E$, $x_2, y_2 \in F$. This is legitimate since

$$< \phi(b)x_2, \phi(< x_1, y_1 >)y_2 > = < x_2, \phi(< x_1b, y_1 >)y_2 >$$

and

$$< x_2, \phi(< x_1, y_1 >)\phi(b)y_2 > = < x_2, \phi(< x_1, y_1b >)y_2 >$$

for all $b \in B$.

Set $N_{EF} = \{z \in E \otimes_B F : < z, z >= 0\}$. Then N_{EF} is an A-submodule, and we can consider the quotient $E \otimes_B F/N_{EF}$ and the quotient map $q : E \otimes_B F \to E \otimes_B F/N_{EF}$. Then $E \otimes_B F/N_{EF}$ is a right A-module : $q(x)a = q(xa), x \in E \otimes_B F, a \in A$, and we can define an A-valued "inner product" on $E \otimes_B F/N_{EF}$ by $< q(x), q(y) > = < x, y >, x, y \in E \otimes_B F$. This makes $E \otimes_B F/N_{EF}$ into a pre-Hilbert A-module, as the reader is urged to check. The completed Hilbert A-module is denoted by $E \otimes_\phi F$, and the image of $x \otimes_B y, x \in E, y \in F$, in $E \otimes_\phi F$ will be denoted by $x \otimes_\phi y$. Note that there is a *-homomorphism $j : \mathcal{L}_B(E) \to \mathcal{L}_A(E \otimes_\phi F)$ given by $j(m)(x \otimes_\phi y) = m(x) \otimes_\phi y, x \in E, y \in F, m \in \mathcal{L}_B(E)$. In certain important cases j maps $\mathcal{K}_B(E)$ into $\mathcal{K}_A(E \otimes_\phi F)$ (cf. Lemma 1.2.8). Since it is a much more suggestive notation we shall write $m \otimes_\phi id$ for the operator $j(m)$.

1.2.4.   *External tensor product.*

Let E be a Hilbert B-module and F a Hilbert C-module. The algebraic tensor product $E \otimes_{\mathbb{C}} F$ is a right module over the algebraic tensor product $B \otimes_{\mathbb{C}} C$ such that $(e \otimes_{\mathbb{C}} f)b \otimes_{\mathbb{C}} c = eb \otimes_{\mathbb{C}} fc, e \in E$, $f \in F, b \in B, c \in C$. By considering $B \otimes_{\mathbb{C}} C$ as a dense *-subalgebra of the spatial tensor product $B \otimes C$, we can define a $B \otimes C$-valued "inner product" on $E \otimes_{\mathbb{C}} F$ as the map $< \cdot, \cdot > : E \otimes_{\mathbb{C}} F \times E \otimes_{\mathbb{C}} F \to B \otimes C$ which is linear in the second variable, conjugate linear in the first and satisfies that

$$< e \otimes_{\mathbb{C}} f, e_1 \otimes_{\mathbb{C}} f_1 > = < e, e_1 > \otimes < f, f_1 >, \ e, e_1 \in E, \ f, f_1 \in F.$$

Then $E \otimes_{\mathbb{C}} F$ is almost a pre-Hilbert $B \otimes C$-module, the difference being that it is only a right module over the dense *-subalgebra $B \otimes_{\mathbb{C}} C$ of $B \otimes C$, not over $B \otimes C$ itself, and that condition (iv) in Definition 1.1.1 needs not hold. In particular condition (i) in Definition 1.1.1 only makes sense for $b \in B \otimes_{\mathbb{C}} C$. To check these assertions, the only difficult point is (iii) : $< x, x > \geq 0$. To see that this holds let $x = \sum_{i=1}^k e_i \otimes_{\mathbb{C}} f_i$. Since

$$\sum_{ij} b_i^* < e_i, e_j > b_j = < \sum_i e_i b_i, \sum_i e_i b_i > \geq 0$$

for all k-tuples $(b_1, b_2, \ldots, b_k)$ in B, it follows that the matrix $\{\pi(< e_i, e_j >)\}$ is a positive element of $M_k(\pi(B))$ for every cyclic representation π of B. Consequently the matrix $\{< e_i, e_j >\}$ is a positive element in $M_k(B)$, so there is an element $\{a_{ij}\} \in M_k(B)$ such that

$$< e_i, e_j > = \sum_{n=1}^{k} a_{ni}^* a_{nj}, \quad i, j = 1, 2, 3, \ldots, k.$$

Similarly there is an element $\{b_{ij}\} \in M_k(C)$ such that

$$< f_i, f_j > = \sum_{n=1}^{k} b_{ni}^* b_{nj}, \quad i, j = 1, 2, 3, \ldots, k.$$

It follows that

$$< x, x > = \sum_{ij} < e_i, e_j > \otimes < f_i, f_j > = \sum_{ij} \sum_{mn} a_{ni}^* a_{nj} \otimes b_{mi}^* b_{mj}$$

$$= \sum_{mn} \left(\sum_i a_{ni} \otimes b_{mi} \right)^* \left(\sum_i a_{ni} \otimes b_{mi} \right) \geq 0.$$

One can still prove a version of Lemma 1.1.2, and consider the $B \otimes_{\mathbb{C}} C$-submodule $N = \{x \in E \otimes_{\mathbb{C}} F :< x, x > = 0\}$. Let $q : E \otimes_{\mathbb{C}} F \to E \otimes_{\mathbb{C}} F/N$ be the quotient map. Then $E \otimes_{\mathbb{C}} F/N$ is equipped with a $B \otimes_{\mathbb{C}} C$-valued "inner product" defined by $< q(x), q(y) > = < x, y >$, $x, y \in E \otimes_{\mathbb{C}} F$. Let $E \hat{\otimes} F$ denote the completion of $E \otimes_{\mathbb{C}} F/N$ in the norm $\| < \cdot, \cdot > \|^{\frac{1}{2}}$. $E \otimes_{\mathbb{C}} F/N$ is a right $B \otimes_{\mathbb{C}} C$-module:

$$q(x)b \otimes_{\mathbb{C}} c = q(x(b \otimes_{\mathbb{C}} c)), x \in E \otimes_{\mathbb{C}} F, \ b \in B, \ c \in C,$$

and we have the inequality $\|zb\| \leq \|z\| \|b\|$ for all $z \in E \otimes_{\mathbb{C}} F/N$, $b \in B \otimes_{\mathbb{C}} C$. Therefore we can extend the right $B \otimes_{\mathbb{C}} C$-module structure by continuity in two steps to obtain a right $B \otimes C$-module structure on $E \hat{\otimes} F$. In this way $E \hat{\otimes} F$ becomes a Hilbert $B \otimes C$-module. For $e \in E$, $f \in F$, we let $e \hat{\otimes} f$ denote the image of $e \otimes_{\mathbb{C}} f$ in $E \hat{\otimes} F$, and observe that the $B \otimes C$-module structure satisfies $(e \hat{\otimes} f)b \otimes c = eb \hat{\otimes} fc$, $b \in B$, $c \in C$.

Let $m \in \mathcal{L}_B(E)$, $n \in \mathcal{L}_C(F)$. Then $m \otimes_{\mathbb{C}} n$ is the linear map on $E \otimes_{\mathbb{C}} F$ given on simple tensors as $m \otimes_{\mathbb{C}} n(e \otimes_{\mathbb{C}} f) = m(e) \otimes_{\mathbb{C}} n(f)$, $e \in E$, $f \in F$. Let $x = \sum_{n=1}^{k} e_i \otimes_{\mathbb{C}} f_i$, $e_i \in E$, $f_i \in F$, $i = 1, 2, 3, \ldots, k$. We now assert that $< m \otimes_{\mathbb{C}} n(x), m \otimes_{\mathbb{C}} n(x) > \leq \|m\|^2 \|n\|^2 < x, x >$. To see this note that the argument used above gives matrices $\{a_{ij}\}$ and $\{b_{ij}\}$ in $M_k(B)$ and $M_k(C)$, respectively, such that

$$\|m\|^2 < e_i, e_j > \; - \; < m(e_i), m(e_j) > \; = \sum_l a_{li}^* a_{lj}$$

and

$$< n(f_i), n(f_j) > \; = \sum_l b_{li}^* b_{lj}, \quad i,j = 1,2,3,\ldots,k.$$

Thus

$$< m \otimes_C n(x), m \otimes_C n(x) > \; = \sum_{ij} < m(e_i), m(e_j) > \otimes < n(f_i), n(f_j) >$$

$$= \sum_{ij} \|m\|^2 < e_i, e_j > \otimes < n(f_i), n(f_j) >$$

$$- \sum_{ij} \sum_l a_{li}^* a_{lj} \otimes < n(f_i), n(f_j) >$$

$$= \sum_{ij} \|m\|^2 < e_i, e_j > \otimes < n(f_i), n(f_j) >$$

$$- \sum_{lk} \left(\sum_i a_{li} \otimes b_{ki} \right)^* \left(\sum_i a_{li} \otimes b_{ki} \right)$$

$$\leq \sum_{ij} \|m\|^2 < e_i, e_j > \otimes < n(f_i), n(f_j) >$$

$$= (*).$$

There are also matrices $\{c_{ij}\}$ and $\{d_{ij}\}$ in $M_k(C)$ and $M_k(B)$, respectively, such that

$$\|n\|^2 < f_i, f_j > \; - \; < n(f_i), n(f_j) > \; = \sum_l c_{li}^* c_{lj}$$

and

$$\|m\|^2 < e_i, e_j > \; = \sum_l d_{li}^* d_{lj}, \quad i,j = 1,2,3,\ldots,k.$$

Thus

$$(*) = \sum_{ij} \|m\|^2 \|n\|^2 < e_i, e_j > \otimes < f_i, f_j >$$

$$- \sum_{ij} \sum_l |m|^2 < e_i, e_j > \otimes c_{li}^* c_{lj}$$

$$= \sum_{ij} \|m\|^2 |n|^2 < e_i, e_j > \otimes < f_i, f_j >$$

$$- \sum_{lk} \left(\sum_i d_{ki} \otimes c_{li} \right)^* \left(\sum_i d_{ki} \otimes c_{li} \right)$$

$$\leq \|m\|^2 \|n\|^2 < x, x > .$$

This proves the assertion, and gives straightforwardly that there is a *-homomorphism $j : \mathcal{L}_B(E) \otimes \mathcal{L}_C(F) \to \mathcal{L}_{B \otimes C}(E \hat{\otimes} F)$ such that

$$j(m \otimes n)(e \hat{\otimes} f) = m(e) \hat{\otimes} n(f), m \in \mathcal{L}_B(E), n \in \mathcal{L}_C(F), e \in E, f \in F.$$

We shall use the notation $m \hat{\otimes} n = j(m \otimes n)$ for the values of j on the simple tensors. We remark next that j is injective and maps the subalgebra $\mathcal{K}_B(E) \otimes \mathcal{K}_C(F)$ of $\mathcal{L}_B(E) \otimes \mathcal{L}_C(F)$ onto $\mathcal{K}_{B \otimes C}(E \hat{\otimes} F)$ (here and in the rest of the book we consider only spatial tensor products of C^*-algebras). To prove these assertions, consider a product state $\psi \otimes \phi$ of $B \otimes C$. Using the notation of Remark 1.1.8 we find that $\mathcal{H}_{\psi \otimes \phi} = \mathcal{H}_\psi \otimes \mathcal{H}_\phi$ and $\pi_{\psi \otimes \phi} \circ j = \pi_\psi \otimes \pi_\phi$. Let π denote the direct sum of all representations π_λ of $\mathcal{L}_{B \otimes C}(E \hat{\otimes} F)$ where λ vary over all product states of $B \otimes C$. π_E is the representation of $\mathcal{L}_B(E)$ obtained as the direct sum of all π_λ where λ vary over all states of B. π_F is defined similarly. Then $\pi \circ j$ is unitarily equivalent to $\pi_E \otimes \pi_F$. Since π_E and π_F are faithful it follows that $\pi_E \otimes \pi_F$ is too. Hence j is injective. To show that j maps $\mathcal{K}_B(E) \otimes \mathcal{K}_C(F)$ onto $\mathcal{K}_{B \otimes C}(E \hat{\otimes} F)$ it suffices to check that $j(\Theta_{e_1, e_2} \otimes \Theta_{f_1, f_2}) = \Theta_{e_1 \hat{\otimes} f_1, e_2 \hat{\otimes} f_2}$ for all $e_1, e_2 \in E, f_1, f_2 \in F$. This follows immediately by checking on a simple tensor.

We shall need a few simple facts about the constructions 1.2.1–1.2.4. Some are proved in the following and we leave others as exercises.

Lemma 1.2.5. *Let E be a Hilbert B-module and $f : B \to A$ a surjective *-homomorphism. Then $E_f \approx E \otimes_f A$.*

Proof. We use the notation from 1.2.2 and 1.2.3. There is a unique linear map $U : E \otimes_B A \to E'_f$ given on simple tensors by $U(x \otimes_B a) = q(x)a$, $x \in E, a \in A$. This gives a linear map from $E \otimes_B A$ to E'_f because $q(x)f(b)a = q(xb)a$, $x \in E, b \in B, a \in A$. It is then clearly an A-module map. Note that

$$\begin{aligned}
< x \otimes_B a, y \otimes_B b > &= a^* f(< x, y >)b \\
&= < q(x)a, q(y)b > \\
&= < U(x \otimes_B a), U(y \otimes_B b) >, \ x, y \in E, \ a, b \in A.
\end{aligned}$$

It follows that U induces a right A-module map $U' : E \otimes_B A / N_{EA} \to E'_f$ which preserves the "inner products". U' extends by continuity to a right

A-module map from $E \otimes_f A$ into E_f, a map which still preserves the "inner products". It now suffices to see that the extension of U' is surjective. If A has a unit 1, this is clear since $U(x \otimes_B 1) = q(x)$, $x \in E$. In general it follows from Lemma 1.1.4 by using an approximate unit.   □

Lemma 1.2.6. *Let $s : \mathbb{C} \to M(B)$ be the *-homomorphism which let $\mathbb{C}$ act by scalar multiplication on B, i.e. $s(\lambda)b = \lambda b$, $\lambda \in \mathbb{C}$, $b \in B$. Then $H_B \approx \mathcal{H} \otimes_s B$.*

Proof. There is a unique linear map $U : \mathcal{H} \otimes_{\mathbb{C}} B \to H_B$ given on simple tensors $\psi \otimes_{\mathbb{C}} b$, $\psi = (\lambda_1, \lambda_2, \ldots) \in \mathcal{H} = l^2$, $b \in B$, by $U(\psi \otimes_{\mathbb{C}} B) = (\lambda_1 b, \lambda_2 b, \ldots)$. Since U preserves the "inner products", it induces a B-module map from $\mathcal{H} \otimes_s B$ to H_B which also preserves the "inner products" and must be surjective since already U has dense range.   □

Lemma 1.2.7. *There is a *-isomorphism, $\mathcal{L}_B(H_B) \simeq M(\mathcal{K} \otimes B)$, mapping $\mathcal{K}_B(H_B)$ onto $\mathcal{K} \otimes B$.*

Proof. By Lemma 1.1.14 it suffices to show that $\mathcal{K}_B(H_B) \simeq \mathcal{K} \otimes B$. Thus, by Lemmas 1.1.19 and 1.2.6, it suffices to show that $\mathcal{K}_B(\mathcal{H} \otimes_s B) \simeq \mathcal{K} \otimes B$. Define $\pi_1 : \mathcal{K} \to \mathcal{L}_B(\mathcal{H} \otimes_s B)$ and $\pi_2 : B \to \mathcal{L}_B(\mathcal{H} \otimes_s B)$ by $\pi_1(k)\psi \otimes_s b = k(\psi) \otimes_s b$ and $\pi_2(c)\psi \otimes_s b = \psi \otimes_s cb$, $k \in \mathcal{K}$, $\psi \in \mathcal{H}$, $c, b \in B$. Then π_1 and π_2 are commuting representations so they induce a *-homomorphism $\lambda : \mathcal{K} \otimes B \to \mathcal{L}_B(\mathcal{H} \otimes_s B)$. Note that $\lambda(\Theta_{\psi,\phi} \otimes bc^*) = \Theta_{\psi \otimes_s b, \phi \otimes_s c}$, $\psi, \phi \in \mathcal{H}$, $b, c \in B$, so that λ maps onto $\mathcal{K}_B(\mathcal{H} \otimes_s B)$.

To show that λ is injective we use the representations introduced in Remark 1.1.8. Consider a state ϕ of B and let π_ϕ be the corresponding representation of $\mathcal{L}_B(\mathcal{H} \otimes_s B)$ on $\mathcal{H}_\phi$. Note that $\mathcal{H}_\phi = \mathcal{H} \otimes \mathcal{H}'_\phi$, where $\mathcal{H}'_\phi$ is the Hilbert space of the GNS-representation π'_ϕ of B associated to ϕ. Note also that $\pi_\phi \circ \lambda = id_\mathcal{K} \otimes \pi'_\phi$. Let $\pi = \sum_\phi^\oplus \pi_\phi$ and $\pi' = \sum_\phi^\oplus \pi'_\phi$. Then the representation $\pi \circ \lambda$ of $\mathcal{K} \otimes B$ is unitarily equivalent to $id_\mathcal{K} \otimes \pi'$. Since $id_\mathcal{K} \otimes \pi'$ is a faithful representation of $\mathcal{K} \otimes B$, this shows that λ is injective.

We remark that this argument is identical with the one used in the end of 1.2.4. An alternative proof is obtained by combining Lemma 1.1.14 with the result of E 1.2.5 (i).   □

Lemma 1.2.8. *Let E be a Hilbert B-module, $f : B \to A$ a *-homomorphism. Then $m \otimes_f id \in \mathcal{K}_A(E \otimes_f A)$ when $m \in \mathcal{K}_B(E)$.*

Proof. Let $x, y \in E$. Then $\Theta_{x,y} \otimes_f id(e \otimes_f a) = x < y, e > \otimes_f a$, $e \in E$, $a \in A$. Let $\{u_i\}$ be an approximate unit in B and set $v_i = \sqrt{f(u_i)}$. Then

$$\Theta_{x \otimes_f v_i, y \otimes_f v_i}(e \otimes_f a) = x \otimes_f v_i < y \otimes_f v_i, e \otimes_f a >$$

$$= (x \otimes_f v_i) v_i f(< y, e >) a$$

$$= x \otimes_f f(u_i < y, e >) a$$

$$= x < y u_i, e > \otimes_f a$$

$$= \Theta_{x, y u_i} \otimes_f id(e \otimes_f a), \quad e \in E, a \in A.$$

Thus $\Theta_{x, y u_i} \otimes_f id = \Theta_{x \otimes_f v_i, y \otimes_f v_i}$ for all i. From Lemma 1.1.4 it follows that $\Theta_{x, y u_i} \to \Theta_{x,y}$ in norm. Hence $\Theta_{x,y} \otimes_f id$ is the norm limit of $\{\Theta_{x \otimes_f v_i, y \otimes_f v_i}\}$, showing that $\Theta_{x,y} \otimes_f id$ is in $\mathcal{K}_B(E \otimes_f A)$. $\qquad\square$

Definition 1.2.9. A *graded* C^*-algebra is a C^*-algebra B equipped with an order two *-automorphism β_B, i.e β_B is a *-automorphism such that $\beta_B^2 = id$. We call β_B the *grading automorphism* for B and say that B is graded by β_B.

When B is graded by β_B, B decomposes into the eigenspaces for β_B, i.e. $B = B_0 \oplus B_1$ where $B_0 = \{b \in B : \beta_B(b) = b\}$ and $B_1 = \{b \in B : \beta_B(b) = -b\}$. This is a Banach space decomposition, not a decomposition of C^*-algebras; while B_0 is a C^*-algebra, B_1 is not. An element which is in either B_0 or B_1 is called *homogeneous*, an element x of B_0 is said to be of *degree* 0 and we write $\deg(x) = 0$. Similarly, an element x of B_1 is said to be of *degree* 1 and we write $\deg(x) = 1$.

A *graded homomorphism* $\phi : A \to B$ between graded C^*-algebras is a *-homomorphism satisfying $\phi \circ \beta_A = \beta_B \circ \phi$. Unless otherwise indicated a *-homomorphism between graded C^*-algebras is assumed to be graded.

The *graded commutator* of B is then defined to be the unique bilinear map $[\cdot, \cdot] : B \times B \to B$ satisfying $[a, b] = ab - (-1)^{ij} ba$, $a \in B_i$, $b \in B_j$, $i, j \in \{0, 1\}$. It is a trivial and tedious task to check that the graded commutator satisfies the following relations for all $x, y, z \in B$:

$$(1.2.1) \qquad [x, y] = -(-1)^{deg(x)deg(y)}[y, x],$$

$$(1.2.2) \qquad [x, yz] = [x, y]z + (-1)^{deg(x)deg(y)}y[x, z],$$

$$(1.2.3) \qquad (-1)^{deg(x)deg(z)}[[x, y], z] + (-1)^{deg(x)deg(y)}[[y, z], x]$$

$$+ (-1)^{deg(y)deg(z)}[[z, x], y] = 0.$$

Here and in the following we have adopted the habit of writing formulas for the graded commutator only for homogeneous elements. Their effect for general elements in $B = B_0 + B_1$ follows from the linearity in both entries.

Definition 1.2.10. A *graded Hilbert B*-module over the graded C^*-algebra B is a Hilbert B-module equipped with a linear bijection $S_E : E \rightarrow E$, called the *grading operator*, satisfying

$$(1.2.4) \qquad S_E(\psi b) = S_E(\psi)\beta_B(b), \quad \psi \in E, b \in B,$$

$$(1.2.5) \qquad < S_E\psi, S_E\phi > = \beta_B(< \psi, \phi >), \quad \psi, \phi \in E, \text{ and}$$

$$(1.2.6) \qquad S_E^2 = id.$$

Then we have $E = E_0 \oplus E_1$ with $E_0 = \{\psi \in E : S_E\psi = \psi\}$ and $E_1 = \{\psi \in E : S_E\psi = -\psi\}$. As in the case of graded C^*-algebras, we call the elements of $E_0 \cup E_1$ *homogeneous*, and say that the elements of E_0 (resp. E_1) have *degree 0* (resp. *degree 1*).

Note that (1.2.4) and (1.2.5) imply

$$(1.2.7) \qquad E_i B_j \subseteq E_{i+j}, \qquad \text{and}$$

$$(1.2.8) \qquad < E_i, E_j > \subseteq B_{i+j}, \quad i, j \in \{0, 1\}.$$

Here and in the following we consider $\{0, 1\}$ as an abelian group with 0 as a neutral element and composition $+$, i.e. as $\mathbb{Z}_2$. Note that condition (1.2.5) implies that $\|S_E\| \leq 1$.

A *countably generated* graded Hilbert B-module is a graded Hilbert B-module which is countably generated as a Hilbert B-module.

Examples 1.2.11.

(a)   Any C^*-algebra B can be graded by taking $\beta_B = id$, *the trivial grading.* Any Hilbert B-module can then be graded by taking $S_E = id$.

(b)   Given any C^*-algebra B, we can define an order two $*$-auto-morphism $\beta_{B \oplus B}$ of $B \oplus B$ by $\beta_{B \oplus B}(a, b) = (b, a)$, $a, b \in B$. This grading of $B \oplus B$ is called *the odd grading*. We let $B_{(1)}$ denote the graded C^*-algebra obtained this way.

(c)   If B is a graded C^*-algebra then B is at the same time a graded Hilbert B-module with $S_B = \beta_B$.

(d)   If E is a graded Hilbert B-module, then $T \to S_E T S_E^{-1}$, $T \in \mathcal{L}_B(E)$, defines an order two automorphism β_E of $\mathcal{L}_B(E)$, and in this way E induces a grading on $\mathcal{L}_B(E)$, *the induced grading*. Since $\beta_E(\mathcal{K}_B(E)) = \mathcal{K}_B(E)$, we obtain a grading of $\mathcal{K}_B(E)$ also. Unless otherwise indicated a grading on $\mathcal{K}_B(E)$ or $\mathcal{L}_B(E)$ will always be the grading induced from the grading of E in this way. When $E = B$ the induced grading on $\mathcal{M}(B)$ is given by the extension, $\underline{\beta}_B \in \mathrm{Aut}\,(\mathcal{M}(B))$, of β_B.

(e)   If E and F are graded Hilbert B-modules, the direct sum $E \oplus F$ is graded by the direct sum operator $S_E \oplus S_F$ defined by $S_E \oplus S_F(e, f) = (S_E(e), S_F(f))$, $(e, f) \in E \oplus F$. Unless something else is explicitly indicated $E \oplus F$ will always be assumed to be graded in this way.

In the development of KK-theory we shall actually only make use of the trivial and the odd grading.

We shall need the graded version of Kasparov's stabilization theorem, Theorem 1.1.24. For this observe that the Hilbert B-module H_B carries two natural but different gradings when B is graded by β_B. The first is given by the symmetry S defined by $S(b_1, b_2, b_3, \ldots) = (\beta_B(b_1), \beta_B(b_2), \beta_B(b_3), \ldots)$, for $(b_1, b_2, b_3, \ldots) \in H_B$, and the second is given by $-S$. We let $\hat{H}_B$ denote the Hilbert B-module $H_B \oplus H_B$ graded by $S \oplus -S$. Two graded Hilbert B-modules E, F are called *isomorphic* when there is an isomorphism $\psi : E \to F$ of Hilbert B-modules such that $\psi \circ S_E = S_F \circ \psi$. As in the ungraded case we write $E \approx F$ in this situation. With these conventions Kasparov's stabilization theorem takes the following form in the graded case.

Theorem 1.2.12.   *Let E be a countably generated graded Hilbert B-module. Then $E \oplus \hat{H}_B \approx \hat{H}_B$.*

Proof.   The proof is just a minor modification of the proof of Theorem 1.1.24. The reduction to the unital case is done as in that proof; the grading of $\hat{B}$ (and hence of $\hat{H}_{\hat{B}}$) is given by the unique automorphic extension of β_B. The generating set $\{\eta_i\}$ for E is chosen in the same way and then decomposed $\eta_i = \eta_i^0 + \eta_i^1$ where $deg(\eta_i^j) = j$, $j = 0, 1$. When $\{\epsilon_i\}$ is the basis for H_B described in the proof of Theorem 1.1.24, then we set $\epsilon_i^0 = (\epsilon_i, 0) \in \hat{H}_B$ and $\epsilon_i^1 = (0, \epsilon_i) \in \hat{H}_B$. We define $T : \hat{H}_B \to E \oplus \hat{H}_B$ to be the element in $\mathcal{L}_B(\hat{H}_B, E \oplus \hat{H}_B)$ given by

$$T(\epsilon_i^j) = (2^{-i}\eta_i^j, 4^{-i}\epsilon_i^j), \; i \in \mathbb{N}, \; j = 0, 1.$$

The proof that the unitary part in the polar decomposition of T gives an isomorphism of Hilbert B-modules can then proceed essentially unchanged. Since T maps degree j elements of $\hat{H}_B$ into degree j elements of $E \oplus \hat{H}_B$, $j \in \{0, 1\}$, it follows that the unitary part V of T intertwines the grading operators. Hence V is an isomorphism of graded Hilbert B-modules. $\qquad\square$

We shall need the following graded version of Lemma 1.2.6.

Lemma 1.2.13. *Let T be a symmetry (i.e., a selfadjoint unitary) acting on $\mathcal{H}$. Assume that both eigenspaces of T are infinite dimensional. Let $\mathcal{H} \otimes_s B$ be graded by the grading operator $T \otimes_s \beta_B$ given on simple tensors by $T \otimes_s \beta_B(\psi \otimes_s b) = T\psi \otimes_s \beta_B(b), \psi \in \mathcal{H}, b \in B$.*
Then $\hat{H}_B \approx \mathcal{H} \otimes_s B$.

Proof. Let $\psi_1, \psi_2, \psi_3, \ldots$ be an orthonormal basis for $\mathcal{H}$ such that $T\psi_i = (-1)^i \psi_i, \; i \in \mathbb{N}$. Define $U : \mathcal{H} \otimes_s B \to \hat{H}_B = H_B \oplus H_B$ on simple tensors by $U(\sum_i \lambda_i \psi_i \otimes b) = ((\lambda_2 b, \lambda_4 b, \ldots), (\lambda_1 b, \lambda_3 b, \ldots))$, $b \in B, \lambda_i \in \mathbb{C}, \sum_i \|\lambda_i\|^2 < \infty$. U gives an isomorphism of Hilbert B-modules and since $U \circ (T \otimes_s \beta_B) = (S \oplus -S) \circ U$, the proof is complete. $\qquad\square$

1.2.14. *Notes and remarks.*

Practically all the material in this section originates from Kasparov's paper [19].

Exercise 1.2

E 1.2.1

Let $E_1, E_2, E_3, \ldots, E_n$ be Hilbert B-modules and let $a_{ij}, \; i, j = 1, 2, \ldots, n$, be an $n \times n$ tuple of operators, $a_{ij} \in \mathcal{L}_B(E_j, E_i)$.

Define $A \in \mathcal{L}_B(E_1 \oplus E_2 \oplus \cdots \oplus E_n)$ by

$$A(e_1, e_2, \ldots, e_n) = \Big(\sum_j a_{1j}e_j, \sum_j a_{2j}e_j, \ldots, \sum_j a_{nj}e_j\Big),$$

$$e_i \in E_i, i = 1, 2, \ldots, n.$$

It is natural to consider A as an n by n matrix whose ij-th entry is a_{ij}.

(i) Show that every $A \in \mathcal{L}_B(E_1 \oplus E_2 \oplus \cdots \oplus E_n)$ is obtained in this way.

(ii) Show that $A \in \mathcal{K}_B(E_1 \oplus E_2 \oplus \cdots \oplus E_n)$ if and only if $a_{ij} \in \mathcal{K}_B(E_j, E_i)$ for all $i, j = 1, 2, \ldots, n$.

(iii) Describe the product and involution $*$ in $\mathcal{L}_B(E_1 \oplus E_2 \oplus \cdots \oplus E_n)$ in terms of $\mathcal{L}_B(E_i, E_j)$, $i, j = 1, 2, \ldots, n$.

(iv) Show that there is a commutative diagram

$$\begin{array}{ccc} M_2(\mathcal{M}(B)) & \simeq & \mathcal{L}_B(B \oplus B) \\ \cup & & \cup \\ M_2(B) & \simeq & \mathcal{K}_B(B \oplus B). \end{array}$$

E 1.2.2

Let J be a selfadjoint subset of B. Prove that $\mathcal{K}_B(\overline{JB}) \simeq \overline{JBJ}$ as C^*-algebras, that $\mathcal{K}_B(B, \overline{JB}) \simeq \overline{JB}$ as Banach spaces and that

$$\begin{bmatrix} \overline{JBJ} & \overline{JB} \\ \overline{BJ} & B \end{bmatrix} \simeq \mathcal{K}_B(\overline{JB} \oplus B)$$

as C^*-algebras.

E 1.2.3

Describe all possible gradings of $M_2(\mathbb{C})$ up to $*$-isomorphism of graded C^*-algebras. Do the same with $M_2(\mathbb{C}) \oplus M_2(\mathbb{C})$.

E 1.2.4

Let $B_{(1)} = B \oplus B$ have the odd grading. Show that the grading automorphism on $\mathcal{M}(B \oplus B)$ for the induced grading is described by $(m_1, m_2) \rightarrow (m_2, m_1)$, $m_1, m_2 \in \mathcal{M}(B)$, on $\mathcal{M}(B) \oplus \mathcal{M}(B)$ under the isomorphism $\mathcal{M}(B \oplus B) \simeq \mathcal{M}(B) \oplus \mathcal{M}(B)$ established in E 1.1.2(i).

E 1.2.5

For $b \in B, i \in \mathbb{N}$, let $\tilde{b}_i \in H_B$ denote the element

$$\tilde{b}_i = (0, 0, 0, \ldots, 0, b, 0, 0, \ldots)$$
$$\uparrow$$
$$i\text{-th coordinate}$$

Fix a full system $\{e_{ij} : i, j \in \mathbb{N}\}$ of matrix units in $\mathcal{K}$. Thus $e_{ij}^* = e_{ji}$, $e_{ij}e_{kl} = \delta(j,k)e_{il}$, $i, j, k, l \in \mathbb{N}$, and the span of $\{e_{ij}\}$ is dense in $\mathcal{K}$.

(i) Prove that there is a unique *-isomorphism $\Psi_B : \mathcal{K} \otimes B \to \mathcal{K}_B(H_B)$ such that $\Psi_B(e_{ij} \otimes bc^*) = \Theta_{\tilde{b}_i, \tilde{c}_j}$, $b, c \in B$, $i, j \in \mathbb{N}$.

Set $IB = C[0,1] \otimes B$ and let $\pi_t : IB \to B$ be evaluation at $t \in [0,1]$. Define $\phi_t : H_{IB} \to H_B$ by

$$\phi_t(f_1, f_2, \ldots) = (\pi_t(f_1), \pi_t(f_2), \ldots), \qquad (f_1, f_2, \ldots) \in H_{IB}.$$

Let $\psi_t : \mathcal{K}_{IB}(H_{IB}) \to \mathcal{K}_B(H_B)$ be the *-homomorphism given by Proposition 1.1.28.

(ii) Show that the diagram

$$
\begin{array}{ccc}
 & \Psi_{IB} & \\
\mathcal{K} \otimes IB & \longrightarrow & \mathcal{K}_{IB}(H_{IB}) \\
{\scriptstyle id_{\mathcal{K}} \otimes \pi_t} \downarrow & & \downarrow {\scriptstyle \psi_t} \\
\mathcal{K} \otimes B & \longrightarrow & \mathcal{K}_B(H_B) \\
 & \Psi_B &
\end{array}
$$

commutes.

1.3. Stable C^*-Algebras

Definition 1.3.1. A C^*-algebra is *stable* when $B \otimes \mathcal{K}$ is *-isomorphic to B.

Lemma 1.3.2. *If B is stable, then $H_B \approx B$.*

Proof. We assert that there is a sequence $V_i, i = 1, 2, 3, \ldots$, of isometries in $\mathcal{L}_B(H_B)$ such that $V_i^* V_j = 0$, $i \neq j$, and $\sum_i V_i V_i^* = 1$ in the strict topology. To see this we proceed exactly as if $B = \mathbb{C}$. Let N_i, $i = 1, 2, 3, \ldots$ be a partition of $\mathbb{N}$ into infinite subsets and let $\phi_i : N_i \to \mathbb{N}$ be bijections. Define $V_i : H_B \to H_B$ by

$$V_i(b_1, b_2, b_3, \ldots)_j = \begin{cases} b_{\phi_i(j)} & \text{if } j \in N_i \\ 0 & \text{otherwise} \end{cases} \qquad (b_1, b_2, b_3, \ldots) \in H_B.$$

We leave the reader to check that $V_i \in \mathcal{L}_B(H_B)$ are isometries with the desired properties.

By combining Lemmas 1.1.17 and 1.2.7 we get isometries U_i, $i \in \mathbb{N}$, in $\mathcal{M}(\mathcal{K} \otimes B)$ such that $U_i^* U_j = 0$, $i \neq j$, and $\sum_k U_k U_k^* = 1$ in the strict topology of $\mathcal{M}(\mathcal{K} \otimes B)$. Since B is stable, Lemma 1.1.14 gives us isometries W_i, $i = 1, 2, \ldots$, in $\mathcal{M}(B)$ with the same properties. Define now $\phi : H_B \to B$ by

$$\phi(b_1, b_2, b_3, \ldots) = \sum_i W_i b_i, \qquad (b_1, b_2, b_3, \ldots) \in H_B.$$

We leave the reader to check that ϕ is an isomorphism of Hilbert B-modules.
$\square$

We also have an "odd graded" version of Lemma 1.3.2.

Lemma 1.3.3. *Assume that B is a stable C^*-algebra. Then $\hat{H}_{B_{(1)}} \approx B_{(1)}$ (as graded Hilbert $B_{(1)}$-modules). Furthermore, when $\mathcal{M}(B) \oplus \mathcal{M}(B)$ is given the odd grading, there is a *-isomorphism $\phi : \mathcal{M}(B)_{(1)} = \mathcal{M}(B) \oplus \mathcal{M}(B) \to \mathcal{M}(B_{(1)})$ of graded C^*-algebras such that the diagram*

$$
\begin{array}{ccc}
& \phi & \\
\mathcal{M}(B) \oplus \mathcal{M}(B) & \to & \mathcal{M}(B_{(1)}) \\
\cup & & \cup \\
B \oplus B & = & B_{(1)}
\end{array}
$$

commutes.

Proof.   As in the proof of Lemma 1.3.2 we can find a sequence W_i, $i = 1, 2, 3, \ldots$ of isometries in $\mathcal{M}(B)$ such that $W_i^* W_j = 0$ $i \neq j$, and $\sum_i W_i W_i^* = 1$ in the strict topology. Let

$$T : \hat{H}_{B_{(1)}} = H_{B \oplus B} \oplus H_{B \oplus B} \to B_{(1)} \oplus B_{(1)} = (B \oplus B) \oplus (B \oplus B)$$

be defined by

$$T((a_1, b_1), (a_2, b_2), \ldots; \ (c_1, d_1), (c_2, d_2), \ldots) =$$

$$= \left((\sum_i W_i a_i, \sum_i W_i b_i); (\sum_i W_i c_i, \sum_i W_i d_i) \right).$$

It is easy to check that T is an isomorphism of graded Hilbert $B_{(1)}$-modules when the first summand of $B_{(1)} \oplus B_{(1)}$ is graded by $\beta_{B \oplus B}$ and the second by $-\beta_{B \oplus B}$.

Take isometries V_1, V_2 in $\mathcal{M}(B)$ such that $V_1 V_1^* + V_2 V_2^* = 1$ and $V_1^* V_2 = 0$. V_1 and V_2 can be found in the same way as the W_i's were found in the proof of Lemma 1.3.2. Define $S : B_{(1)} \oplus B_{(1)} \to B_{(1)}$ by $S((a, b), (c, d)) = (V_1 a + V_2 c, V_1 b - V_2 d)$. It is straightforward to check that S defines an isomorphism of graded Hilbert $B_{(1)}$ -modules. Thus ST gives the desired isomorphism $\hat{H}_{B_{(1)}} \approx B_{(1)}$.

Define $\phi : \mathcal{M}(B) \oplus \mathcal{M}(B) \to \mathcal{M}(B_{(1)})$ by $\phi(m_1, m_2)(a, b) = (m_1(a), m_2(b))$, $m_1, m_2 \in \mathcal{M}(B)$, $a, b \in B$. Then ϕ is clearly an injective *-homomorphism. Let $m \in \mathcal{M}(B_{(1)})$. Then $m(B \oplus 0) \subseteq B \oplus 0$ and $m(0 \oplus B) \subseteq 0 \oplus B$ since m is a $B \oplus B$-module map. So there are multipliers, $m_1, m_2 \in \mathcal{M}(B)$ such that $m(b, 0) = (m_1 b, 0)$ and $m(0, b) = (0, m_2 b)$, $b \in B$. Then $m = \phi(m_1, m_2)$, showing that ϕ is also surjective. Since $\phi(B \oplus B) = B_{(1)}$ we have the desired commutative diagram.   □

Definition 1.3.4.   Let B be a C^*-algebra and $p \in \mathcal{M}(B)$ a projection with complement $p^\perp = 1 - p$. Then p is called *fully complemented* if $p^\perp B \approx B$.

Theorem 1.3.5.   *Let B be a stable C^*-algebra and E a countably generated Hilbert B-module. Then there is a fully complemented projection $p \in \mathcal{M}(B)$ such that $E \approx pB$.*

If $q \in \mathcal{M}(B)$ is another fully complemented projection such that $E \approx qB$, then $q = upu^*$ for some unitary $u \in \mathcal{M}(B)$.

Proof. By Kasparov's stabilization theorem, Theorem 1.1.24, and Lemma 1.3.2 we have that $E \oplus B \approx B$. Let $\psi : E \oplus B \to B$ be an isomorphism, and let $f \in \mathcal{L}_B(E \oplus B)$ be the projection onto E, i.e. $f(e, b) = (e, 0)$, $e \in E$, $b \in B$. Set $p = \psi f \psi^{-1}$. Then $pB = \psi f(E \oplus B) = \psi(E \oplus 0) \approx E$ and $p^\perp B = \psi f^\perp(E \oplus B) = \psi(0 \oplus B) \approx B$. This proves the existence part. If $q \in \mathcal{M}(B)$ is another fully complemented projection such that $qB \approx E$, then $pB \approx qB$ and $p^\perp B \approx q^\perp B$. Consequently there is an automorphism u of $B = pB + p^\perp B = qB + q^\perp B$ (as a Hilbert B-module) such that $u(pB) = qB$. u is a unitary in $\mathcal{M}(B)$ such that $upu^* = q$. $\qquad\square$

When B is σ-unital with strictly positive element $h \in B$ and $p \in \mathcal{M}(B)$ is a projection, then $\{ph\}$ is a set of generators for pB by Lemma 1.1.21. In particular, pB is countably generated. Thus if B is both stable and σ-unital, Theorem 1.3.5 gives a complete description and classification of all countably generated Hilbert B-modules in terms of projections in $\mathcal{M}(B)$.

Lemma 1.3.6. *Let B be a stable C^*-algebra. Then there is a path $\{v_t : t \in]0, 1]\}$ of isometries in $\mathcal{M}(B)$ such that*

(i) *the map $t \to v_t$ is strictly continuous,*

(ii) $v_1 = 1$, *and*

(iii) $\lim_{t \to 0} v_t v_t^* = 0$ *in the strict topology.*

Proof. There is a strong * continuous path w_t, $t \in]0, 1]$, of isometries on a separable infinite dimensional Hilbert space $\mathcal{H}$ such that $w_1 = 1$ and $w_t w_t^* \to 0$ strongly as $t \to 0$. For example such a path can be defined on $\mathcal{H} = L^2[0, 1]$ by

$$
w_t f(s) = \begin{cases} t^{-\frac{1}{2}} f(s/t), & s \in [0, t] \\ 0 & s \in]t, 1] \end{cases} \qquad f \in L^2[0, 1], t \in]0, 1].
$$

Set $u_t = w_t \otimes_s 1 \in \mathcal{L}_B(\mathcal{H} \otimes_s B)$, $t \in]0, 1]$. Then u_t, $t \in]0, 1]$, is a strictly continuous path of isometries in $\mathcal{L}_B(\mathcal{H} \otimes_s B)$ such that $u_1 = 1$ and $u_t u_t^* \to 0$ strictly as $t \to 0$. This is easily seen by checking on simple tensors in $\mathcal{H} \otimes_s B$.

By combining Lemmas 1.2.6, 1.1.19 and 1.2.7 we get a commutative diagram

$$\begin{array}{ccc} \mathcal{L}_B(\mathcal{H} \otimes_s B) & \simeq & \mathcal{M}(\mathcal{K} \otimes B) \\ \cup & & \cup \\ \mathcal{K}_B(\mathcal{H} \otimes_s B) & \simeq & \mathcal{K} \otimes B. \end{array}$$

Since B is stable Lemma 1.1.14 gives a commutative diagram

$$\begin{array}{ccc} \mathcal{M}(\mathcal{K} \otimes B) & \simeq & \mathcal{M}(B) \\ \cup & & \cup \\ \mathcal{K} \otimes B & \simeq & B. \end{array}$$

In combination this gives *-isomorphisms $\psi_1 : \mathcal{L}_B(\mathcal{H} \otimes_s B) \to \mathcal{M}(B)$ and $\psi_2 : \mathcal{K}_B(\mathcal{H} \otimes_s B) \to B$ such that

$$\begin{array}{ccc} & \psi_1 & \\ \mathcal{L}_B(\mathcal{H} \otimes_s B) & \to & \mathcal{M}(B) \\ \cup & & \cup \\ \mathcal{K}_B(\mathcal{H} \otimes_s B) & \to & B \\ & \psi_2 & \end{array}$$

commutes. Set $v_t = \psi_1(u_t), t \in {]0,1]}$. Then the path $\{v_t : t \in {]0,1]}\}$ satisfies (i), (ii) and (iii) by Lemma 1.1.17. □

Lemma 1.3.7. *Let B be a stable C^*-algebra and $w \in \mathcal{M}(B)$ an isometry (a unitary). Then there is a strictly continuous path w_t, $t \in [0,1]$, of isometries (unitaries) such that $w_0 = 1, w_1 = w$.*

Proof. Let $\{v_t : t \in {]0,1]}\}$ be a path of isometries in $\mathcal{M}(B)$ with the properties stated in Lemma 1.3.6. Set $w_t = v_t w v_t^* + 1 - v_t v_t^*, t \in {]0,1]}$, and $w_0 = 1$. We leave the reader to check that this path has the desired properties, both when w is a unitary and when it is an isometry. □

Definition 1.3.8. Let B be a stable C^*-algebra. A *-isomorphism $\Theta : M_n(B) \to B$ is called *inner* when there are isometries $w_1, w_2, \ldots, w_n \in \mathcal{M}(B)$ such that $w_i^* w_j = 0$, $i \neq j$, $\sum_i w_i w_i^* = 1$ and

$$(1.3.1) \qquad \Theta(\{b_{ij}\}) = \sum_{i,j} w_i b_{ij} w_j^*, \qquad \{b_{ij}\} \in M_n(B).$$

If Θ is an inner *-isomorphism $\Theta : M_n(B) \to B$ given by the isometries $w_1, w_2, \ldots, w_n$, then there is also a *-isomorphism $M_n(\mathcal{M}(B)) \to \mathcal{M}(B)$ given by the same formula as (1.3.1). Since this *-isomorphism is an extension of Θ, we denote it by Θ again.

Lemma 1.3.9. *Let B be a stable C*-algebra. For each n there are inner *-isomorphisms, $M_n(B) \to B$, and if $\Theta_1, \Theta_2 : M_n(B) \to B$ are two inner *-isomorphisms, then there is a unitary $u \in \mathcal{M}(B)$ such that $\Theta_1 = Adu \circ \Theta_2$.*

Proof. By the same procedure as in the proof of Lemma 1.3.2 we can construct isometries $w_1, w_2, \ldots, w_n$ in $\mathcal{M}(B)$ satisfying the conditions of Definition 1.3.8. Thus inner *-isomorphisms, $M_n(B) \to B$, exist for all $n \in \mathbb{N}$. If Θ_1 is the inner *-isomorphism given by $w_1, w_2, w_3, \ldots, w_n$ and Θ_2 is another given $v_1, v_2, \ldots, v_n$, then $u = \sum_k w_k v_k^*$ is a unitary such that $\Theta_1 = Adu \circ \Theta_2$. $\square$

Definition 1.3.10.   Let A and B be C^*-algebras.   Two *-homomorphisms $\phi, \psi : A \to B$ are called *homotopic* when there is a path λ_t, $t \in [0, 1]$, of *-homomorphisms $\lambda_t : A \to B$ such that

(i) $t \to \lambda_t(a)$ is continuous for all $a \in A$, and

(ii) $\lambda_0 = \phi, \lambda_1 = \psi$.

The path λ_t, $t \in [0, 1]$, is called a homotopy from ϕ to ψ and we write $\phi \sim \psi$ when ϕ and ψ are homotopic.

Let $\mathrm{Hom}\,(A, B)$ denote the set of *-homomorphisms from A to B. It is clear that homotopy defines an equivalence relation on $\mathrm{Hom}\,(A, B)$. We denote the homotopy classes in $\mathrm{Hom}\,(A, B)$ by $[A, B]$, i.e. $[A, B] = \mathrm{Hom}\,(A, B)/\sim$. For $\phi \in \mathrm{Hom}\,(A, B)$ we let $[\phi]$ denote the homotopy class in $[A, B]$ containing ϕ. In the following we shall investigate $[A, B]$ in the particular case where B is stable.

Lemma 1.3.11.   *Assume that B is a stable C^*-algebra. Let Θ : $M_2(B) \to B$ be an inner *-isomorphism and let $j : B \to M_2(B)$ be the embedding into the upper left-hand corner, i.e. $j(x) = \begin{bmatrix} x & 0 \\ 0 & 0 \end{bmatrix}$, $x \in B$.*

Then $\Theta \circ j$ is homotopic to the identity map id_B on B.

Proof. By the definition of an inner *-isomorphism, there is an isometry $w \in \mathcal{M}(B)$ such that $\Theta \circ j(x) = wxw^*$, $x \in B$. By Lemma 1.3.7 there is a strictly continuous path w_t, $t \in [0, 1]$, of isometries in $\mathcal{M}(B)$ such that $w_0 = 1$ and $w_1 = w$. Define $\lambda_t \in \mathrm{Hom}\,(B, B)$ by $\lambda_t(x) = w_t x w_t^*$, $x \in B, t \in [0, 1]$. Then λ_t, $t \in [0, 1]$, is a homotopy from id_B to $\Theta \circ j$. $\square$

Let now B be a stable C^*-algebra and $\Theta_B : M_2(B) \to B$ an inner isomorphism. We can then define a composition $+$ in $[A, B]$ by

$$(1.3.2) \qquad [\phi] + [\psi] = \left[\Theta_B \circ \begin{bmatrix} \phi & 0 \\ 0 & \psi \end{bmatrix}\right], \quad \phi, \psi \in \text{Hom}\,(A, B).$$

Here $\begin{bmatrix} \phi & 0 \\ 0 & \psi \end{bmatrix} : A \to M_2(B)$ is the *-homomorphism sending $a \in A$ to $\begin{bmatrix} \phi(a) & 0 \\ 0 & \psi(a) \end{bmatrix} \in M_2(B)$.

Lemma 1.3.12. *Let B be a stable C^*-algebra and A any C^*-algebra. Then $[A, B]$ is an abelian semigroup with a 0 element represented by the zero homomorphism.*

Proof. That the zero homomorphism represents a neutral element for $+$ follows from Lemma 1.3.11.

Let

$$(1.3.3) \qquad R_t = \begin{bmatrix} \cos\frac{\pi}{2}t & \sin\frac{\pi}{2}t \\ -\sin\frac{\pi}{2}t & \cos\frac{\pi}{2}t \end{bmatrix} \in M_2(\mathcal{M}(B)), \quad t \in [0, 1].$$

When $\phi, \psi \in \text{Hom}\,(A, B)$ we can define $\lambda_t \in \text{Hom}\,(A, M_2(B))$ by

$$\lambda_t(a) = AdR_t \circ \begin{bmatrix} \phi(a) & 0 \\ 0 & \psi(a) \end{bmatrix}, \quad a \in A, t \in [0, 1].$$

Then $\lambda_t, t \in [0, 1]$, is a homotopy from $\begin{bmatrix} \phi & 0 \\ 0 & \psi \end{bmatrix}$ to $\begin{bmatrix} \psi & 0 \\ 0 & \phi \end{bmatrix}$. It follows that $[\phi] + [\psi] = [\psi] + [\phi]$.

To prove the associativity of $+$ we use that homomorphisms "on the diagonal" can be permuted freely by using rotation matrices similar to R_t, $t \in [0, 1]$. By using this and Lemma 1.3.11, we get the following string of homotopies with $\phi, \psi, \lambda \in \text{Hom}\,(A, B)$:

$$\Theta_B \circ \begin{bmatrix} \Theta_B \circ \begin{bmatrix} \phi & 0 \\ 0 & \psi \end{bmatrix} & 0 \\ 0 & \lambda \end{bmatrix} \sim \Theta_B \circ \begin{bmatrix} \Theta_B \circ \begin{bmatrix} \phi & 0 \\ 0 & \psi \end{bmatrix} & 0 \\ 0 & \Theta_B \circ \begin{bmatrix} \lambda & 0 \\ 0 & 0 \end{bmatrix} \end{bmatrix}$$

$$= \Theta_B \circ (id_{M_2(\mathbb{C})} \otimes \Theta_B) \circ \begin{bmatrix} \phi & 0 & 0 & 0 \\ 0 & \psi & 0 & 0 \\ 0 & 0 & \lambda & 0 \\ 0 & 0 & 0 & 0 \end{bmatrix}$$

$$\sim \Theta_B \circ (id_{M_2(\mathbb{C})} \otimes \Theta_B) \circ \begin{bmatrix} \phi & 0 & 0 & 0 \\ 0 & 0 & 0 & 0 \\ 0 & 0 & \psi & 0 \\ 0 & 0 & 0 & \lambda \end{bmatrix}$$

$$= \Theta_B \circ \begin{bmatrix} \Theta_B \circ \begin{bmatrix} \phi & 0 \\ 0 & 0 \end{bmatrix} & 0 \\ 0 & \Theta_B \circ \begin{bmatrix} \psi & 0 \\ 0 & \lambda \end{bmatrix} \end{bmatrix}$$

$$\sim \Theta_B \circ \begin{bmatrix} \phi & 0 \\ 0 & \Theta_B \circ \begin{bmatrix} \psi & 0 \\ 0 & \lambda \end{bmatrix} \end{bmatrix}.$$

Thus $([\phi] + [\psi]) + [\lambda] = [\phi] + ([\psi] + [\lambda])$.   □

It will be essential for us later to be able to extend certain *-homo-morphisms $A \to B$ to *-homomorphisms $\mathcal{M}(A) \to \mathcal{M}(B)$. In view of Corollary 1.1.15 it is therefore natural to consider the following class of *-homomorphisms.

Definition 1.3.13. A *-homomorphism $\phi : A \to B$ is called *quasi-unital* when there is a projection $p \in \mathcal{M}(B)$ such that $\overline{\phi(A)B} = pB$. We denote the set of quasi-unital *-homomorphisms in $\mathrm{Hom}\,(A, B)$ by $\mathrm{Hom}\,_q(A, B)$.

By Corollary 1.1.15 all quasi-unital *-homomorphisms $\phi : A \to B$ ad-mit a unique strictly continuous extension $\underline{\phi} : \mathcal{M}(A) \to \mathcal{M}(B)$. The pro-jection p appearing in Definition 1.3.13 is then $\underline{\phi}(1)$. We call it *the relative unit* for ϕ and denote it by p_ϕ. It is important to notice that the composi-tion of quasi-unital *-homomorphisms yields a quasi-unital homomorphism (cf. E 1.3.2). To introduce the notion of homotopy in $\mathrm{Hom}\,_q(A, B)$ we let IB denote the C^*-algebra $IB = C[0, 1] \otimes B$, identified as the C^*-algebra of continuous B-valued functions on $[0, 1]$, and let $\pi_t : IB \to B$ denote the *-homomorphisms obtained by evaluation at t, $t \in [0, 1]$.

Definition 1.3.14.   Two quasi-unital *-homomorphisms $\phi, \psi \in \mathrm{Hom}\,_q(A, B)$ are called *strongly homotopic* when there is a *-homomorphism $\lambda \in \mathrm{Hom}\,_q(A, IB)$ such that $\pi_0 \circ \lambda = \phi$ and $\pi_1 \circ \lambda = \psi$. We write $\phi \overset{\sim}{\to} \psi$ in this case and call λ a strong homotopy from ϕ to ψ.

This definition is justified by E 1.3.1. In contrast to the case of or-dinary homotopy in $\mathrm{Hom}\,(A, B)$, it is not obvious at a glance that strong homotopy defines an equivalence relation in $\mathrm{Hom}_q(A,B)$. This important fact follows from the following alternative description of $\overset{\sim}{}$.

Lemma 1.3.15. *Let* $f, g \in \mathrm{Hom}_q(A,B)$. *Then* $f \overset{\sim}{} g$ *if and only if there is a path* λ_t, $t \in [0,1]$, *in* $\mathrm{Hom}\,(\mathcal{M}(A), \mathcal{M}(B))$ *such that*

(i) λ_t *is strictly continuous for all* t,

(ii) $t \to \lambda_t(m)$ *is strictly continuous for all* $m \in \mathcal{M}(A)$,

(iii) $t \to \lambda_t(a)$ *is norm continuous from* $[0,1]$ *into* B *for all* $a \in A$, *and*

(iv) $\lambda_0 = \underline{f}$, $\lambda_1 = \underline{g}$.

Proof.   If such a path exists we can define $\lambda \in \mathrm{Hom}\,(A, IB)$ by $\lambda(a)(t) = \lambda_t(a)$, $a \in A$, $t \in [0,1]$. This is possible by condition (iii). We assert that λ is quasi-unital. To see this observe that by (ii) we have a projection $p \in \mathcal{M}(IB)$ given by $(pf)(t) = \lambda_t(1)f(t)$, $f \in IB$, $t \in [0,1]$. It is clear that $\overline{\lambda(A)IB} \subseteq pIB$. Let $z \in pIB$. For any $t_0 \in [0,1]$ we have that $\pi_{t_0}(z) \in \lambda_{t_0}(1)B$. Since λ_{t_0} is strictly continuous by condition (i), we have that $\lambda_{t_0}(1)B = \overline{\lambda_{t_0}(A)B}$. Thus we can find finite sets $\{a_i\} \subseteq A$ and $\{b_i\} \subseteq B$ such that $\left\| \sum_i \lambda_{t_0}(a_i)b_i - \pi_{t_0}(z) \right\| < \epsilon$.

But then, by (iii), this inequality extends to an open neighborhood of t_0 and, as $[0,1]$ is compact, it follows that we can find a finite open cover $\{U_j : j = 1, 2, \dots, N\}$ of $[0,1]$ and for each j finite subsets $\{a_i^j\} \subseteq A$ and $\{b_i^j\} \subseteq B$ such that $\left\| \sum_i \lambda_t(a_i^j)b_i^j - \pi_t(z) \right\| < \epsilon$ for all $j = 1, 2, \dots, N$. Let $\{g_j\}$ be a partition of unity subordinate to this cover of $[0,1]$. Define $c_i^j \in IB$ by $c_i^j(t) = g_j(t)b_i^j$, $t \in [0,1]$. Then $\left\| \sum_i \lambda(a_i^j)c_i^j - z \right\| < \epsilon$. It follows that $\overline{\lambda(A)IB} = pIB$. Thus λ is quasi-unital as asserted. Since $\pi_0 \circ \lambda = f$ and $\pi_1 \circ \lambda = g$, we have shown that $f \overset{\sim}{} g$.

To prove the converse, let $\lambda \in \mathrm{Hom}_q(A, IB)$ such that $\pi_0 \circ \lambda = f$, $\pi_1 \circ \lambda = g$. Since the π_t's are surjective they are in particular quasi-unital. Then $\lambda_t = \underline{\pi_t} \circ \underline{\lambda}$, $t \in [0,1]$, defines a path in $\mathrm{Hom}\,(\mathcal{M}(A), \mathcal{M}(B))$ having all the stated properties. $\square$

Let $[A,B]_q$ denote the strong homotopy classes in $\mathrm{Hom}_q(A,B)$ and when $\phi \in \mathrm{Hom}_q(A,B)$ we let $\{\phi\}$ denote the class in $[A,B]_q$ containing ϕ. Since strong homotopy is stronger that homotopy (cf. E 1.3.1), we can define a map $\Lambda : [A,B]_q \to [A,B]$ by $\Lambda\{\phi\} = [\phi]$, $\phi \in \mathrm{Hom}_q(A,B)$. We shall prove the following.

Theorem 1.3.16. *Let A and B be σ-unital C^*-algebras. If B is stable the map $\Lambda : [A, B]_q \to [A, B]$ is a bijection.*

The proof of this theorem requires a considerable amount of preparations, some of which are interesting in their own right.

Lemma 1.3.17. *Let B be a stable σ-unital C^*-algebra and $J \subseteq B$ a σ-unital C^*-subalgebra. Then there is a fully complemented projection $p \in \mathcal{M}(B)$ and a *-isomorphism*

$$\psi : \begin{bmatrix} \overline{JBJ} & \overline{JB} \\ \overline{BJ} & B \end{bmatrix} \to \begin{bmatrix} pBp & pB \\ Bp & B \end{bmatrix}$$

such that

$$\psi \begin{bmatrix} 0 & \overline{JB} \\ 0 & 0 \end{bmatrix} = \begin{bmatrix} 0 & pB \\ 0 & 0 \end{bmatrix} \text{ and } \psi \begin{bmatrix} 0 & 0 \\ 0 & b \end{bmatrix} = \begin{bmatrix} 0 & 0 \\ 0 & b \end{bmatrix}, \quad b \in B.$$

Proof.   It is easy to check that it suffices to find an isomorphism $\lambda : \overline{JB} \to pB$ of Hilbert B-modules for some fully complemented projection $p \in \mathcal{M}(B)$, and a *-isomorphism $\phi : \overline{JBJ} \to pBp$ such that

(a) $\phi(xy^*) = \lambda(x)\lambda(y)^*$, $x, y \in \overline{JB}$ and

(b) $\lambda(az) = \phi(a)\lambda(z)$, $a \in \overline{JBJ}, z \in \overline{JB}$.

For in that case the map

$$\begin{bmatrix} x_1 & x_2 \\ x_3 & x_4 \end{bmatrix} \to \begin{bmatrix} \phi(x_1) & \lambda(x_2) \\ \lambda(x_3^*)^* & x_4 \end{bmatrix},$$

$$x_1 \in \overline{JBJ}, x_2 \in \overline{JB}, x_3 \in \overline{BJ}, x_4 \in B,$$

can serve as ψ.

Let $p \in \mathcal{M}(B)$ be a fully complemented projection and $\lambda : \overline{JB} \to pB$ an isomorphism of Hilbert B-modules. p and λ exist by Theorem 1.3.5 since $\overline{JB}$ is countably generated; in fact $\overline{JB}$ is generated by $\{kh\}$ where k is strictly positive in J and h is strictly positive in B. Now $\mathcal{K}_B(\overline{JB})$ is *-isomorphic to $\overline{JBJ}$ through an isomorphism sending a generating element $\Theta_{x,y}$, $x, y \in \overline{JB}$, to xy^* (cf. E 1.2.2). In the same way $\mathcal{K}_B(pB)$ is *-isomorphic to pBp through $\Theta_{x,y} \to xy^*$, $x, y \in pB$. On the other hand, λ induces a *-isomorphism $\mathcal{K}_B(\overline{JB}) \simeq \mathcal{K}_B(pB)$ by sending $\Theta_{x,y}$ to

$\Theta_{\lambda(x),\lambda(y)}$, $x, y \in \overline{JB}$. Thus we get a *-isomorphism $\phi : \overline{JBJ} \to pBp$ which satisfies (a) by construction.

To check (b) it suffices to check for $a = xy^*$, $x, y \in \overline{JB}$, in which case we get $\lambda(az) = \lambda(xy^*z) = \lambda(x)y^*z$ since λ is a B-module map. Using (a) and that λ preserves the inner product we get $\lambda(x)y^*z = \lambda(x)\lambda(y)^*\lambda(z) = \phi(xy^*)\lambda(z) = \phi(a)\lambda(z)$. $\qquad\square$

Lemma 1.3.18. *Let B be a stable σ-unital C^*-algebra and $J \subseteq B$ be a σ-unital C^*-subalgebra. Then there is a *-isomorphism*

$$\phi : \begin{bmatrix} \overline{JBJ} & \overline{JB} \\ \overline{BJ} & B \end{bmatrix} \to B$$

such that $\phi \circ j_1 \sim id_B$, where $j_1 : B \to \begin{bmatrix} \overline{JBJ} & \overline{JB} \\ \overline{BJ} & B \end{bmatrix}$ is the embedding into the lower right-hand corner.

Proof. Let p and ψ be as in Lemma 1.3.17. Note that B and pB are countably-generated as Hilbert B-modules because B is σ-unital. Therefore

$$\begin{bmatrix} p & 0 \\ 0 & 1 \end{bmatrix}(B \oplus B) = pB \oplus B \approx B \approx \begin{bmatrix} 0 & 0 \\ 0 & 1 \end{bmatrix}(B \oplus B)$$

by Theorem 1.1.24 and Lemma 1.3.2. The same theorem and lemma show that $B \oplus B \approx B$. Let $\delta : B \oplus B \to B$ be an isomorphism. Set $q_1 = \delta \begin{bmatrix} p & 0 \\ 0 & 1 \end{bmatrix} \delta^{-1}, q_2 = \delta \begin{bmatrix} 0 & 0 \\ 0 & 1 \end{bmatrix} \delta^{-1}$. Then q_1 and q_2 are projections in $\mathcal{M}(B)$ by Lemma 1.1.19. Note that $q_1^\perp B = \delta(p^\perp B \oplus 0) \approx p^\perp B \approx B$ and $q_2^\perp B = \delta(B \oplus 0) \approx B$, so that q_1 and q_2 are both fully complemented. Since $q_1 B = \delta(pB \oplus B) \approx pB \oplus B \approx B \approx \delta(0 \oplus B) \approx q_2 B$, we conclude from Theorem 1.3.5 that q_1 and q_2 are unitarily equivalent in $\mathcal{M}(B)$. It follows now from Lemma 1.1.19 that $\begin{bmatrix} p & 0 \\ 0 & 1 \end{bmatrix}$ is unitarily equivalent to $\begin{bmatrix} 0 & 0 \\ 0 & 1 \end{bmatrix}$ in $\mathcal{L}_B(B \oplus B) \simeq M_2(\mathcal{M}(B))$. Since $\begin{bmatrix} 0 & 0 \\ 0 & 1 \end{bmatrix}$ is equivalent to $\begin{bmatrix} 1 & 0 \\ 0 & 1 \end{bmatrix}$, we conclude that $WW^* = \begin{bmatrix} p & 0 \\ 0 & 1 \end{bmatrix}$ for some isometry W in $M_2(\mathcal{M}(B))$. Let $\Theta_B : M_2(\mathcal{M}(B)) \to \mathcal{M}(B)$ be an inner *-isomorphism. Set $e = \Theta_B \begin{bmatrix} p & 0 \\ 0 & 1 \end{bmatrix}$ and $V = \Theta_B(W)$. Then

$$\Theta_B \circ \psi : \begin{bmatrix} \overline{JBJ} & \overline{JB} \\ \overline{BJ} & B \end{bmatrix} \to eBe$$

is a *-isomorphism, and we define the desired ϕ by $\phi(\cdot) = V^*\Theta_B \circ \psi(\cdot)V$. ϕ is then a *-isomorphism. To check that $\phi \circ j_1$ is homotopic to id, let U_t, $t \in [0,1]$, be a strictly continuous path of isometries in $\mathcal{M}(B)$ connecting V to 1 (cf. Lemma 1.3.7). Then $U_t\phi(\cdot)U_t^*$, $t \in [0,1]$, is a continuous path of *-homomorphisms connecting ϕ to $\Theta_B \circ \psi$. Note that $\Theta_B \circ \psi \circ j_1(b) = wbw^*$, $b \in B$, for some isometry $w \in \mathcal{M}(B)$ used to define the inner *-isomorphism Θ_B. By connecting w to 1 through a strictly continuous path of isometries in $\mathcal{M}(B)$, we conclude that $\Theta_B \circ \psi \circ j_1 \sim id_B$. Thus $\phi \circ j_1 \sim id_B$.   □

Lemma 1.3.19. *Assume that A and B are σ-unital and that B is stable. Then any *-homomorphism $\mu : A \to B$ is homotopic to a quasi-unital *-homomorphism $\lambda : A \to B$ such that $1 - p_\lambda = WW^*$ for some isometry $W \in \mathcal{M}(B)$.*

Proof.   Set $J = \mu(A)$. Then J is a σ-unital C^*-subalgebra of B and Lemmas 1.3.17 and 1.3.18 apply. Let ψ be the *-isomorphism and $p \in \mathcal{M}(B)$ the projection given by Lemma 1.3.17, and

$$\phi : \begin{bmatrix} \overline{JBJ} & \overline{JB} \\ \overline{BJ} & B \end{bmatrix} \to B$$

the *-isomorphism given by Lemma 1.3.18.

Then $\mu \sim \phi \circ j_1 \circ \mu$ where $j_1 : B \to \begin{bmatrix} \overline{JBJ} & \overline{JB} \\ \overline{BJ} & B \end{bmatrix}$ is the imbedding into the lower right-hand corner. Note that

$$j_1 \circ \mu(A) \subseteq \begin{bmatrix} J & J \\ J & J \end{bmatrix} \subseteq \begin{bmatrix} \overline{JBJ} & \overline{JB} \\ \overline{BJ} & B \end{bmatrix}.$$

So by considering the rotation matrices R_t, $t \in [0,1]$, from (1.3.3) as elements of $M_2(\mathcal{M}(J))$, we can define $\lambda_t \in \text{Hom}(A,B)$ by

$$\lambda_t = \phi \circ AdR_t \circ j_1 \circ \mu, \ t \in [0,1].$$

This gives a homotopy showing that $\phi \circ j_1 \circ \mu \sim \phi \circ j_2 \circ \mu$, where j_2 is the imbedding of J into the upper left-hand corner. Note that

$$\overline{j_2 \circ \mu(A)} \begin{bmatrix} \overline{JBJ} & \overline{JB} \\ \overline{BJ} & B \end{bmatrix} = \begin{bmatrix} \overline{JBJ} & \overline{JB} \\ 0 & 0 \end{bmatrix}.$$

Thus if we set $\mu_0 = \psi \circ j_2 \circ \mu \in \text{Hom}(A, \begin{bmatrix} pBp & pB \\ Bp & B \end{bmatrix})$, the properties of ψ (cf. Lemma 1.3.17), ensure that

$$\overline{\mu_0(A)} \begin{bmatrix} pBp & pB \\ Bp & B \end{bmatrix} = \begin{bmatrix} pBp & pB \\ 0 & 0 \end{bmatrix}.$$

Note that

$$\begin{bmatrix} pBp & pB \\ Bp & B \end{bmatrix} = \begin{bmatrix} p & 0 \\ 0 & 1 \end{bmatrix} M_2(B) \begin{bmatrix} p & 0 \\ 0 & 1 \end{bmatrix}$$

so that we can make the identification:

$$\mathcal{M}\left(\begin{bmatrix} pBp & pB \\ Bp & B \end{bmatrix}\right) = \begin{bmatrix} p & 0 \\ 0 & 1 \end{bmatrix} M_2(\mathcal{M}(B)) \begin{bmatrix} p & 0 \\ 0 & 1 \end{bmatrix}$$

(cf. E 1.1.8). Then the preceding calculations show that μ_0 is quasi-unital with relative unit $p_{\mu_0} = \begin{bmatrix} p & 0 \\ 0 & 0 \end{bmatrix}$. Set $\lambda = \phi \circ \psi^{-1} \circ \mu_0 \in \mathrm{Hom}\,(A, B)$. Then $\lambda = \phi \circ j_2 \circ \mu \sim \mu$ by the property of ϕ (cf. Lemma 1.3.18). $\phi \circ \psi^{-1}$ extends to a *-isomorphism

$$\phi \circ \psi^{-1} : \begin{bmatrix} p & 0 \\ 0 & 1 \end{bmatrix} M_2(\mathcal{M}(B)) \begin{bmatrix} p & 0 \\ 0 & 1 \end{bmatrix} \to \mathcal{M}(B)$$

such that

$$\overline{\lambda(A)B} = \phi \circ \psi^{-1} \overline{\begin{bmatrix} pBp & pB \\ 0 & 0 \end{bmatrix}} = \phi \circ \psi^{-1}(\begin{bmatrix} p & 0 \\ 0 & 0 \end{bmatrix})B.$$

Thus λ is quasi-unital with relative unit $p_\lambda = \phi \circ \psi^{-1} \begin{bmatrix} p & 0 \\ 0 & 0 \end{bmatrix}$. Then

$$1 - p_\lambda = \phi \circ \psi^{-1} \begin{bmatrix} 0 & 0 \\ 0 & 1 \end{bmatrix}.$$

By the proof of Lemma 1.3.18 there is a partial isometry V in $M_2(\mathcal{M}(B))$ such that $V^*V = \begin{bmatrix} p & 0 \\ 0 & 1 \end{bmatrix}$ and $VV^* = \begin{bmatrix} 0 & 0 \\ 0 & 1 \end{bmatrix}$. Then V is an isometry in $\begin{bmatrix} p & 0 \\ 0 & 1 \end{bmatrix} M_2(\mathcal{M}(B)) \begin{bmatrix} p & 0 \\ 0 & 1 \end{bmatrix}$ and thus $W = \phi \circ \psi^{-1}(V)$ is an isometry in $\mathcal{M}(B)$ with $WW^* = 1 - p_\lambda$. $\qquad\square$

We are now ready for

Proof of Theorem 1.3.16. By Lemma 1.3.19, $\Lambda : [A, B]_q \to [A, B]$ is surjective. To see that it is also injective, let $\lambda_1, \lambda_2 \in \mathrm{Hom}\,_q(A, B)$ be such that $\lambda_1 \sim \lambda_2$. It is not difficult to see (cf. E 1.3.1), that it means that there is a *-homomorphism $\alpha \in \mathrm{Hom}\,(A, IB)$ such that $\pi_0 \circ \alpha = \lambda_1$ and $\pi_1 \circ \alpha = \lambda_2$. Set $J = \alpha(A)$. Since both A and B are σ-unital we can combine Lemma 1.3.2 with Kasparov's stabilization theorem, Theorem 1.1.24, to conclude that the Hilbert IB-modules $\overline{JIB} \oplus IB$ and $IB \oplus IB$ are isomorphic. Let $W : \overline{JIB} \oplus IB \to IB \oplus IB$ be an isomorphism. Identify $\mathcal{K}_{IB}(\overline{JIB} \oplus IB)$ with

$$B_J = \begin{bmatrix} \overline{JIBJ} & \overline{JIB} \\ \overline{IBJ} & IB \end{bmatrix}$$

and $\mathcal{K}_{IB}(IB \oplus IB)$ with $M_2(IB)$ (cf. E 1.2.1 and E 1.2.2). Set $\beta = W \begin{bmatrix} \alpha & 0 \\ 0 & 0 \end{bmatrix} W^*$. Note that $\begin{bmatrix} \alpha & 0 \\ 0 & 0 \end{bmatrix} \in \mathrm{Hom}_q(A, B_J)$ so that $\beta \in \mathrm{Hom}_q(A, M_2(IB))$.

Now apply Proposition 1.1.28 with $A = IB$, $C = B$, $\pi = \pi_t$, $E_1 = \overline{JIB} \oplus IB$, $E_2 = IB \oplus IB$, $F_1 = \overline{\pi_t(J)B} \oplus B$, $F_2 = B \oplus B$, $\phi_2 = \pi_t \oplus \pi_t$ and $\phi_1 = \phi_2|_{\overline{JIB} \oplus IB}$, $t \in [0, 1]$. This gives us linear maps

$$\psi_t^3 : \mathcal{L}_{IB}(\overline{JIB} \oplus IB, IB \oplus IB) \to \mathcal{L}_B(\overline{\pi_t(J)B} \oplus B, B \oplus B)$$

such that

$$\psi_t^3(m)(\pi_t(b_1), \pi_t(b_2)) = \pi_t \oplus \pi_t(m(b_1, b_2)), \quad b_1 \in \overline{JIB}, b_2 \in IB.$$

In particular, it is easy to see that $\psi_t^3(W) : \overline{\pi_t(J)B} \oplus B \to B \oplus B$ is an isomorphism of Hilbert B-modules, for any $t \in [0, 1]$. Apply next Proposition 1.1.27 twice, in both cases with $A = IB, C = B, E_1 = E_2 = E$ and $F_1 = F_2 = F, \phi_1 = \phi_2 = \phi$; first with $E = \overline{JIB} \oplus IB, F = \overline{\pi_t(J)B} \oplus B, \phi = \pi_t \oplus \pi_t|_{\overline{JIB} \oplus IB}$ and secondly with $E = IB \oplus IB, F = B \oplus B$ and $\phi = \pi_t \oplus \pi_t$. This gives us *-homomorphisms $\psi_t^1 : \mathcal{L}_{IB}(\overline{JIB} \oplus IB) \to \mathcal{L}_B(\overline{\pi_t(J)B} \oplus B)$ and $\psi_t^2 : \mathcal{L}_{IB}(IB \oplus IB) \to \mathcal{L}_B(B \oplus B)$ such that

$$(1.3.4) \qquad \psi_t^1(m)(\pi_t(b_1), \pi_t(b_2)) = \pi_t \oplus \pi_t(m(b_1, b_2)),$$

$$b_1 \in \overline{JIB}, b_2 \in IB,$$

and $\psi_t^2(m)(\pi_t(b_1), \pi_t(b_2)) = \pi_t \oplus \pi_t(m(b_1, b_2))$, $b_1, b_2 \in IB$. When we identify $\mathcal{L}_{IB}(IB \oplus IB) = M_2(\mathcal{M}(IB))$ and $\mathcal{L}_B(B \oplus B) = M_2(\mathcal{M}(B))$ (cf. E 1.2.1), we see that $\psi_t^2 = id_{M_2(\mathbb{C})} \otimes \underline{\pi}_t$, where $\underline{\pi}_t : \mathcal{M}(IB) \to \mathcal{M}(B)$ is the strictly continuous *-homomorphism extending π_t.

Since $\psi_t^3(W) \circ (\pi_t \oplus \pi_t) = (\pi_t \oplus \pi_t) \circ W$ the uniqueness statement of Proposition 1.1.28 implies that the diagram

$$
\begin{array}{ccc}
 & W \cdot W^* & \\
B_J & \longrightarrow & M_2(IB) \\
\psi_t^1 \downarrow & & \downarrow id_{M_2(\mathbb{C})} \otimes \pi_t \\
\mathcal{K}_B(\overline{\pi_t(J)B} \oplus B) & \longrightarrow & M_2(B) \\
 & \psi_t^3(W) \cdot \psi_t^3(W)^* &
\end{array}
$$

commutes.

Thus $(id_{M_2(\mathbb{C})} \otimes \pi_t) \circ \beta(a) = \psi_t^3(W)\psi_t^1\left(\begin{bmatrix} \alpha(a) & 0 \\ 0 & 0 \end{bmatrix}\right)\psi_t^3(W)^*$,
$t \in [0,1]$, $a \in A$. So by identifying $M_2(IB) = IM_2(B)$ we see that $\beta \in \mathrm{Hom}_q(A, IM_2(B))$ is a homotopy connecting the two quasi-unital *-homomorphisms $\psi_0^3(W)\psi_0^1\left(\begin{bmatrix} \alpha(\cdot) & 0 \\ 0 & 0 \end{bmatrix}\right)\psi_0^3(W)^*$ and $\psi_1^3(W)\psi_1^1\left(\begin{bmatrix} \alpha(\cdot) & 0 \\ 0 & 0 \end{bmatrix}\right)\psi_1^3(W)^*$. Observe that (1.3.4) gives the equalities
$$\psi_0^1 \circ \begin{bmatrix} \alpha & 0 \\ 0 & 0 \end{bmatrix} = \begin{bmatrix} \lambda_1 & 0 \\ 0 & 0 \end{bmatrix} \text{ and } \psi_1^1 \circ \begin{bmatrix} \alpha & 0 \\ 0 & 0 \end{bmatrix} = \begin{bmatrix} \lambda_2 & 0 \\ 0 & 0 \end{bmatrix}.$$

Since $\pi_0(J) = \lambda_1(A)$ we have that $\overline{\pi_0(J)B} = p_{\lambda_1}B$. Thus $\psi_0^3(W)$: $p_{\lambda_1}B \oplus B \to B \oplus B$ is an isomorphism of Hilbert B-modules and
$$V_0 = \psi_0^3(W)\begin{bmatrix} p_{\lambda_1} & 0 \\ 0 & 1 \end{bmatrix}$$

is a coisometry in $\mathcal{L}_B(B \oplus B) = M_2(\mathcal{M}(B))$. Similarly
$$V_1 = \psi_1^3(W)\begin{bmatrix} p_{\lambda_2} & 0 \\ 0 & 1 \end{bmatrix}$$

is a coisometry in $\mathcal{L}_B(B \oplus B) = M_2(\mathcal{M}(B))$. As shown above
$$V_0\begin{bmatrix} \lambda_1(\cdot) & 0 \\ 0 & 0 \end{bmatrix}V_0^* \sim V_1\begin{bmatrix} \lambda_2(\cdot) & 0 \\ 0 & 0 \end{bmatrix}V_1^*$$

in $\mathrm{Hom}_q(A, IM_2(B))$. By Lemma 1.3.7, there is a strictly continuous path S_t, $t \in [0,1]$, of isometries in $M_2(\mathcal{M}(B)) = \mathcal{M}(M_2(B))$ such that $S_0 = 1$ and $S_1 = V_0^*$. Define $\lambda : A \to IM_2(B)$ by
$$\lambda(a)(t) = S_t V_0\begin{bmatrix} \lambda_1(a) & 0 \\ 0 & 0 \end{bmatrix}V_0^* S_t^*, \quad t \in [0,1], a \in A.$$

This provides us with a homotopy showing that
$$V_0\begin{bmatrix} \lambda_1(\cdot) & 0 \\ 0 & 0 \end{bmatrix}V_0^* \sim \begin{bmatrix} \lambda_1(\cdot) & 0 \\ 0 & 0 \end{bmatrix}.$$

Similarly, one shows that
$$V_1\begin{bmatrix} \lambda_2(\cdot) & 0 \\ 0 & 0 \end{bmatrix}V_1^* \sim \begin{bmatrix} \lambda_2(\cdot) & 0 \\ 0 & 0 \end{bmatrix}$$

in $\mathrm{Hom}_q(A, M_2(B))$. Putting these homotopies together leads to the conclusion that
$$\begin{bmatrix} \lambda_1 & 0 \\ 0 & 0 \end{bmatrix} \sim \begin{bmatrix} \lambda_2 & 0 \\ 0 & 0 \end{bmatrix}.$$

Applying an inner isomorphism $\Theta_B : M_2(B) \to B$ we get that $U\lambda_1(\cdot)U^* \overset{\sim}{\sim} U\lambda_2(\cdot)U^*$ for some isometry $U \in \mathcal{M}(B)$. Connecting U to 1 through a strictly continuous path of isometries (cf. Lemma 1.3.7), we reach the desired conclusion: $\lambda_1 \overset{\sim}{\sim} \lambda_2$.   □

1.3.20.   *Notes and remarks.*

The basic construction in Lemma 1.3.6 goes back to Dixmier and Douady [9]. As seen from Lemma 1.3.7 it has the almost immediate consequence that the unitary group of a stable C^*-algebra is connected in the strict topology. For stable and σ-unital C^*-algebras, this group is now known to be contractible in norm (cf. [7]). It was J. Cuntz who first pointed out the fundamental importance of the property of stable C^*-algebras described in Lemma 1.3.11. In fact he took it as the definition of stability in [5]. Theorem 1.3.16 comes from [31].

Exercise 1.3

E 1.3.1

Let $\phi, \psi \in \mathrm{Hom}\,(A, B)$. Show that $\phi \sim \psi$, i.e. that ϕ is homotopic to ψ, if and only if there is a *-homomorphism $\alpha : A \to IB$ such that $\pi_0 \circ \alpha = \phi$ and $\pi_1 \circ \alpha = \psi$.

E 1.3.2

Let $\phi \in \mathrm{Hom}\,_q(A, B)$ and $\psi \in \mathrm{Hom}\,_q(B, C)$. Show that $\psi \circ \phi$ is quasi-unital with relative unit $\underline{\psi}(p_\phi)$.

E 1.3.3

Let B be an arbitrary C^*-algebra. Show that $\mathcal{K} \otimes B$ is stable.

E 1.3.4

Let B, C be stable C^*-algebras and $\Theta_C : M_2(C) \to C$, $\Theta_B : M_2(B) \to B$ inner *-isomorphisms. Prove that

$$\Theta_C \circ \begin{bmatrix} \psi \circ \phi_1 & 0 \\ 0 & \psi \circ \phi_2 \end{bmatrix} \sim \psi \circ \Theta_B \circ \begin{bmatrix} \phi_1 & 0 \\ 0 & \phi_2 \end{bmatrix}$$

for all $\psi \in \mathrm{Hom}\,(B, C)$, $\phi_1, \phi_2 \in \mathrm{Hom}\,(A, B)$.

E 1.3.5

Let A, B, C, D be stable C^*-algebras. By Lemma 1.3.12, $[A, B]$ is an abelian semi-group.

(i) Show that there is a well-defined map $\circ : [A, B] \times [B, C] \to [A, C]$ given by $[\phi] \circ [\psi] = [\psi \circ \phi]$, $\psi \in \text{Hom}\,(B, C), \phi \in \text{Hom}\,(A, B)$.

(ii) Show that

$$x \circ (y + z) = x \circ y + x \circ z, \quad x \in [A, B], \ y, \ z \in [B, C],$$

and by using E 1.3.4 that

$$(x + y) \circ z = x \circ z + y \circ z, \quad x, y \in [A, B], \ z \in [B, C].$$

(iii) Show that $(x \circ y) \circ z = x \circ (y \circ z)$, $x \in [A, B]$, $y \in [B, C]$, $z \in [C, D]$.

CHAPTER 2

The Kasparov Approach to KK-Theory

2.1. The KK-Groups

In this chapter, **we impose the restriction on all** C^*-algebras $A, B, C, D, \ldots$ **that they are** σ-unital.

Let A, B be graded C^*-algebras.

Definition 2.1.1. A *Kasparov* $A - B$-*module* is a triple $\mathcal{E} = (E, \phi, F)$ where E is a countably generated graded Hilbert B-module, $\phi : A \to \mathcal{L}_B(E)$ is a *-homomorphism and $F \in \mathcal{L}_B(E)$ is an element of degree 1 such that

(i) $\phi \circ \beta_A = \beta_E \circ \phi$ (i.e. ϕ is a graded homomorphism),

(ii) $[F, \phi(a)] \in \mathcal{K}_B(E)$, $\qquad a \in A$,

(iii) $(F^2 - 1)\phi(a) \in \mathcal{K}_B(E)$, $\qquad a \in A$,

(iv) $(F^* - F)\phi(a) \in \mathcal{K}_B(E)$, $\qquad a \in A$.

We remind the reader that the commutator occurring in (ii) is graded. In the following we let $\mathbb{E}(A, B)$ denote the set of Kasparov $A - B$-modules.

The constructions with Hilbert C^*-modules described in 1.2.1–1.2.4 can all be performed with Kasparov modules. We will consider them below, leaving a lot of details to the readers own considerations.

2.1.2. *Direct sum.*

Let $\mathcal{E}_i = (E_i, \phi_i, F_i)$, $i = 1, 2, \ldots, n$, be Kasparov $A - B$-modules. We can then form the Hilbert B-module $E_1 \oplus E_2 \oplus \cdots \oplus E_n$ (cf. 1.2.1). Given $T_i \in \mathcal{L}_B(E_i)$, $i = 1, 2, \ldots, n$, we can define an element $T_1 \oplus T_2 \oplus \cdots \oplus T_n \in \mathcal{L}_B(E_1 \oplus E_2 \oplus \cdots \oplus E_n)$ by

$$T_1 \oplus T_2 \oplus \cdots \oplus T_n(e_1, e_2, \ldots, e_n) =$$
$$(T_1 e_1, T_2 e_2, \ldots, T_n e_n), \quad e_i \in E_i, \quad i = 1, 2, \ldots, n.$$

It is easy to see that $T_1 \oplus T_2 \oplus \cdots \oplus T_n \in \mathcal{K}_B(E_1 \oplus E_2 \oplus \cdots \oplus E_n)$ if and only if $T_i \in \mathcal{K}_B(E_i)$, $i = 1, 2, \ldots, n$ (cf. E 1.2.1).

To make $E_1 \oplus E_2 \oplus \cdots \oplus E_n$ graded define the grading operator $S_{E_1 \oplus E_2 \oplus \cdots \oplus E_n}$ by

$$S_{E_1 \oplus E_2 \oplus \cdots \oplus E_n}(e_1, e_2, \ldots, e_n) = (S_{E_1} e_1, S_{E_2} e_2, \ldots, S_{E_n} e_n).$$

Then $E_1 \oplus E_2 \oplus \cdots \oplus E_n$ is a graded Hilbert B-module.

Define $\phi_1 \oplus \phi_2 \oplus \cdots \oplus \phi_n : A \to \mathcal{L}_B(E_1 \oplus E_2 \oplus \cdots \oplus E_n)$ by

$$\phi_1 \oplus \phi_2 \oplus \cdots \oplus \phi_n(a) = \phi_1(a) \oplus \phi_2(a) \oplus \cdots \oplus \phi_n(a), \quad a \in A.$$

Then $(E_1 \oplus E_2 \oplus \cdots \oplus E_n, \phi_1 \oplus \phi_2 \oplus \cdots \oplus \phi_n, F_1 \oplus F_2 \oplus \cdots \oplus F_n)$ is a Kasparov $A - B$-module which we denote by $\mathcal{E}_1 \oplus \mathcal{E}_2 \oplus \cdots \oplus \mathcal{E}_n$ and call the direct sum of $\mathcal{E}_1, \mathcal{E}_2, \ldots, \mathcal{E}_n$.

2.1.3.   *Pullback.*

Let $\mathcal{E} = (E, \phi, F) \in \mathbb{E}(A, B)$ and let $\psi : C \to A$ be a *-homomorphism of graded C^*-algebras. Then $(E, \phi \circ \psi, F)$ is a Kasparov $C - B$-module which we denote by $\psi^*(\mathcal{E})$.

2.1.4.   *Internal tensor product.*

Let $\mathcal{E} = (E, \phi, F) \in \mathbb{E}(A, B)$ and $\psi : B \to C$ be a *-homomorphism of graded C^*-algebras. We can then form the Hilbert C-module $E \otimes_\psi C$ as a special case of the internal tensor product of Hilbert B-modules, 1.2.3. From the construction of $E \otimes_\psi C$ and the properties of S_E it follows that there is a linear bijection $S_{E \otimes_\psi C}$ on $E \otimes_\psi C$ with the property that

$$S_{E \otimes_\psi C}(e \otimes_\psi c) = S_E(e) \otimes_\psi \beta_C(c), \quad e \in E, c \in C.$$

In fact, it suffices to observe that for $a, b \in C, x, y \in E$ we have

$$< \beta_C(a), \psi(< S_E(x), S_E(y) >)\beta_C(b) >$$
$$= < \beta_C(a), \psi \circ \beta_B(< x, y >)\beta_C(b) >$$
$$= < \beta_C(a), \beta_C \circ \psi(< x, y >)\beta_C(b) >$$
$$= \beta_C(< a, \psi(< x, y >)b >).$$

It follows that $S_{E \otimes_\psi C}$ is the grading operator for a grading of $E \otimes_\psi C$.

As noted in 1.2.3, there is a *-homomorphism

$$j : \mathcal{L}_B(E) \to \mathcal{L}_C(E \otimes_\psi C)$$

given by $j(m) = m \otimes_\psi id$, $m \in \mathcal{L}_B(E)$. Define $\phi \otimes_\psi id : A \to \mathcal{L}_C(E \otimes_\psi C)$ by $\phi \otimes_\psi id(a) = \phi(a) \otimes_\psi id$, $a \in A$. By Lemma 1.2.8, $m \otimes_\psi id \in \mathcal{K}_C(E \otimes_\psi C)$ when $m \in \mathcal{K}_B(E)$. Using this one sees that $(E \otimes_\psi C, \phi \otimes_\psi id, F \otimes_\psi id)$ is a Kasparov $A - C$-module which we denote by $\psi_*(\mathcal{E})$. In particular, $E \otimes_\psi C$ is countably generated because, when $\{e_i\}$ is a set of generators for E and $h \in C$ a strictly positive element, then $\{e_i \otimes_\psi h\}$ is a set of generators for $E \otimes_\psi C$.

2.1.5.   *Pushout.*

In the same setting as in 2.1.4, assume that ψ is surjective. Then we can define E_ψ as in 1.2.2. Since

$$\psi(< S_E(x), S_E(x) >) = \psi \circ \beta_B(< x, x >) = \beta_C \circ \psi(< x, x >), \quad x \in E,$$

it follows from the construction of E_ψ that there is a grading operator S_{E_ψ} on E_ψ given by $S_{E_\psi}(q(x)) = q(S_E(x))$, $x \in E$, on E_ψ' and extended to E_ψ by continuity. In a similar way one sees that there is, for each $T \in \mathcal{L}_B(E)$, an element $T_\psi \in \mathcal{L}_C(E_\psi)$ which is given by $T_\psi(q(x)) = q(T(x))$, $x \in E$, on E_ψ' and extended to E_ψ by continuity. The map $T \to T_\psi$ is a *-homomorphism from $\mathcal{L}_B(E)$ into $\mathcal{L}_C(E_\psi)$, and by checking for $T = \Theta_{x,y}$, $x, y \in E$, one sees that $T_\psi \in \mathcal{K}_C(E_\psi)$ when $T \in \mathcal{K}_B(E)$. Define $\phi_\psi : A \to \mathcal{L}_C(E_\psi)$ by $\phi_\psi(a) = \phi(a)_\psi$, $a \in A$. Then $(E_\psi, \phi_\psi, F_\psi)$ is a Kasparov $A - C$-module which we denote by $\mathcal{E}_\psi$.

2.1.6.   *External tensor product.*

Let $\mathcal{E} = (E, \phi, F) \in \mathbb{E}(A, B)$ and let C be a graded C^*-algebra. Then we can form the Hilbert $B \otimes C$-module $E \hat{\otimes} C$, a special case of 1.2.4. From the construction of $E \hat{\otimes} C$ it follows that there is a linear bijection $S_{E \hat{\otimes} C}$ on $E \hat{\otimes} C$ given by $S_{E \hat{\otimes} C}(e \hat{\otimes} c) = S_E(e) \hat{\otimes} \beta_C(c)$, $e \in E$, $c \in C$. It is then clear that $S_{E \hat{\otimes} C}$ is the grading operator for a grading of $E \hat{\otimes} C$. Define

$$\phi \hat{\otimes} id : A \otimes C \to \mathcal{L}_{B \otimes C}(E \hat{\otimes} C)$$

by

$$\phi \hat{\otimes} id(a \otimes c) = \phi(a) \hat{\otimes} c = j(\phi(a) \otimes c), \quad a \in A, c \in C.$$

Using what was proved about j in 1.2.4 it follows that $(E \hat{\otimes} C, \phi \hat{\otimes} id, F \hat{\otimes} id)$ is a Kasparov $A \otimes C - B \otimes C$-module. We denote it by $\tau_C(\mathcal{E})$.

Definition 2.1.7. Two Kasparov $A - B$-modules $\mathcal{E}_1 = (E_1, \phi_1, F_1)$, $\mathcal{E}_2 = (E_2, \phi_2, F_2)$ are *isomorphic* when there is an isomorphism $\psi : E_1 \to E_2$ of Hilbert B-modules such that $S_{E_2} \circ \psi = \psi \circ S_{E_1}$, $F_2 \circ \psi = \psi \circ F_1$ and $\phi_2(a) \circ \psi = \psi \circ \phi_1(a)$, $a \in A$. We write $\mathcal{E}_1 \simeq \mathcal{E}_2$ in this case.

It is clear that $\simeq$ defines an equivalence relation on $\mathbb{E}(A, B)$.

Lemma 2.1.8. *Let $\mathcal{E}_1, \mathcal{E}_2, \ldots, \mathcal{E}_n \in \mathbb{E}(A, B)$ and let 0 denote the trivial Kasparov $A - B$-module $(0, 0, 0)$. Then*

$$\mathcal{E}_{\sigma(1)} \oplus \mathcal{E}_{\sigma(2)} \oplus \cdots \oplus \mathcal{E}_{\sigma(n)} \simeq \mathcal{E}_1 \oplus \mathcal{E}_2 \cdots \oplus \mathcal{E}_n$$

for every permutation σ of $\{1, 2, \ldots, n\}$, $(\mathcal{E}_1 \oplus \mathcal{E}_2) \oplus \mathcal{E}_3 \simeq \mathcal{E}_1 \oplus (\mathcal{E}_2 \oplus \mathcal{E}_3) \simeq \mathcal{E}_1 \oplus \mathcal{E}_2 \oplus \mathcal{E}_3$ and $\mathcal{E}_1 \oplus 0 \simeq \mathcal{E}_1$.

Proof. The proof is left as an exercise, E 2.1.1. $\square$

We shall also now consider two weaker equivalence relations on $\mathbb{E}(A, B)$. To define the first, recall that π_t, $t \in [0, 1]$, denotes the surjection $IB = B \otimes C[0, 1] \to B$ obtained by evaluation at t. When B is graded by the automorphism β_B, we consider $IB = B \otimes C[0, 1]$ to be graded by the automorphism $\beta_B \otimes id$, and each π_t is then a graded homomorphism. If $\mathcal{E}$ is a Kasparov $A - IB$-module we obtain a Kasparov $A - B$-module for each $t \in [0, 1]$, namely the pushout $\mathcal{E}_{\pi_t}$ (cf. 2.1.5).

Definition 2.1.9. Two Kasparov $A - B$-modules $\mathcal{E}, \mathcal{F} \in \mathbb{E}(A, B)$ are called *homotopic* when there is a Kasparov $A - IB$-module $\mathcal{G} \in \mathbb{E}(A, IB)$ such that $\mathcal{G}_{\pi_0} \simeq \mathcal{E}$ and $\mathcal{G}_{\pi_1} \simeq \mathcal{F}$. We write $\mathcal{E} \sim \mathcal{F}$ when there is a finite set $\{\mathcal{E}_1, \mathcal{E}_2, \ldots, \mathcal{E}_n\} \subseteq \mathbb{E}(A, B)$ such that $\mathcal{E}_1 = \mathcal{E}, \mathcal{E}_n = \mathcal{F}$ and $\mathcal{E}_i$ is homotopic to $\mathcal{E}_{i+1}$, $i = 1, 2, \ldots, n - 1$.

At this point it is not obvious that $\sim$ is an equivalence relation on $\mathbb{E}(A, B)$. To show that it is we need some lemmas of independent interest.

Lemma 2.1.10. *Let $\mathcal{E} \in \mathbb{E}(A, B)$ and let $f : B \to C$ and $g : C \to D$ be *-homomorphisms of graded C^*-algebras. Then $g_*(f_*(\mathcal{E})) \simeq (g \circ f)_*(\mathcal{E})$ in $\mathbb{E}(A, D)$.*

Proof. Let $\mathcal{E} = (E, \phi, F)$. By construction there is a D-module map $U : (E \otimes_f C) \otimes_g D \to E \otimes_{g \circ f} D$ given on simple tensors by

$$U((e \otimes_f c) \otimes_g d) = e \otimes_{g \circ f} g(c)d, \quad e \in E, \, c \in C, \, d \in D.$$

We leave the reader to check that U does the job. The only point that may cause trouble is the surjectivity of U. For this point take an approximate unit $\{u_i\}$ for B and note that

$$U((e \otimes_f f(u_i)) \otimes_g d) = e \otimes_{g \circ f} g(f(u_i))d = eu_i \otimes_{g \circ f} d \to e \otimes_{g \circ f} d. \quad \square$$

Lemma 2.1.11. *Let $\mathcal{E} \in \mathbb{E}(A, B)$ and let $f : B \to C$ be a surjective homomorphism of graded C^*-algebras. Then $\mathcal{E}_f \simeq f_*(\mathcal{E})$.*

Proof. Let $\mathcal{E} = (E, \phi, F)$. In Lemma 1.2.5 we constructed an isomorphism of Hilbert C-modules between E_f and $E \otimes_f C$. We leave the reader to check that this isomorphism has all the properties we need to reach the conclusion: $\mathcal{E}_f \simeq f_*(\mathcal{E})$. $\quad \square$

Lemma 2.1.12. $\sim$ *is an equivalence relation on $\mathbb{E}(A, B)$.*

Proof. By definition $\sim$ is transitive. It suffices to show that $\mathcal{E} \sim \mathcal{E}$, $\mathcal{E} \in \mathbb{E}(A, B)$, and that $\mathcal{E}_1 \sim \mathcal{E}_2$ implies $\mathcal{E}_2 \sim \mathcal{E}_1$.

To prove the last thing first, let $\psi' : C[0,1] \to C[0,1]$ be the *-auto-morphism induced by the homeomorphism $t \to 1 - t$ of $[0,1]$ and set $\psi = id \otimes \psi' \in \mathrm{Aut}(IB)$. Then $\pi_0 \circ \psi = \pi_1$ and $\pi_0 = \pi_1 \circ \psi$. So if $\mathcal{E} \in \mathbb{E}(A, IB)$ satisfies that $\mathcal{E}_{\pi_0} \simeq \mathcal{E}_1$ and $\mathcal{E}_{\pi_1} \simeq \mathcal{E}_2$, then by Lemmas 2.1.10 and 2.1.11, $\psi_*(\mathcal{E})_{\pi_0} \simeq \pi_{0*}(\psi_*(\mathcal{E})) \simeq (\pi_0 \circ \psi)_*(\mathcal{E}) = \pi_{1*}(\mathcal{E}) \simeq \mathcal{E}_2$ and similarly $\psi_*(\mathcal{E})_{\pi_1} \simeq \mathcal{E}_1$.

To prove that $\mathcal{E} \sim \mathcal{E}$, let $\phi : B \to IB$ be the *-homomorphism which identifies $b \in B$ with the corresponding constant B-valued function on $[0,1]$. Then $\pi_t \circ \phi = id$ for all $t \in [0,1]$. Thus $\phi_*(\mathcal{E}) \in \mathbb{E}(A, IB)$ and $\phi_*(\mathcal{E})_{\pi_0} \simeq (\pi_0 \circ \phi)_*(\mathcal{E}) = id_*(\mathcal{E})$ and $\phi_*(\mathcal{E})_{\pi_1} \simeq id_*(\mathcal{E})$. As it is easy to check that $id_*(\mathcal{E}) \simeq \mathcal{E}$, this completes the proof. $\quad \square$

Lemma 2.1.13. *Let $\mathcal{E}_1, \mathcal{E}_2 \in \mathbb{E}(A, B)$ and let $f : B \to C$ be a homomorphism of graded C^*-algebras. Then $f_*(\mathcal{E}_1 \oplus \mathcal{E}_2) \simeq f_*(\mathcal{E}_1) \oplus f_*(\mathcal{E}_2)$.*

Proof. Let $\mathcal{E}_1 = (E_1, \phi_1, F_1), \mathcal{E}_2 = (E_2, \phi_2, F_2)$. There is an isomorphism $U : (E_1 \oplus E_2) \otimes_f C \to (E_1 \otimes_f C) \oplus (E_2 \otimes_f C)$ of Hilbert C-modules given by $U((e_1, e_2) \otimes_f c) = (e_1 \otimes_f c, e_2 \otimes_f c)$, $e_1 \in E_1$, $e_2 \in E_2$, $c \in C$. We leave the reader to check that U does the job. $\square$

Corollary 2.1.14.   *Let $\mathcal{E}_1, \mathcal{E}_2, \mathcal{E}_3, \mathcal{E}_4 \in \mathbb{E}(A, B)$ and now assume that $\mathcal{E}_1 \sim \mathcal{E}_2$ and $\mathcal{E}_3 \sim \mathcal{E}_4$. Then $\mathcal{E}_1 \oplus \mathcal{E}_3 \sim \mathcal{E}_2 \oplus \mathcal{E}_4$.*

Proof. Let $\mathcal{E}, \mathcal{F} \in \mathbb{E}(A, IB)$ such that

$$\pi_{0*}(\mathcal{E}) \simeq \mathcal{E}_1, \pi_{1*}(\mathcal{E}) \simeq \mathcal{E}_2, \pi_{0*}(\mathcal{F}) \simeq \mathcal{E}_3$$

and $\pi_{1*}(\mathcal{F}) \simeq \mathcal{E}_4$. Then $\pi_{0*}(\mathcal{E} \oplus \mathcal{F}) \simeq \mathcal{E}_1 \oplus \mathcal{E}_3$ and $\pi_{1*}(\mathcal{E} \oplus \mathcal{F}) \simeq \mathcal{E}_2 \oplus \mathcal{E}_4$ by Lemma 2.1.13. The corollary follows immediately from this. $\square$

Definition 2.1.15.   An element $\mathcal{E} = (E, \phi, F) \in \mathbb{E}(A, B)$ is called *degenerate* when $[F, \phi(a)] = (F^2 - 1)\phi(a) = (F^* - F)\phi(a) = 0$ for all $a \in A$. The class of degenerate Kasparov $A - B$-modules is denoted $\mathbb{D}(A, B)$.

Definition 2.1.16.   Two Kasparov $A - B$-modules $\mathcal{E}_1, \mathcal{E}_2 \in \mathbb{E}(A, B)$ are said to be *operator homotopic* when there are a graded Hilbert B-module E, a graded homomorphism $\phi : A \to \mathcal{L}_B(E)$ and a norm continuous path F_t, $t \in [0, 1]$, in $\mathcal{L}_B(E)$ such that

(i) $\mathcal{F}_t = (E, \phi, F_t) \in \mathbb{E}(A, B)$ for all $t \in [0, 1]$,

(ii) $\mathcal{F}_0 \simeq \mathcal{E}_1$ and $\mathcal{F}_1 \simeq \mathcal{E}_2$.

We write $\mathcal{E}_1 \approx \mathcal{E}_2$ when there are degenerate Kasparov $A - B$-modules $\mathcal{F}_1, \mathcal{F}_2 \in \mathbb{D}(A, B)$ such that $\mathcal{E}_1 \oplus \mathcal{F}_1$ is operator homotopic to $\mathcal{E}_2 \oplus \mathcal{F}_2$.

Lemma 2.1.17. $\approx$ *defines an equivalence relation on $\mathbb{E}(A, B)$.*

Proof. Only the transitivity condition is not trivial. So let $\mathcal{E}_1, \mathcal{E}_2, \mathcal{E}_3 \in \mathbb{E}(A, B), \mathcal{F}_1, \mathcal{F}_2, \mathcal{F}_3, \mathcal{F}_4 \in \mathbb{D}(A, B)$ and assume that $\mathcal{E}_1 \oplus \mathcal{F}_1$ is operator homotopic to $\mathcal{E}_2 \oplus \mathcal{F}_2$ and that $\mathcal{E}_2 \oplus \mathcal{F}_3$ is operator homotopic to $\mathcal{E}_3 \oplus \mathcal{F}_4$. Then it is easy to see that $\mathcal{E}_1 \oplus \mathcal{F}_1 \oplus \mathcal{F}_3$ is operator homotopic to $\mathcal{E}_2 \oplus \mathcal{F}_2 \oplus \mathcal{F}_3$, and $\mathcal{E}_2 \oplus \mathcal{F}_2 \oplus \mathcal{F}_3$ to $\mathcal{E}_3 \oplus \mathcal{F}_2 \oplus \mathcal{F}_4$.

Let $\mathcal{E}'_t = (E, \phi, F_t) \in \mathbb{E}(A, B)$ and $\mathcal{E}''_t = (E', \phi', F'_t) \in \mathbb{E}(A, B)$, $t \in [0, 1]$, be the respective operator homotopies. Then $\mathcal{E}'_1 \simeq \mathcal{E}''_0$. Let $\psi : E \to E'$ be an isomorphism of Hilbert B-modules implementing this last isomorphism of Kasparov $A - B$-modules. Define $G_t \in \mathcal{L}_B(E)$, $t \in [0, 1]$, by $G_t = F_{2t}, t \in [0, \frac{1}{2}]$, and $G_t = \psi^{-1} \circ F'_{2t-1} \circ \psi$, $t \in [\frac{1}{2}, 1]$. Then $\mathcal{G}_t = (E, \phi, G_t)$, $t \in [0, 1]$, gives an operator homotopy between $\mathcal{E}_1 \oplus \mathcal{F}_1 \oplus \mathcal{F}_3$

and $\mathcal{E}_3 \oplus \mathcal{F}_2 \oplus \mathcal{F}_4$. Since the direct sum of two degenerate Kasparov $A - B$-modules is again degenerate, we see that $\mathcal{E}_1 \approx \mathcal{E}_3$. □

The following sufficient condition for two Kasparov $A - B$ modules to be operator homotopic will be very useful later on.

Lemma 2.1.18. *Let* $\mathcal{E} = (E, \phi, F)$, $\mathcal{E}' = (E, \phi, F') \in \mathbb{E}(A, B)$. *If* $\phi(a)[F, F']\phi(a)^* \geq 0 \mod \mathcal{K}_B(E)$, $a \in A$, *then* $\mathcal{E}$ *and* $\mathcal{E}'$ *are operator homotopic.*

(To be completely precise, the assumption of the lemma is that $q(\phi(a)[F, F']\phi(a)^*) \geq 0$ in $\mathcal{L}_B(E)/\mathcal{K}_B(E)$ for all $a \in A$, where $q : \mathcal{L}_B(E) \to \mathcal{L}_B(E)/\mathcal{K}_B(E)$ is the quotient map.)

Proof. Let $\mathcal{A} = \{T \in \mathcal{L}_B(E) : [T, \phi(a)] \in \mathcal{K}_B(E), a \in A\}$. Then $\mathcal{A}$ is a C^*-algebra and $J = \{T \in \mathcal{A} : T\phi(a) \in \mathcal{K}_B(E), a \in A\}$ is an ideal in $\mathcal{A}$. Using equation (1.2.3) one sees that $[F, F'] \in \mathcal{A}$. We assert that $[F, F'] \geq 0 \mod J$. To see this we introduce for each state ω of $\mathcal{L}_B(E)/\mathcal{K}_B(E)$ and each $a \in A$ the positive linear functional $T \to \omega(q(\phi(a)T\phi(a)^*))$ on $\mathcal{A}$, where $q : \mathcal{L}_B(E) \to \mathcal{L}_B(E)/\mathcal{K}_B(E)$ is the quotient map. Since the functional annihilates J, we get a positive functional $\mu(\omega, a)$ on $\mathcal{A}/J$. Then $\{\mu(\omega, a)\}$ is a faithful family of positive linear functionals on $\mathcal{A}/J$, so the corresponding direct sum of GNS-representations is faithful too. It follows that $[F, F'] \geq 0 \mod J$ if and only if $\sum_{ij} q(\phi(a)S_i[F, F']S_j^*\phi(a)^*) \geq 0$ for all finite sets $\{S_i\} \subseteq \mathcal{A}$ and all $a \in A$. Since $[F, F']$ has degree zero and $\mathcal{A}$ is left globally invariant by the grading operator of $\mathcal{L}_B(E)$, we may assume that each S_i is of degree 0 too. But then each S_i commutes with $\phi(a) \mod \mathcal{K}_B(E)$, so that

$$\sum_{ij} q(\phi(a)S_i[F, F']S_j^*\phi(a)^*) = q((\sum_i S_i)\phi(a)[F, F']\phi(a)^*(\sum_j S_j^*)).$$

This is positive by assumption, so that $[F, F']$ is indeed positive mod J. Thus $[F, F'] = P + K$, where P is positive in $\mathcal{A}$ and $K \in J$. Since $[F, F']$ has degree 0, we can assume both P and K have degree 0 too. Set

$$F_t = (1 + \sin t \cos t P)^{-\frac{1}{2}}(\cos t F + \sin t F'), \quad t \in [0, \frac{\pi}{2}].$$

Then F_t, $t \in [0, \frac{\pi}{2}]$, is a norm continuous path of degree 1 elements in $\mathcal{L}_B(E)$. Since $F_0 = F$ and $F_{\frac{\pi}{2}} = F'$, the proof can be completed by checking that $(E, \phi, F_t) \in \mathbb{E}(A, C)$ for each t. Condition (ii) of Definition

2.1.1 follows from the observation that $F_t \in \mathcal{A}$ and the conditions (iii) and (iv) are equivalent to $F_t^2 - 1$, $F_t - F_t^* \in J$. Observe that $F^2 = 1 \bmod J$ and $F'^2 = 1 \bmod J$. Thus

$$FP = F(FF' + F'F) = F^2F' + FF'F = F' + FF'F \bmod J.$$

Similarly,

$$PF = (FF' + F'F)F = FF'F + F'F^2 = FF'F + F' \bmod J.$$

Hence P and F commute mod J. In the same way one sees that so do P and F'. Thus

$$
\begin{aligned}
F_t^2 &= (\cos t F + \sin t F')^2 (1 + \sin t \cos t P)^{-1} \\
&= (\cos t^2 F^2 + \sin t^2 F'^2 + \cos t \sin t [F, F'])(1 + \cos t \sin t P)^{-1} \\
&= (1 + \cos t \sin t P)(1 + \cos t \sin t P)^{-1} = 1 \bmod J.
\end{aligned}
$$

Similarly,

$$F_t - F_t^* = (1 + \sin t \cos t P)^{-\frac{1}{2}}(\cos t(F - F^*) + \sin t(F' - F'^*)) = 0 \bmod J$$

since $F - F^*$, $F' - F'^* \in J$.   □

Definition 2.1.19. $KK(A, B)$ is the equivalence classes in $\mathbb{E}(A, B)$ under the equivalence relation $\sim$, i.e. $KK(A, B) = \mathbb{E}(A, B)/\sim$. We let $[\mathcal{E}]$ denote the element in $KK(A, B)$ represented by $\mathcal{E} \in \mathbb{E}(A, B)$.

$K\hat{K}(A, B)$ is the equivalence classes in $\mathbb{E}(A, B)$ under the equivalence relation $\approx$, i.e. $K\hat{K}(A, B) = \mathbb{E}(A, B)/\approx$. We let $\{\mathcal{E}\}$ denote the element in $K\hat{K}(A, B)$ represented by $\mathcal{E} \in \mathbb{E}(A, B)$.

We start to relate $KK(A, B)$ to $K\hat{K}(A, B)$. Ultimately we shall see that when A is separable, we have $KK(A, B) = K\hat{K}(A, B)$, i.e. the two equivalence relations $\approx$ and $\sim$ agree on $\mathbb{E}(A, B)$. The following lemmas give the easy part of this assertion.

Lemma 2.1.20. *Every degenerate Kasparov $A-B$-module $\mathcal{E} \in \mathbb{D}(A, B)$ is homotopic to 0.*

Proof. Let $\mathcal{E} = (E, \phi, F)$ and consider the ideal $C_0[0, 1)$ in $C[0, 1]$ consisting of the functions f with the property that $f(1) = 0$. $C_0[0, 1)$ is a Hilbert $C[0, 1]$-module, so we can form the external tensor product $E \hat{\otimes} C_0[0, 1)$ (cf. 1.2.4). $E \hat{\otimes} C_0[0, 1)$ can be graded by a symmetry

S given on simple tensors by the formula $S(e\hat{\otimes}f) = S_E(e)\hat{\otimes}f$, $e \in E$, $f \in C_0[0,1)$. In the same way we can define $\hat{\phi} : A \to \mathcal{L}_{IB}(E\hat{\otimes}C_0[0,1))$ by $\hat{\phi}(a) = \phi(a)\hat{\otimes}id$, $a \in A$ (cf. 1.2.4). Since $\mathcal{E}$ is degenerate it follows that not only is $\mathcal{F} = (E\hat{\otimes}C_0[0,1), \hat{\phi}, F\hat{\otimes}id)$ a Kasparov $A - IB$-module, but in fact $\mathcal{F} \in \mathbb{D}(A, IB)$.

There is a unique map $U : (E\hat{\otimes}C_0[0,1))_{\pi_0} \to E$ given (in the notation of 1.2.2 and 1.2.4) by $U(q(e\hat{\otimes}f)) = f(0)e$, $e \in E$, $f \in C_0[0,1)$. It is easy to check that U defines an isomorphism of Kasparov $A - B$-modules, i.e. $\mathcal{E} \simeq \pi_{0*}(\mathcal{F})$. Since $\pi_1(< x,x >) = 0$, $x \in E\hat{\otimes}C_0[0,1)$, it follows that $\pi_{1*}(\mathcal{F}) = 0$. $\qquad\square$

Lemma 2.1.21. *Let $\mathcal{E}_1, \mathcal{E}_2 \in \mathbb{E}(A, B)$. Then $\mathcal{E}_1 \approx \mathcal{E}_2$ implies that $\mathcal{E}_1 \sim \mathcal{E}_2$.*

Proof. By the preceding lemma it suffices to show the following: Let E be a graded Hilbert B-module, $\phi : A \to \mathcal{L}_B(E)$ a graded homomorphism and F_t, $t \in [0,1]$, a norm continuous path in $\mathcal{L}_B(E)$ such that $\mathcal{F}_t = (E, \phi, F_t) \in \mathbb{E}(A, B)$, $t \in [0,1]$. Then $\mathcal{F}_0$ is homotopic to $\mathcal{F}_1$. To see this consider the vector space IE consisting of continuous functions $f : [0,1] \to E$. IE is a right IB-module: $(xf)(t) = x(t)f(t)$, $x \in IE$, $f \in IB$, $t \in [0,1]$, and is a Hilbert IB-module in the "inner product" :

$$< f, g > (t) = < f(t), g(t) >, \quad f, g \in IE, \, t \in [0,1].$$

IE is graded by the grading operator S_{IE} given by $S_{IE}(f)(t) = S_E(f(t))$, $t \in [0,1]$, $f \in IE$. Recall that the grading automorphism of IB is $\beta_B \otimes id$.

There is a graded *-homomorphism $\pi : C[0,1] \otimes \mathcal{L}_B(E) \to \mathcal{L}_{IB}(IE)$ given by $(\pi(T)f)(t) = T_t(f(t))$, $t \in [0,1]$, $f \in IE$. Let F denote the element in $C[0,1] \otimes \mathcal{L}_B(E)$ corresponding to the path F_t, $t \in [0,1]$, and define $1 \otimes \phi : A \to IB$ by $(1\otimes\phi)(a) = 1\otimes\phi(a)$, $a \in A$. We assert that $\mathcal{F} = (IE, \pi \circ (1 \otimes \phi), \pi(F))$ is a Kasparov $A - IB$-module. The only nontrivial things to check in Definition 2.1.1 are the conditions (ii), (iii) and (iv). They will follow if we can show that $\pi(T) \in \mathcal{K}_{IB}(IE)$ when $T_t \in \mathcal{K}_B(E)$ for all $t \in [0,1]$. This is equivalent to $\pi(C[0,1]\otimes\mathcal{K}_B(E)) \subseteq \mathcal{K}_{IB}(IE)$, so it suffices to check that $\pi(f \otimes \Theta_{x,y}) \in \mathcal{K}_{IB}(IE)$ when $x, y \in E$ and $f \in C[0,1]$ is positive. Define $\tilde{x}, \tilde{y} \in IE$ by $\tilde{x}(t) = \sqrt{f(t)}x$ and $\tilde{y}(t) = \sqrt{f(t)}y$, $t \in [0,1]$, and note that $\pi(f \otimes \Theta_{x,y}) = \Theta_{\tilde{x},\tilde{y}}$. Thus $\mathcal{F} \in \mathbb{E}(A, IB)$ as asserted. This can also be seen by using 1.2.4 and the fact that $IE \simeq E\hat{\otimes}C[0,1]$.

We leave the reader to complete the proof by showing that $\mathcal{F}_{\pi_0} \simeq \mathcal{F}_0$ and $\mathcal{F}_{\pi_1} \simeq \mathcal{F}_1$.   □

By Lemma 2.1.21 there is a map $\mu : K\hat{K}(A, B) \to KK(A, B)$ given by $\mu\{\mathcal{E}\} = [\mathcal{E}]$, $\mathcal{E} \in \mathbb{E}(A, B)$. μ is obviously surjective.

Definition 2.1.22. Now let us define compositions, $+$, in $KK(A, B)$ and $K\hat{K}(A, B)$ by

$$[\mathcal{E}_1] + [\mathcal{E}_2] = [\mathcal{E}_1 \oplus \mathcal{E}_2]$$

and

$$\{\mathcal{E}_1\} + \{\mathcal{E}_2\} = \{\mathcal{E}_1 \oplus \mathcal{E}_2\}, \quad \mathcal{E}_1, \mathcal{E}_2 \in \mathbb{E}(A, B),$$

respectively.

Corollary 2.1.14 ensures that the composition is well-defined in $KK(A, B)$. It is easy to see that the corresponding corollary also holds for $\approx$, so that the composition is also well-defined in $K\hat{K}(A, B)$. By Lemma 2.1.8 both $KK(A, B)$ and $K\hat{K}(A, B)$ are commutative semigroups with neutral elements $[0]$ and $\{0\}$, respectively. However, more is true :

Theorem 2.1.23. $KK(A, B)$ and $K\hat{K}(A, B)$ are both abelian groups and $\mu : K\hat{K}(A, B) \to KK(A, B)$ is a surjective group homomorphism.

Proof. It is clear that μ is a surjective homomorphism of abelian semigroups with neutral elements. It therefore suffices to show that inverses exist in $K\hat{K}(A, B)$. So let $\mathcal{E} = (E, \phi, F) \in \mathbb{E}(A, B)$. Define $-E$ to be the graded Hilbert B-module which, as a Hilbert B-module, is identical to E but is graded by $-S_E$. Define $\phi_- : A \to \mathcal{L}_B(E) = \mathcal{L}_B(-E)$ by $\phi_- = \phi \circ \beta_A$. (Recall that β_A is the grading automorphism for A.) Then ϕ_- is a graded homomorphism. Define $G_t \in \mathcal{L}_B(E \oplus -E)$ to be the element given by the matrix:

$$\begin{bmatrix} c_t F & s_t \\ s_t & -c_t F \end{bmatrix}$$

where $c_t = \cos\frac{\pi}{2}t$ and $s_t = \sin\frac{\pi}{2}t$, $t \in [0, 1]$. G_t is clearly norm-continuous in t and $(E \oplus -E, \phi \oplus \phi_-, F \oplus -F) = (E \oplus -E, \phi \oplus \phi_-, G_0)$. Note that G_t

is of degree one since

$$(S_E \oplus -S_E)G_t = \begin{bmatrix} c_t S_E F & s_t S_E \\ -s_t S_E & c_t S_E F \end{bmatrix}$$

$$= \begin{bmatrix} -c_t F S_E & s_t S_E \\ -s_t S_E & -c_t F S_E \end{bmatrix}$$

$$= -G_t(S_E \oplus -S_E), \ t \in [0,1].$$

We assert that $\mathcal{F}_t = (E \oplus -E, \phi \oplus \phi_-, G_t) \in \mathbb{E}(A, B)$ for all $t \in [0,1]$. This amounts to the assertions:

(a) $[G_t, \phi(a) \oplus \phi \circ \beta_A(a)] \in \mathcal{K}_B(E \oplus -E)$,

(b) $(G_t^2 - 1)(\phi(a) \oplus \phi \circ \beta_A(a)) \in \mathcal{K}_B(E \oplus -E)$,

(c) $(G_t^* - G_t)(\phi(a) \oplus \phi \circ \beta_A(a)) \in \mathcal{K}_B(E \oplus -E)$, $\quad a \in A$, $\quad t \in [0,1]$.

To check (a) observe that it suffices to check when $a \in A$ has degree 0 or 1. Thus (a) decomposes into

(a') $G_t(\phi(a) \oplus \phi(a)) - (\phi(a) \oplus \phi(a))G_t \in \mathcal{K}_B(E \oplus -E)$, $\deg(a) = 0$,

(a'') $G_t(\phi(a) \oplus \phi(-a)) + (\phi(a) \oplus \phi(-a))G_t \in \mathcal{K}_B(E \oplus -E)$, $\deg(a) = 1$.

We find

$$G_t(\phi(a) \oplus \phi(a)) + (\phi(a) \oplus \phi(a))G_t =$$

$$\begin{bmatrix} c_t(F\phi(a) - \phi(a)F) & 0 \\ 0 & c_t(\phi(a)F - F\phi(a)) \end{bmatrix} \in \mathcal{K}_B(E \oplus -E)$$

for $\deg(a) = 0$ and

$$G_t(\phi(a) \oplus \phi(-a)) - (\phi(a) \oplus \phi(-a))G_t =$$

$$\begin{bmatrix} c_t(F\phi(a) + \phi(a)F) & 0 \\ 0 & c_t(F\phi(a) + \phi(a)F) \end{bmatrix} \in \mathcal{K}_B(E \oplus -E)$$

for $\deg(a) = 1$.

To check (b) above it suffices to observe that

$$(G_t^2 - 1) = c_t^2(F^2 - 1) \oplus c_t^2(F^2 - 1).$$

Similarly (c) follows since we have $G_t^* - G_t = c_t(F^* - F) \oplus c_t(F - F^*)$.

Note that $G_1 = \begin{bmatrix} 0 & 1 \\ 1 & 0 \end{bmatrix}$, that $G_1^* = G_1, G_1^2 = 1$. Inserting $t = 1$ in the preceding calculations gives that $[G_1, \phi(a) \oplus \phi_-(a)] = 0$. Thus

$\mathcal{F}_1 = (E \oplus -E, \phi \oplus \phi_-, G_1)$ is a degenerate Kasparov $A - B$-module. It follows that $\mathcal{F}_t$, $t \in [0,1]$, gives an operator homotopy between $\mathcal{E} \oplus -\mathcal{E}$, where $-\mathcal{E} = (-E, \phi_-, -F) \in \mathbb{E}(A,B)$, and an element in $\mathbb{D}(A,B)$. Thus $\{\mathcal{E}\} + \{-\mathcal{E}\} = 0$. $\qquad\qquad\qquad\qquad\qquad\qquad\qquad\qquad\qquad\qquad\qquad\square$

Next we consider the functorial properties of the KK-groups.

Lemma 2.1.24.   *Let $\mathcal{E}_1, \mathcal{E}_2 \in \mathbb{E}(A,B)$ and let $f : D \to A$ and $g : B \to C$ be graded homomorphisms. Then $f^*(g_*(\mathcal{E}_1)) = g_*(f^*(\mathcal{E}_1))$. If $\mathcal{E}_1 \simeq \mathcal{E}_2$, then $f^*(\mathcal{E}_1) \simeq f^*(\mathcal{E}_2)$, $g_*(\mathcal{E}_1) \simeq g_*(\mathcal{E}_2)$ and $\tau_C(\mathcal{E}_1) \simeq \tau_C(\mathcal{E}_2)$.*

Proof.   The proof is an easy exercise in manipulations with the constructions of 2.1.2 - 2.1.6. We leave it as an exercise (cf. E 2.1.1). $\qquad\square$

Lemma 2.1.25. *Let $f : A \to B$ be a homomorphism of graded C^*-algebras. Then for every C^*-algebra C there are group homomorphisms $f^* : KK(B,C) \to KK(A,C)$ and $f^* : K\hat{K}(B,C) \to K\hat{K}(A,C)$ given by $f^*[\mathcal{E}] = [f^*(\mathcal{E})]$ and $f^*\{\mathcal{E}\} = \{f^*(\mathcal{E})\}$, $\mathcal{E} \in \mathbb{E}(A,B)$, respectively.*

Proof.   To show that f^* is well-defined on $KK(B,C)$, note first that by Lemma 2.1.24 it suffices to take $\mathcal{F} \in \mathbb{E}(A, IC)$ and show that $f^*(\pi_{0*}(\mathcal{F})) \sim f^*(\pi_{1*}(\mathcal{F}))$. But $f^*(\pi_{0*}(\mathcal{F})) = \pi_{0*}(f^*(\mathcal{F}))$ and $f^*(\pi_{1*}(\mathcal{F})) = \pi_{1*}(f^*(\mathcal{F}))$ by Lemma 2.1.24, so $f^*(\mathcal{F})$ gives the desired homotopy.

To see that f^* is well-defined on $K\hat{K}(B,C)$, assume that $\mathcal{E}_1 \approx \mathcal{E}_2 \in \mathbb{E}(B,C)$. Let $\mathcal{F}_1, \mathcal{F}_2 \in \mathbb{D}(B,C)$ and let $\mathcal{G}_t = (E, \phi, F_t)$, $t \in [0,1]$, be an operator homotopy connecting $\mathcal{E}_1 \oplus \mathcal{F}_1$ to $\mathcal{E}_2 \oplus \mathcal{F}_2$. Then $f^*(\mathcal{G}_t) = (E, \phi \circ f, F_t)$, $t \in [0,1]$, gives an operator homotopy between $f^*(\mathcal{E}_1 \oplus \mathcal{F}_1)$ and $f^*(\mathcal{E}_2 \oplus \mathcal{F}_2)$. Since $f^*(\mathcal{F}_i) \in \mathbb{D}(A,C)$ it follows from Lemma 2.1.13 that $f^*(\mathcal{E}_i \oplus \mathcal{F}_i) \simeq f^*(\mathcal{E}_i) \oplus f^*(\mathcal{F}_i)$, $i = 1,2$, and we conclude that f^* is well-defined on both KK and $K\hat{K}$. Both maps are obviously group homomorphisms. $\qquad\qquad\qquad\qquad\qquad\qquad\qquad\qquad\qquad\qquad\qquad\square$

Lemma 2.1.26.   *Let $f : B \to C$ be a homomorphism of graded C^*-algebras. Then for every graded C^*-algebra A there are group homomorphisms $f_* : KK(A,B) \to KK(A,C)$ and $f_* : K\hat{K}(A,B) \to K\hat{K}(A,C)$ given by $f_*[\mathcal{E}] = [f_*(\mathcal{E})]$ and $f_*\{\mathcal{E}\} = \{f_*(\mathcal{E})\}$, $\mathcal{E} \in \mathbb{E}(A,B)$, respectively.*

Proof. By Lemma 2.1.13 it suffices to check that both maps are well-defined. To check the KK-case it suffices by Lemma 2.1.24 to take $\mathcal{F} \in \mathbb{E}(A, IB)$ and show that $f_*(\pi_{0*}(\mathcal{F})) \sim f_*(\pi_{1*}(\mathcal{F}))$. But $f_*(\pi_{i*}(\mathcal{F})) \simeq (f \circ \pi_i)_*(\mathcal{F})$, $i = 0, 1$, by Lemma 2.1.10, so it suffices to show that when $\mathcal{E} \in \mathbb{E}(A, B)$ and g_t, $t \in [0, 1]$, is a path of graded homomorphisms $g_t : B \to C$ which is continuous in the sense that $t \to g_t(b)$ is continuous for all $b \in B$, then $g_{0*}(\mathcal{E})$ is homotopic to $g_{1*}(\mathcal{E})$. The path g_t, $t \in [0, 1]$, defines a graded homomorphism $g : IB \to IC$ by $g(x)(t) = g_t(x(t))$, $t \in [0, 1]$, $x \in IB$. Let $f : B \to IB$ be the graded homomorphism which identifies $b \in B$ with the corresponding constant function in IB. Then $(g \circ f)_*(\mathcal{E}) \in \mathbb{E}(A, IC)$ and an easy application of Lemma 2.1.10 gives that $(g \circ f)_*(\mathcal{E})$ defines a homotopy between $g_{0*}(\mathcal{E})$ and $g_{1*}(\mathcal{E})$.

To check for $K\hat{K}$, assume that $\mathcal{E}_1 \approx \mathcal{E}_2$ in $\mathbb{E}(A, B)$. Let $\mathcal{F}_1, \mathcal{F}_2 \in \mathbb{D}(A, B)$ and let $\mathcal{G}_t = (E, \phi, G_t) \in \mathbb{E}(A, B)$, $t \in [0, 1]$, be an operator homotopy connecting $\mathcal{E}_1 \oplus \mathcal{F}_1$ to $\mathcal{E}_2 \oplus \mathcal{F}_2$. Then

$$f_*(\mathcal{G}_t) = (E \otimes_f C, \phi \otimes_f id, G_t \otimes_f id), \ t \in [0, 1],$$

is an operator homotopy connecting $f_*(\mathcal{E}_1 \oplus \mathcal{F}_1)$ to $f_*(\mathcal{E}_2 \oplus \mathcal{F}_2)$. Since $f_*(\mathcal{F}_i) \in \mathbb{D}(A, C)$, $i = 1, 2$, Lemma 2.1.13 tells us that $f_*(\mathcal{E}_1) \approx f_*(\mathcal{E}_2)$. $\square$

Lemma 2.1.25 says essentially that both $KK(\cdot, B)$ and $K\hat{K}(\cdot, B)$ are contravariant functors from the category of graded (σ-unital) C^*-algebras to the category of abelian groups for all fixed graded C^*-algebras B. Similarly, Lemma 2.1.26 says essentially that $KK(A, \cdot)$ and $K\hat{K}(A, \cdot)$ are covariant functors between the same categories for all fixed A.

Lemma 2.1.27. *Let A, B, D be graded C^*-algebras. Then there are group homomorphisms $\tau_D : KK(A, B) \to KK(A \otimes D, B \otimes D)$ and $\tau_D : K\hat{K}(A, B) \to K\hat{K}(A \otimes D, B \otimes D)$ given by $\tau_D[\mathcal{E}] = [\tau_D(\mathcal{E})]$ and $\tau_D\{\mathcal{E}\} = \{\tau_D(\mathcal{E})\}, \mathcal{E} \in \mathbb{E}(A, B)$, respectively.*

Proof. First we observe that $\tau_D(\mathcal{E}_1 \oplus \mathcal{E}_2) \simeq \tau_D(\mathcal{E}_1) \oplus \tau_D(\mathcal{E}_2)$, $\mathcal{E}_1, \mathcal{E}_2 \in \mathbb{E}(A, B)$. Thus it suffices to check that τ_D is well-defined in both cases.

To show this in the KK-case it suffices by Lemma 2.1.24 to take $\mathcal{E} \in \mathbb{E}(A, IB)$ and prove that $\tau_D(\mathcal{E}_{\pi_0}) \sim \tau_D(\mathcal{E}_{\pi_1})$. We leave the reader to show that $\tau_D(\mathcal{E}_{\pi_t}) \simeq (\tau_D(\mathcal{E}))_{\pi_t}$ for all $t \in [0, 1]$, which is certainly sufficient to reach the desired conclusion.

To show that τ_D is well-defined on $K\hat{K}$, let $\mathcal{E}_1, \mathcal{E}_2 \in \mathbb{E}(A, B)$, $\mathcal{F}_1, \mathcal{F}_2 \in \mathbb{D}(A, B)$ and let $\mathcal{G}_t = (E, \phi, G_t)$, $t \in [0, 1]$, be an operator homotopy connecting $\mathcal{E}_1 \oplus \mathcal{F}_1$ to $\mathcal{E}_2 \oplus \mathcal{F}_2$. Then $\tau_D(\mathcal{G}_t) = (E\hat{\otimes}D, \phi\hat{\otimes}id, G_t\hat{\otimes}id)$, $t \in [0, 1]$, is an operator homotopy connecting $\tau_D(\mathcal{E}_1 \oplus \mathcal{F}_1)$ to $\tau_D(\mathcal{E}_2 \oplus \mathcal{F}_2)$. Since $\tau_D(\mathcal{F}_i) \in \mathbb{D}(A \otimes D, B \otimes D)$, $i = 1, 2$, the proof is complete. $\qquad\square$

In concluding this section, we remark that although we have restricted attention to σ-unital C^*-algebras, the whole development up to now has made no application of this assumption and it can be avoided all the way simply by deleting the word "countably generated" in the definition of the Kasparov modules. But now we turn to the construction of the Kasparov product which is a bihomomorphism $KK(A, B) \times KK(B, C) \to KK(A, C)$, and for this purpose the restriction to σ-unital C^*-algebras is unavoidable.

2.1.28.   *Notes and remarks.*

All the constructions described in this sections appeared in [19]. The exposition given here follows [28] closely.

Exercise 2.1

E 2.1.1

Prove Lemma 2.1.8 and Lemma 2.1.24.

E 2.1.2

The purpose of this exercise is to show that $K\hat{K}(\mathbb{C}, \mathbb{C}) \simeq \mathbb{Z}$ and it uses the theory of Fredholm operators as developed for example in [10].

(a) Show that every element in $K\hat{K}(\mathbb{C}, \mathbb{C})$ can be represented by a triple $\left(\mathcal{H} \oplus \mathcal{H}, \phi, \begin{bmatrix} 0 & S \\ T & 0 \end{bmatrix}\right) \in \mathbb{E}(\mathbb{C}, \mathbb{C})$, where $\mathcal{H}$ is a Hilbert space and $\mathcal{H} \oplus \mathcal{H}$ is graded by the symmetry U given by $U(\psi, \chi) = (\psi, -\chi)$, $\psi, \chi \in \mathcal{H}$, $\phi : \mathbb{C} \to \mathcal{B}(\mathcal{H} \oplus \mathcal{H})$ is a unital *-homomorphism and $S, T \in \mathcal{B}(\mathcal{H})$ are unitary mod $\mathcal{K}(\mathcal{H})$ such that $T^* = S$ mod $\mathcal{K}(\mathcal{H})$.

Such a triple will be called a standard triple.

(b) Show that there is a map $K\hat{K}(\mathbb{C}, \mathbb{C}) \to \mathbb{Z}$ determined by the condition that an element in $K\hat{K}(\mathbb{C}, \mathbb{C})$ represented by a standard triple $\left(\mathcal{H} \oplus \mathcal{H}, \phi, \begin{bmatrix} 0 & S \\ T & 0 \end{bmatrix}\right)$ is mapped to index(T), the Fredholm index of the Fredholm operator T.

(c) Show that the map constructed in (b) is a group isomorphism.

E 2.1.3

Let E be a countably generated graded Hilbert B-module over the graded C^*-algebra B and $\phi_t : A \to \mathcal{L}_B(E)$, $t \in [0, 1]$, a path of graded *-homomorphisms such that $t \to \phi_t(a)$ is continuous in norm for all $a \in A$. Let F_t, $t \in [0, 1]$, be a norm continuous path in $\mathcal{L}_B(E)$ such that $(E, \phi_t, F_t) \in \mathbb{E}(A, B)$, $t \in [0, 1]$. Show that (E, ϕ_0, F_0) and (E, ϕ_1, F_1) represent the same element in $KK(A, B)$.

E 2.1.4

Let A, B and C be graded σ-unital C^*-algebras.

(i) Let $\phi_t : C \to A$, $t \in [0, 1]$, be a graded homotopy, i.e. $t \to \phi_t(c)$ is continuous for all $c \in C$ and ϕ_t is a graded *-homomorphism for all $t \in [0, 1]$. Show that $\phi_0^* = \phi_1^* : KK(A, B) \to KK(C, B)$.

(ii) Let $\psi_t : B \to C$, $t \in [0, 1]$, be a graded homotopy. Show that $\psi_{0*} = \psi_{1*} : KK(A, B) \to KK(A, C)$.

2.2. The Kasparov Product

The main technical result used to define the Kasparov product and establish its properties is the following:

Theorem 2.2.1 [Kasparov's technical theorem]. *Let B be a graded σ-unital C^*-algebra, let E_1, E_2 be σ-unital subalgebras of $\mathcal{M}(B)$ and let $\mathcal{F}$ be a separable closed linear subspace of $\mathcal{M}(B)$. Assume that*

(i) $\underline{\beta}_B(E_i) = E_i$, $i = 1, 2$, *and* $\underline{\beta}_B(\mathcal{F}) = \mathcal{F}$,

(ii) $E_1 E_2 \subseteq B$

(iii) $[\mathcal{F}, E_1] \subseteq E_1$.

Then there exist elements $M, N \in \mathcal{M}(B)$ of degree 0 such that $N + M = 1$, $N, M \geq 0$, $ME_1 \subseteq B$, $NE_2 \subseteq B$ and $[N, \mathcal{F}] \subseteq B$.

For the proof of this we need the following lemma.

Lemma 2.2.2. *For every $\epsilon > 0$ there exists $\delta(\epsilon) > 0$ such that if x and y are elements of a graded C^*-algebra, $\|x\| \leq 1$, $\|y\| \leq 1$, $x \geq 0$, $\deg(x) = 0$ and $\|[x, y]\| < \delta(\epsilon)$, then $\|[\sqrt{x}, y]\| < \epsilon$.*

Proof. First choose a polynomial p such that $|\sqrt{t} - p(t)| < \frac{\epsilon}{4}$ for all $t \in [0, 1]$. We can assume that $p(0) = 0$ so that

$$p(t) = c_1 t + c_2 t^2 + \cdots + c_m t^m.$$

Let a be any element in a C^*-algebra of norm less than one. Then $D(x) = ax - xa$ is a linear operator of norm less than 2 and, since D is a derivation: $D(xy) = D(x)y + xD(y)$, we find that $\|D(x^n)\| \leq n\|D(x)\|$ for all x of norm less than 1. Hence for all positive x, $\|x\| \leq 1$, we have $\|D(p(x))\| \leq \sum_1^m |c_k| k \|D(x)\|$. Therefore

$$\|D(\sqrt{x})\| \leq \|D(p(x))\| + 2\|\sqrt{x} - p(x)\| \leq \sum_1^m |c_k| k \|D(x)\| + \frac{\epsilon}{2}.$$

Thus $\delta(\epsilon) = \epsilon(2\sum_1^m |c_k| k)^{-1}$ will work because $[\cdot, \cdot]$ reduces to the usual commutator when one of the entries is of degree 0, and because $\deg(x) = 0$ implies $\deg(\sqrt{x}) = 0$. $\qquad\square$

are the E_2-tensor operators for E, then $\widetilde{U}T_x = T'_{U(x)}$ and $(T'_{U(x)})^* = T^*_x \widetilde{U}^{-1}$, $x \in E_1$.

(iii)   If $P \in \mathcal{L}_A(E_1)$ is a projection of degree zero and $F \in \mathcal{L}_B(E_{12})$ is an F_2-connection for E_1 then the operator $(P \otimes_f id)F(P \otimes_f id)$ is an F_2-connection for PE_1. To see this it suffices to observe that

$$(P \otimes_f id)T_x = T_x,$$
$$(P \otimes_f id)T_{S_{E_1}(x)} = T_{S_{E_1}(x)}, \ x \in PE_1,$$

and that

$$(P \otimes_f id)\mathcal{K}_B(E_2, E_{12}) \subseteq \mathcal{K}_B(E_2, PE_1 \otimes_f E_2),$$
$$\mathcal{K}_B(E_{12}, E_2)(P \otimes_f id) \subseteq \mathcal{K}_B(PE_1 \otimes_f E_2, E_2).$$

For example, for $x \in PE_1$, we find

$$T_x F_2 - (P \otimes_f id)F(P \otimes_f id)T_{S_{E_1}(x)}$$
$$= T_x F_2 - (P \otimes_f id)F T_{S_{E_1}(x)}$$
$$= (P \otimes_f id)(T_x F_2 - F T_{S_{E_1}(x)}) \in (P \otimes_f id)\mathcal{K}_B(E_2, E_{12})$$
$$\subseteq \mathcal{K}_B(E_2, PE_1 \otimes_f E_2).$$

(iv)   Assume that $E_1 = \mathcal{H} \otimes_s A$ as described in Lemma 1.2.13. Then there is an F_2-connection for E_1 of degree 1. To see this observe that there is a linear B-module map $W : (\mathcal{H} \otimes_s A) \otimes_f E_2 \to \mathcal{H} \otimes_s E_2$ given on simple tensors by $W(\phi \otimes_s a \otimes_f e_2) = \phi \otimes_s f(a)e_2$, $\phi \in \mathcal{H}$, $a \in A$, $e_2 \in E_2$. W preserves the "inner products" and intertwines the grading operators. The grading operator on $\mathcal{H}$ is any symmetry U with infinite dimensional eigenspaces. Since we can assume that f is unital (by (i)) it is also clear that W is surjective. Hence W is an isomorphism of graded Hilbert B-modules. Define $F \in \mathcal{L}_B(E_1 \otimes_f E_2)$ by $F = W^{-1}(U \otimes_s F_2)W$, where $U \otimes_s F_2$ is given by $U \otimes_s F_2(\phi \otimes_s e_2) = U(\phi) \otimes_s F_2(e_2)$, $\phi \in \mathcal{H}$, $e_2 \in E_2$. Clearly, F has degree 1 since F_2 has. If $x = \phi \otimes_s 1 \in E_1$, $\phi \in \mathcal{H}$, an easy calculation shows that $T_x F_2 - F T_{S_{E_1}(x)} = 0 = F_2 T^*_x - T^*_{S_{E_1}(x)}F$. Thus (2.2.1)–(2.2.2) hold for such $x \in E_1$. Since $[F_2, f(b)] \in \mathcal{K}_B(E_2)$, $b \in A$, by assumption, it is a simple calculation to show that if (2.2.1) and (2.2.2) hold for $x \in E_1$, then it holds for xb, for all $b \in A$. (Check for $\deg(b) = 0$ and $\deg(b) = 1$). Thus (2.2.1) and (2.2.2) hold for vectors of the form $\phi \otimes_s b$, $\phi \in \mathcal{H}$, $b \in B$. These span a dense set in E_1, so that it holds for all $x \in E_1$ by linearity and continuity.

Now the proof is completed by using (ii), (iii) and (iv). Recall that since E_1 is countably generated a combination of Theorem 1.2.12 and

Lemma 1.2.13 gives that there is a projection $P \in \mathcal{L}_B(\mathcal{H} \otimes_s A)$ of degree zero such that $E_1 \simeq P(\mathcal{H} \otimes_s A)$. By (ii) and (iii) it is therefore sufficient to prove the lemma when $E_1 = \mathcal{H} \otimes_s A$. This case was handled in (iv). $\square$

Lemma 2.2.6. *In the setting above the following hold.*

(a) *If $F_2 \in \mathcal{L}_B(E_2)$ is of degree 1, then an element F of degree 1 in $\mathcal{L}_B(E_{12})$ is an F_2-connection for E_1 if and only if F^* is an F_2^*-connection for E_1.*

(b) *Let $F_2, F_2' \in \mathcal{L}_B(E_2)$ and $F, F' \in \mathcal{L}_B(E_{12})$ all have degree 1. If F and F' are F_2 and F_2'-connections for E_1, respectively, then $F + F'$ is an $F_2 + F_2'$-connection for E_1.*

(c) *The set of 0-connections for E_1 is*

$$\{V \in \mathcal{L}_B(E_{12}) : V(k \otimes_f id), \ (k \otimes_f id)V \in \mathcal{K}_B(E_{12}), \ k \in \mathcal{K}_B(E_1)\}.$$

(d) *If $F_2 \in \mathcal{L}_B(E_{12})$ has degree 1 and F is an F_2-connection for E_1 of degree 1, then $[F, k \otimes_f id] \in \mathcal{K}_B(E_{12})$ for all $k \in \mathcal{K}_B(E_1)$.*

(e) *If $F_2 \in \mathcal{L}_B(E_2)$ has degree 1 and $[F_2, f(a)] = 0$ for all $a \in A$, then there is an operator $S_{E_1} \otimes_f F_2 \in \mathcal{L}_B(E_{12})$ given on simple tensors by*

$$S_{E_1} \otimes_f F_2(x \otimes_f y) = S_{E_1}(x) \otimes_f F_2(y), \ x \in E_1, \ y \in E_2.$$

$S_{E_1} \otimes_f F_2$ is an F_2-connection for E_1 of degree 1. Moreover,

$$[T \otimes_f id, S_{E_1} \otimes_f F_2] = 0$$

for all $T \in \mathcal{L}_A(E_1)$.

(f) *If $\mathcal{E}_2 = (E_2, f, F_2) \in \mathbb{E}(A, B)$ and G is an F_2-connection for E_1 of degree 1, then $G - G^*$ and $G^2 - 1$ are 0-connections for E_1.*

Proof.

(a)   Just take the adjoint of the relations (2.2.1) and (2.2.2).

(b)   Obvious.

(c) $V \in \mathcal{L}_B(E_{12})$ is a 0-connection for E_1 if and only if $VT_x \in \mathcal{K}_B(E_2, E_{12})$ and $T_x^*V \in \mathcal{K}_B(E_{12}, E_2)$ for all $x \in E_1$. Note that for $x, y \in E_1$, $T_x T_y^* = \Theta_{x,y} \otimes_f id$. Thus if V is a 0 connection, then

$$V(\Theta_{x,y} \otimes_\psi id) = (VT_x)T_y^* \in \mathcal{K}_B(E_2, E_{12})\mathcal{L}_B(E_{12}, E_1) \subseteq \mathcal{K}_B(E_2)$$

and similarly $(\Theta_{x,y} \otimes_f id)V \in \mathcal{K}_B(E_2)$. By linearity and continuity V lies in the set described in (c). Conversely, if V lies in this set,

$$VT_x T_x^* V^* = V(\Theta_{x,x} \otimes_\psi id)V^* \in \mathcal{K}_B(E_{12})$$

for all $x \in E_1$. By Lemma 1.1.10, $VT_x \in \mathcal{K}_B(E_2, E_{12})$ for all $x \in E_1$. Similarly one concludes that $T_x^* V \in \mathcal{K}_B(E_{12}, E_2)$ for all $x \in E_1$. Thus V is a 0-connection for E_1.

(d)   It suffices to check for $\deg(k) = 0$ and $\deg(k) = 1$. Note that the degree 0 subspace of $\mathcal{K}_B(E_1)$ is the closed linear span of operators of the form $S_{E_1} \Theta_{x,y} S_{E_1}^{-1} + \Theta_{x,y}$, $x, y \in E_1$, and that the degree 1 subspace is the closed linear span of operators of the form $S_{E_1} \Theta_{x,y} S_{E_1}^{-1} - \Theta_{x,y}$, $x, y \in E_1$. Since $\Theta_{x,y} \otimes_f id = T_x T_y^*$ and $S_{E_1} \Theta_{x,y} S_{E_1}^{-1} \otimes_f id = T_{S_{E_1}(x)} T_{S_{E_1}(y)}^*$ it suffices to check that

$$F(T_{S_{E_1}(x)} T_{S_{E_1}(y)}^* \pm T_x T_y^*) \mp (T_{S_{E_1}(x)} T_{S_{E_1}(y)}^* \pm T_x T_y^*)F \in \mathcal{K}_B(E_{12})$$

for all $x, y \in E_1$. This follows readily from the relations (2.2.1) and (2.2.2).

(e)   The condition that $[F_2, f(a)] = 0$ for all $a \in A$, is precisely what we need to define $S_{E_1} \otimes_A F_2$ on the algebraic tensor product $E_1 \otimes_A E_2$. It is easy to check that this map extends by continuity to the desired element $S_{E_1} \otimes_f F_2$ in $\mathcal{L}_B(E_{12})$. To check that $S_{E_1} \otimes_f F_2$ has the stated properties is left to the reader.

(f)   Note that $(G - G^*)T_{S_{E_1}(x)} = T_x(F_2 - F_2^*)$ and $(G^2 - 1)T_x = T_x(F_2^2 - 1) \bmod \mathcal{K}_B(E_2, E_{12})$ for all $x \in E_1$. Let X denote either $F_2 - F_2^*$ or $F_2^2 - 1$. It suffices to show that $T_x X \in \mathcal{K}_B(E_2, E_{12})$ for an arbitrary $x \in E_1$. Since $T_{x v_i} \to T_x$ when $\{v_i\}$ is an approximate unit for A, it suffices to show that $T_{xa} X \in \mathcal{K}_B(E_2, E_{12})$ for arbitrary $x \in E_1$, $a \in A$. Note that $T_{xa} X(z) = xa \otimes_f X(z) = x \otimes_f f(a)X(z)$. Thus $T_{xa} X = T_x f(a)X$. Since (E_2, f, F_2) is a Kasparov $A - B$ module, $f(a)X \in \mathcal{K}_B(E_2)$. The proof is complete. $\square$

Definition 2.2.7.   Next we let $\mathcal{E}_1 = (E_1, \phi_1, F_1) \in \mathbb{E}(A, B)$, $\mathcal{E}_2 = (E_2, \phi_2, F_2) \in \mathbb{E}(B, C)$, set $E_{12} = E_1 \otimes_{\phi_2} E_2$ and define $\tilde{\phi}_1 : A \to \mathcal{L}_C(E_{12})$ by $\tilde{\phi}_1(a) = \phi_1(a) \otimes_{\phi_2} id$, $a \in A$. A triple $\mathcal{E}_{12} = (E_{12}, \tilde{\phi}_1, F) \in \mathbb{E}(A, C)$ is called a *Kasparov product* of $\mathcal{E}_1$ by $\mathcal{E}_2$ if

(a) F is an F_2-connection for E_1, and

(b) $\tilde{\phi}_1(a)[F_1 \otimes_{\phi_2} id, F]\tilde{\phi}_1(a)^* \geq 0 \bmod \mathcal{K}_C(E_{12})$, $a \in A$.

Recall that E_{12} is a graded Hilbert C-module, graded by the operator given on simple tensors by $x \otimes_{\phi_2} y \to S_{E_1}(x) \otimes_{\phi_2} S_{E_2}(y)$.

Theorem 2.2.8. *Assume that A is separable. Let $\mathcal{E}_1 = (E_1, \phi_1, F_1) \in$ $\mathbb{E}(A, B)$, $\mathcal{E}_2 = (E_2, \phi_2, F_2) \in \mathbb{E}(B, C)$. Then there exists a Kasparov product, $\mathcal{E}_{12} \in \mathbb{E}(A, C)$, of $\mathcal{E}_1$ by $\mathcal{E}_2$. $\mathcal{E}_{12}$ is unique up to operator homotopy.*

Proof. Note first that E_{12} is countably generated since E_1 and E_2 are; in fact if $\{x_n\}$ and $\{y_n\}$ are sets of generators for E_1 and E_2, respectively, then $\{x_n \otimes_{\phi_2} y_m\}$ is a set of generators for E_{12}.

Proposition 2.2.5 guarantees the existence of an F_2-connection G for E_1 of degree 1. Set $A_1 = \mathcal{K}_B(E_1) \otimes_{\phi_2} id + \mathcal{K}_C(E_{12})$ and let A_2 be the C^*-subalgebra of $\mathcal{L}_B(E_{12})$ generated by $G^2 - 1$, $[G, \tilde{\phi}_1(A)]$, $G - G^*$ and $[G, F_1 \otimes_{\phi_2} id]$. Let $\mathcal{F}$ be the closed vector space generated by $F_1 \otimes_{\phi_2} id$, G, and $\tilde{\phi}_1(A)$. We want to apply Corollary 2.2.3 with $E = E_{12}$. To check that all conditions of this corollary are met, observe that A_2 is separable, hence σ-unital, and that A_1 is σ-unital since $h_1 \otimes_{\phi_2} id + h_2$ is strictly positive when $h_1 \in \mathcal{K}_B(E_1)$ and $h_2 \in \mathcal{K}_C(E_{12})$ both are strictly positive. The h_i's exist by Corollary 1.1.25. Since A is separable, so is $\mathcal{F}$, and we see that all the size conditions of Corollary 2.2.3 are satisfied. It is a trivial exercise in handling graded commutators to see that condition (i) is fulfilled. To check (ii) it suffices to take $k \in \mathcal{K}_B(E_1)$ and show that $(k \otimes_{\phi_2} id)T \in \mathcal{K}_C(E_{12})$ for $T = G^2 - 1$, $G - G^*$, $[G, F_1 \otimes_{\phi_2} id]$, $[G, \tilde{\phi}_1(a)]$, $a \in A$. Since $G - G^*$ and $G^2 - 1$ are 0-connections for E_1 by 2.2.6 (f), these choices for T are taken care of by 2.2.6 (c). Note that

$$k \otimes_{\phi_2} id[G, \phi_1(a) \otimes_{\phi_2} id] = (-1)^{deg k}[G, k\phi_1(a) \otimes_{\phi_2} id]$$
$$- (-1)^{deg(k)}[G, k \otimes_{\phi_2} id]\phi_1(a) \otimes_{\phi_2} id, \ a \in A,$$

and

$$k \otimes_{\phi_2} id\, [G, F_1 \otimes_{\phi_2} id] = (-1)^{deg(k)}[G, kF_1 \otimes_{\phi_2} id]$$
$$- (-1)^{deg(k)}[G, k \otimes_{\phi_2} id]F_1 \otimes_{\phi_2} id$$

by (1.2.2). Hence the two remaining possibilities for T are handled by 2.2.6 (d). To check condition (iii) of Corollary 2.2.3 it suffices to check that $[F_1 \otimes_{\phi_2} id, k \otimes_{\phi_2} id]$, $[G, k \otimes_{\phi_2} id]$, $[\tilde{\phi}_1(a), k \otimes_{\phi_2} id]$ is in A_1, for $k \in \mathcal{K}_A(E_1)$. The first and third commutator is in A_1 for trivial reasons, and the second because of 2.2.6 (d). Thus Corollary 2.2.3 applies.

It follows that there are two positive elements M, N in $\mathcal{L}_C(E_{12})$ of degree 0 such that $M + N = 1$, $MA_1 \subseteq \mathcal{K}_C(E_{12})$, $NA_2 \subseteq \mathcal{K}_C(E_{12})$ and

$[\mathcal{F}, M] \subseteq \mathcal{K}_C(E_{12})$. Set $F = M^{\frac{1}{2}}(F_1 \otimes_{\phi_2} id) + N^{\frac{1}{2}}G$. Then F has degree 1. Note that $M^{\frac{1}{2}}A_1 \subseteq \mathcal{K}_C(E_{12})$, so that $M^{\frac{1}{2}}T_x \in \mathcal{K}_C(E_2, E_{12})$ for all $x \in E_1$ since $M^{\frac{1}{2}}T_x T_x^* M^{\frac{1}{2}} = M^{\frac{1}{2}}(\Theta_{x,x} \otimes_{\phi_2} id)M^{\frac{1}{2}} \in \mathcal{K}_C(E_{12})$ (cf. Lemma 1.1.10). Thus $T_x^* M^{\frac{1}{2}} \in \mathcal{K}_C(E_{12}, E_2)$ for all $x \in E_1$. It follows from this that $M^{\frac{1}{2}}(F_1 \otimes_{\phi_2} id)$ is a 0-connection for E_1. Since $MT_x = M^{\frac{1}{2}}M^{\frac{1}{2}}T_x \in \mathcal{K}_C(E_2, E_{12})$, we see that

$$NT_x = (1 - M)T_x = T_x \bmod \mathcal{K}_C(E_2, E_{12}).$$

Hence $N^{\frac{1}{2}}T_x = T_x \bmod \mathcal{K}_C(E_2, E_{12})$ for all $x \in E_1$. Thus $T_x^* N^{\frac{1}{2}} = T_x^* \bmod \mathcal{K}_C(E_{12}, E_2)$ for all $x \in E_1$. It follows from this that $N^{\frac{1}{2}}G$ is an F_2-connection for E_1. Thus F is an F_2-connection for E_1 by 2.2.6 (b).

We assert that $(E_{12}, \tilde{\phi}_1, F) \in \mathbb{E}(A, C)$. So we have to check the conditions in Definition 2.1.1. (i) is trivially satisfied. To check the rest we fix an arbitrary homogeneous element $a \in A$. Then

$$
\begin{aligned}
[F, \tilde{\phi}_1(a)] &= (-1)^{deg(a)}[\tilde{\phi}_1(a), F] \\
&= (-1)^{deg(a)}[\tilde{\phi}_1(a), M^{\frac{1}{2}}(F_1 \otimes_{\phi_2} id)] + (-1)^{deg(a)}[\tilde{\phi}_1(a), N^{\frac{1}{2}}G] \\
&= (-1)^{deg(a)}([\tilde{\phi}_1(a), M^{\frac{1}{2}}](F_1 \otimes_{\phi_2} id) + M^{\frac{1}{2}}[\tilde{\phi}_1(a), F_1 \otimes_{\phi_2} id]) \\
&\quad + (-1)^{deg(a)}([\tilde{\phi}_1(a), N^{\frac{1}{2}}]G + N^{\frac{1}{2}}[\tilde{\phi}_1(a), G]).
\end{aligned}
$$

The first term is in $\mathcal{K}_C(E_{12})$ because $\tilde{\phi}_1(a) \in \mathcal{F}$, the second because $[\tilde{\phi}_1(a), F_1 \otimes_{\phi_2} id] = [\phi_1(a), F_1] \otimes_{\phi_2} id \in A_1$, the third because $\tilde{\phi}_1(a) \in \mathcal{F}$ and $[\tilde{\phi}_1(a), N] \in \mathcal{K}_C(E_{12}) \Rightarrow [\tilde{\phi}_1(a), N^{\frac{1}{2}}] \in \mathcal{K}_C(E_{12})$, and the fourth because $[\tilde{\phi}_1(a), G] \in A_2$. Thus (ii) is satisfied too. To check (iii) observe $[\mathcal{F}, M] \subseteq \mathcal{K}_C(E_{12})$ implies that N and M commute with everything in $\mathcal{F} \bmod \mathcal{K}_C(E_{12})$ since $\deg(N) = \deg(M) = 0$. Hence

$$
\begin{aligned}
F^2 - 1 &= M(F_1^2 \otimes_{\phi_2} id) + NG^2 + N^{\frac{1}{2}}M^{\frac{1}{2}}G(F_1 \otimes_{\phi_2} id) \\
&\quad + N^{\frac{1}{2}}M^{\frac{1}{2}}(F_1 \otimes_{\phi_2} id)G - 1 \\
&= M((F_1^2 - 1) \otimes_{\phi_2} id) + N(G^2 - 1) \\
&\quad + N^{\frac{1}{2}}M^{\frac{1}{2}}[G, F_1 \otimes_{\phi_2} id] \bmod \mathcal{K}_C(E_{12}).
\end{aligned}
$$

Thus

$$
\begin{aligned}
(F^2 - 1)\tilde{\phi}_1(a) &= M((F_1^2 - 1)\phi_1(a) \otimes_{\phi_2} id) + N(G^2 - 1)\tilde{\phi}_1(a) \\
&\quad + N^{\frac{1}{2}}M^{\frac{1}{2}}[G, F_1 \otimes_{\phi_2} id]\tilde{\phi}_1(a) \bmod \mathcal{K}_C(E_{12}).
\end{aligned}
$$

The first term is in $\mathcal{K}_C(E_{12})$ because $(F_1^2 - 1)\phi_1(a) \in \mathcal{K}_C(E_1)$ so that $(F_1^2 - 1)\phi_1(a) \otimes_{\phi_2} id \in A_1$, the second because $G^2 - 1 \in A_2$ and the third because $[G, F_1 \otimes_{\phi_2} id] \in A_2$. To check (iv) observe that

$$(F - F^*)\tilde{\phi}_1(a) = M^{\frac{1}{2}}((F_1 - F_1^*)\phi_1(a) \otimes_{\phi_2} id)$$
$$+ N^{\frac{1}{2}}(G - G^*)\tilde{\phi}_1(a) \bmod \mathcal{K}_C(E_{12}).$$

The first term is in $\mathcal{K}_C(E_{12})$ because $(F_1 - F_1^*)\phi_1(a) \otimes_{\phi_2} id \in A_1$, the second because $G - G^* \in A_2$. This completes the proof that $\mathcal{E}_{12} = (E_{12}, \tilde{\phi}_1, F)$ is a Kasparov $A - C$ module.

To show that $\mathcal{E}_{12}$ is a Kasparov product it now suffices to demonstrate condition (b) of Definition 2.2.7. Observe that

$$[F_1 \otimes_{\phi_2} id, F] = [F_1 \otimes_{\phi_2} id, M^{\frac{1}{2}}](F_1 \otimes_{\phi_2} id) + M^{\frac{1}{2}}[F_1 \otimes_{\phi_2} id, F_1 \otimes_{\phi_2} id]$$
$$+ [F_1 \otimes_{\phi_2} id, N^{\frac{1}{2}}]G + N^{\frac{1}{2}}[F_1 \otimes_{\phi_2} id, G].$$

The first term is in $\mathcal{K}_C(E_{12})$ because $F_1 \otimes_{\phi_2} id \in \mathcal{F}$, the third for the same reason and the fourth because $[F_1 \otimes_{\phi_2} id, G] \in A_2$. Calculating mod $\mathcal{K}_C(E_{12})$ we find

$$\tilde{\phi}_1(a)[F_1 \otimes_{\phi_2} id, F]\tilde{\phi}_1(a)^* = \tilde{\phi}_1(a)M^{\frac{1}{2}}[F_1 \otimes_{\psi_2} id, F_1 \otimes_{\phi_2} id]\tilde{\phi}_1(a)^*$$
$$= 2\tilde{\phi}_1(a)M^{\frac{1}{2}}(F_1^2 \otimes_{\phi_2} id)\tilde{\phi}_1(a)^*$$
$$= 2\tilde{\phi}_1(a)M^{\frac{1}{2}}\tilde{\phi}_1(a)^* \geq 0 \bmod \mathcal{K}_C(E_{12})$$

because

$$M^{\frac{1}{2}}(F_1^2 \otimes_{\phi_2} id)\tilde{\phi}_1(a)^* - M^{\frac{1}{2}}\tilde{\phi}_1(a)^* =$$
$$M^{\frac{1}{2}}((F_1^2 - 1)\phi_1(a^*) \otimes_{\phi_2} id) \in M^{\frac{1}{2}}A_1 \subseteq \mathcal{K}_C(E_{12}).$$

This completes the proof of the existence part.

Assume $\mathcal{E}' = (E_{12}, \tilde{\phi}_1, F') \in \mathbb{E}(A, C)$ is another Kasparov product of $\mathcal{E}_1$ by $\mathcal{E}_2$. Set $A_1 = \mathcal{K}_B(E_1) \otimes_{\phi_2} id + \mathcal{K}_C(E_{12})$, let A_2 be the C^*-subalgebra generated by $[F_1 \otimes_{\phi_2} id, F], [F_1 \otimes_{\phi_2} id, F'], F - F'$ and let $\mathcal{F}$ be the closed linear subspace spanned by $\tilde{\phi}_1(A), F_1 \otimes_{\phi_2} id, F,$ and F'. As above one can check that Corollary 2.2.3 applies to give $M, N \geq 0$ in $\mathcal{L}_C(E_{12})$ of degree 0 such that $M + N = 1$, $MA_1 \subseteq \mathcal{K}_C(E_{12})$, $NA_2 \subseteq \mathcal{K}_C(E_{12})$ and $[\mathcal{F}, M] \subseteq \mathcal{K}_C(E_{12})$. Set $F'' = M^{\frac{1}{2}}(F_1 \otimes_{\phi_2} id) + N^{\frac{1}{2}}F$. As above one can check that $(E_{12}, \tilde{\phi}_1, F'') \in \mathbb{E}(A, C)$. The only difference is that we do not assume that A_2 contains $F^2 - 1, F - F^*$ and $[F, \tilde{\phi}_1(A)]$. In all the places where this was used above it can be replaced by the fact that $(E_{12}, \tilde{\phi}_1, F) \in \mathbb{E}(A, C)$.

Note that

$$[X, F''] = [X, M^{\frac{1}{2}}]F_1 \otimes_{\phi_2} id + M^{\frac{1}{2}}[X, F_1 \otimes_{\phi_2} id] + [X, N^{\frac{1}{2}}]F + N^{\frac{1}{2}}[X, F]$$

for $X = F$ or F'. Corollary 2.2.3 was applied in a way that assures that the first three terms are all in $\mathcal{K}_C(E_{12})$. Thus

$$[F, F''] = N^{\frac{1}{2}}[F, F] = 2N^{\frac{1}{4}}F^2 N^{\frac{1}{4}}$$

and

$$[F', F''] = N^{\frac{1}{2}}[F', F] = N^{\frac{1}{4}}(F'F + F'F)N^{\frac{1}{4}} \bmod \mathcal{K}_C(E_{12})$$

since $F, F' \in \mathcal{F}$. But $N^{\frac{1}{4}}F' = N^{\frac{1}{4}}F \bmod \mathcal{K}_C(E_{12})$ since $F - F' \in A_2$, so that $[F', F''] = [F, F''] = 2N^{\frac{1}{4}}F^2 N^{\frac{1}{4}} \bmod \mathcal{K}_C(E_{12})$. By using that $\tilde{\phi}_1(A) \in \mathcal{F}$ and that $(F^2 - 1)\tilde{\phi}_1(A) \subseteq \mathcal{K}_C(E_{12})$ we see that

$$
\begin{aligned}
\tilde{\phi}_1(a)[F, F'']\tilde{\phi}_1(a)^* &= \tilde{\phi}_1(a)[F', F'']\tilde{\phi}_1(a)^* \\
&= 2N^{\frac{1}{4}}\tilde{\phi}_1(a)F^2\tilde{\phi}_1(a)^* N^{\frac{1}{4}} \\
&= 2N^{\frac{1}{4}}\tilde{\phi}_1(aa^*)N^{\frac{1}{4}} \geq 0 \bmod \mathcal{K}_C(E_{12}).
\end{aligned}
$$

Thus two applications of Lemma 2.1.18 give that $\mathcal{E}_{12}$ is operator homotopic to $\mathcal{E}'$. $\qquad\square$

Now let A, B, C be graded C^*-algebras. For the rest of this chapter, all C^*-algebras denoted by $A, A_1, A_2, \ldots$ are assumed to be separable.

For $\mathcal{E}_1 \in \mathbb{E}(A, B), \mathcal{E}_2 \in \mathbb{E}(B, C)$ we write $\mathcal{E}_{12}$ for a Kasparov product of $\mathcal{E}_1$ by $\mathcal{E}_2$. We want to show that there are maps, $KK(A, B) \times KK(B, C) \to KK(A, C)$ and $K\hat{K}(A, B) \times K\hat{K}(B, C) \to K\hat{K}(A, C)$, given by

$$(2.2.3) \qquad\qquad ([\mathcal{E}_1], [\mathcal{E}_2]) \to [\mathcal{E}_{12}],$$

and

$$(2.2.4) \qquad ([\mathcal{E}_1], [\mathcal{E}_2]) \to \{\mathcal{E}_{12}\}, \qquad \mathcal{E}_1 \in \mathbb{E}(A, B), \mathcal{E}_2 \in \mathbb{E}(B, C),$$

respectively. We will do this by showing it in the $K\hat{K}$-case and then proceed to show that $\mu: K\hat{K}(A, B) \to KK(A, B)$ is an isomorphism.

Lemma 2.2.9. *Let* $\mathcal{E}_1 = (E_1, \phi_1, F_1)$, $\mathcal{E}_3 = (E_3, \phi_3, F_3) \in \mathbb{E}(A, B)$, $\mathcal{E}_2 = (E_2, \phi_2, F_2)$, $\mathcal{E}_4 = (E_4, \phi_4, F_4) \in \mathbb{E}(B, C)$.

If $\mathcal{E}_1 \simeq \mathcal{E}_3$, there are Kasparov products $\mathcal{E}_{12}$ of $\mathcal{E}_1$ by $\mathcal{E}_2$ and $\mathcal{E}_{32}$ of $\mathcal{E}_3$ by $\mathcal{E}_2$ such that $\mathcal{E}_{12} \simeq \mathcal{E}_{32}$.

If $\mathcal{E}_2 \simeq \mathcal{E}_4$, there are Kasparov products $\mathcal{E}_{12}$ of $\mathcal{E}_1$ by $\mathcal{E}_2$ and $\mathcal{E}_{14}$ of $\mathcal{E}_1$ by $\mathcal{E}_4$ such that $\mathcal{E}_{12} \simeq \mathcal{E}_{14}$.

Proof. Let $\psi : \mathcal{E}_1 \to \mathcal{E}_3$ be an isomorphism of Kasparov $A - B$-modules. Then there is an isomorphism

$$\tilde{\psi} : E_{12} = E_1 \otimes_{\phi_2} E_2 \to E_{32} = E_3 \otimes_{\phi_2} E_2$$

of Hilbert C-modules given on simple tensors by

$$\tilde{\psi}(e_1 \otimes_{\phi_2} e_2) = \psi(e_1) \otimes_{\phi_2} e_2, \quad e_i \in E_i, \quad i = 1, 2.$$

If $\mathcal{E}_{12} = (E_{12}, \tilde{\phi}_1, F)$ is a Kasparov product, then $(E_{32}, \tilde{\phi}_3, \tilde{\psi} F \tilde{\psi}^{-1}) = \mathcal{E}_{32}$ is a Kasparov product of $\mathcal{E}_3$ by $\mathcal{E}_2$ such that $\mathcal{E}_{12} \simeq \mathcal{E}_{32}$ via $\tilde{\psi}$. The straightforward verification of this is left to the reader.

Let $\psi : \mathcal{E}_2 \to \mathcal{E}_4$ be an isomorphism of Kasparov $A - B$-modules. Then there is an isomorphism $\tilde{\psi} : E_{12} = E_1 \otimes_{\phi_2} E_2 \to E_{14} = E_1 \otimes_{\phi_4} E_4$ given by $\tilde{\psi}(e_1 \otimes_{\phi_2} e_2) = e_1 \otimes_{\phi_4} \psi(e_2)$. If $\mathcal{E}_{12} = (E_{12}, \tilde{\phi}_1, F)$ is a Kasparov product of $\mathcal{E}_1$ by $\mathcal{E}_2$, then $(E_{14}, \tilde{\phi}_1, \tilde{\psi} \circ F \circ \tilde{\psi}^{-1}) = \mathcal{E}_{14}$ is a Kasparov product of $\mathcal{E}_1$ by $\mathcal{E}_4$. Again the verification is left to the reader. $\square$

Lemma 2.2.10. *Let* $\mathcal{E}_1 \in \mathbb{D}(A, B), \mathcal{E}_2 \in \mathbb{E}(B, C)$. *Then any Kasparov product of* $\mathcal{E}_1$ *by* $\mathcal{E}_2$ *is operator homotopic to a degenerate element, i.e. to an element in* $\mathbb{D}(A, C)$.

Proof. Let $\mathcal{E}_1 = (E_1, \phi_1, F_1), \mathcal{E}_2 = (E_2, \phi_2, F_2)$. Since $\mathcal{E}_1$ is degenerate, $\mathcal{E} = (E_{12}, \tilde{\phi}_1, F_1 \otimes_{\phi_2} id) \in \mathbb{D}(A, C)$. If $\mathcal{F} = (E_{12}, \tilde{\phi}_1, F)$ is any Kasparov product of $\mathcal{E}_1$ by $\mathcal{E}_2$, then $\tilde{\phi}_1(a)[F_1 \otimes_{\phi_2} id, F]\tilde{\phi}_1(a)^* \geq 0 \mod \mathcal{K}_C(E_{12})$ by definition. Hence $\mathcal{F}$ and $\mathcal{E}$ are operator homotopic by Lemma 2.1.18. $\square$

Lemma 2.2.11. *Let* $\mathcal{E}_1 \in \mathbb{E}(A, B), \mathcal{E}_2 \in \mathbb{D}(B, C)$. *Then there is a degenerate Kasparov product* $\mathcal{E}_{12} \in \mathbb{D}(A, C)$ *of* $\mathcal{E}_1$ *by* $\mathcal{E}_2$.

Proof. Let $\mathcal{E}_1 = (E_1, \phi_1, F_1), \mathcal{E}_2 = (E_2, \phi_2, F_2)$. Define $S_{E_1} \otimes_{\phi_2} F_2 \in \mathcal{L}_C(E_1 \otimes_{\phi_2} E_2)$ by $S_{E_1} \otimes_{\phi_2} F_2(e_1 \otimes_{\phi_2} e_2) = S_{E_1}(e_1) \otimes_{\phi_2} F_2(e_2), e_1 \in E_1, e_2 \in E_2$. This can be done because $[F_2, \phi_2(b)] = 0, b \in B$, since $\mathcal{E}_2$ is degenerate (cf. 2.2.6 (e)). We assert that $\mathcal{E}_{12} = (E_{12}, \tilde{\phi}_1, S_{E_1} \otimes_{\phi_2} F_2) \in \mathbb{D}(A, C)$. By Lemma 2.2.6 (e), $\mathcal{E}_{12}$ will then be a Kasparov product of $\mathcal{E}_1$ by

$\mathcal{E}_2$, and the proof will be complete. First observe that $[\tilde{\phi}_1(a), S_{E_1} \otimes_{\phi_2} F_2] = 0$ by Lemma 2.2.6 (e). Fix $a \in A, e_1 \in E_1, e_2 \in E_2$ and let $\{v_i\}$ be an approximate unit for B of degree 0. Then

$$((S_{E_1} \otimes_{\phi_2} F_2)^2 - 1)\tilde{\phi}_1(a))(e_1 \otimes_{\phi_2} e_2) =$$
$$\phi_1(a)(e_1) \otimes_{\phi_2} (F_2^2 - 1)(e_2) =$$
$$\lim_i \phi_1(a)(e_1 v_i) \otimes_{\phi_2} (F_2^2 - 1)(e_2) =$$
$$\lim_i \phi_1(a)(e_1) \otimes_{\phi_2} \phi_2(v_i)(F_2^2 - 1)(e_2) =$$
$$\lim_i \phi_1(a)(e_1) \otimes_{\phi_2} (F_2^2 - 1)\phi_2(v_i)(e_2) = 0.$$

By observing that $(S_{E_1} \otimes_{\phi_2} F_2)^* = S_{E_1} \otimes_{\phi_2} F_2^*$ we get similarly

$$(S_{E_1} \otimes_{\phi_2} F_2 - (S_{E_1} \otimes_{\phi_2} F_2)^*)\tilde{\phi}_1(a)(e_1 \otimes_{\phi_2} e_2) =$$
$$S_{E_1}\phi_1(a)(e_1) \otimes_{\phi_2} (F_2 - F_2^*)(e_2) =$$
$$\lim_i S_{E_1}\phi_1(a)(e_1) \otimes_{\phi_2} (F_2 - F_2^*)\phi_2(v_i)(e_2) = 0. \qquad \square$$

Lemma 2.2.12. *Let $\mathcal{E}_1, \mathcal{E}_2 \in \mathbb{E}(A, B)$, $\mathcal{E}_3 \in \mathbb{E}(B, C)$. If $\mathcal{E}_{i3}$ is a Kasparov product of $\mathcal{E}_i$ by $\mathcal{E}_3$, $i = 1, 2$, then $\mathcal{E}_{13} \oplus \mathcal{E}_{23}$ is isomorphic to a Kasparov product of $\mathcal{E}_1 \oplus \mathcal{E}_2$ by $\mathcal{E}_3$.*

Proof. Let $\mathcal{E}_i = (E_i, \phi_i, F_i)$, $i = 1, 2, 3$, and $\mathcal{E}_{i3} = (E_{i3}, \tilde{\phi}_i, G_i)$, $i = 1, 2$. There is an isomorphism

$$\psi : (E_1 \oplus E_2) \otimes_{\phi_3} E_3 \to E_1 \otimes_{\phi_3} E_3 \oplus E_2 \otimes_{\phi_3} E_3 = E_{13} \oplus E_{23}$$

of graded Hilbert C-modules given by

$$\psi((e_1, e_2) \otimes_{\phi_3} e_3) = (e_1 \otimes_{\phi_3} e_3, e_2 \otimes_{\phi_3} e_3), \quad e_i \in E_i, \; i = 1, 2, 3.$$

Set

$$\mathcal{F} = ((E_1 \oplus E_2) \otimes_{\phi_3} E_3, (\phi_1 \oplus \phi_2) \otimes_{\phi_3} id, \psi^{-1}(G_1 \oplus G_2)\psi).$$

It is clearly sufficient to show that $\mathcal{F}$ is a Kasparov product of $\mathcal{E}_1 \oplus \mathcal{E}_2$ by $\mathcal{E}_3$ since ψ is then an isomorphism of Kasparov $A - C$-modules.

It is easy to see that $\mathcal{F} \in \mathbb{E}(A, C)$ since $\mathcal{E}_{13} \oplus \mathcal{E}_{23} \in \mathbb{E}(A, C)$. Thus it suffices to check the conditions (a) and (b) of Definition 2.2.7. The E_3-tensor operators $T_{(e_1, e_2)}$, $e_1 \in E_1, e_2 \in E_2$, for $E_1 \oplus E_2$ are given by

$T_{(e_1,e_2)} = \psi^{-1}(T'_{e_1} \oplus T''_{e_2})$ where T'_{e_1}, $e_1 \in E_1$, and T''_{e_2}, $e_2 \in E_2$, are the E_3 tensor operators for E_1 and E_2, respectively. Thus

$$\psi(T_{(e_1,e_2)}F_3 - \psi^{-1}(G_1 \oplus G_2)\psi T_{(S_{E_1(e_1)}, S_{E_2(e_2)})}) =$$

$$(T'_{e_1}F_3 - G_1 T'_{S_{E_1}(e_1)}) \oplus (T''_{e_2}F_3 - G_2 T''_{S_{E_2}(e_2)}).$$

It follows that the first condition for $\psi^{-1}(G_1 \oplus G_2)\psi$ to be an F_3-connection for $E_1 \oplus E_2$ is satisfied. The second follows from a similar identity. With $\tilde{\phi}_{12}(a) = (\phi_1(a) \oplus \phi_2(a)) \otimes_{\phi_3} id$ for $a \in A$ we have that

$$\psi(\tilde{\phi}_{12}(a))\left[(F_1 \oplus F_2) \otimes_{\phi_3} id, \psi^{-1}(G_1 \oplus G_2)\psi\right]\tilde{\phi}_{12}(a^*)\psi^{-1} =$$
$$\tilde{\phi}_1(a)[F_1 \otimes_{\phi_3} id, G_1]\tilde{\phi}_1(a^*) \oplus \tilde{\phi}_2(a)[F_2 \otimes_{\phi_3} id, G_2]\tilde{\phi}_2(a^*), \ a \in A.$$

It follows readily from this that $\mathcal{F}$ is a Kasparov product of $\mathcal{E}_1 \oplus \mathcal{E}_2$ by $\mathcal{E}_3$.

$\square$

Lemma 2.2.13. *Let $\mathcal{E}_1 \in \mathbb{E}(A,B)$, $\mathcal{E}_2, \mathcal{E}_3 \in \mathbb{E}(B,C)$. If $\mathcal{E}_{1i}$ is a Kasparov product of $\mathcal{E}_1$ by $\mathcal{E}_i$, $i = 2,3$, then $\mathcal{E}_{12} \oplus \mathcal{E}_{13}$ is isomorphic to a Kasparov product of $\mathcal{E}_1$ by $\mathcal{E}_2 \oplus \mathcal{E}_3$.*

Proof. Let $\mathcal{E}_i = (E_i, \phi_i, F_i)$, $i = 1,2,3$, $\mathcal{E}_{1i} = (E_1 \otimes_{\phi_i} E_i, \tilde{\phi}_1, G_i)$, $i = 2,3$. There is an isomorphism

$$\psi : E_1 \otimes_{\phi_2 \oplus \phi_3} (E_2 \oplus E_3) \to (E_1 \otimes_{\phi_2} E_2) \oplus (E_1 \otimes_{\phi_3} E_3)$$

of graded Hilbert C-modules given by

$$\psi(e_1 \otimes_{\phi_2 \oplus \phi_3} (e_2, e_3)) = (e_1 \otimes_{\phi_2} e_2, e_1 \otimes_{\phi_3} e_3), e_i \in E_i, \ i = 1,2,3.$$

Set

$$\mathcal{F} = (E_1 \otimes_{\phi_2 \oplus \phi_3} (E_2 \oplus E_3), \phi_1 \otimes_{\phi_2 \oplus \phi_3} id, \psi^{-1}(G_2 \oplus G_3)\psi).$$

As in the preceding proof it suffices now to show that $\mathcal{F}$ is a Kasparov product of $\mathcal{E}_1$ by $\mathcal{E}_2 \oplus \mathcal{E}_3$. $\mathcal{F} \in \mathbb{E}(A,C)$ because $\mathcal{E}_{12} \oplus \mathcal{E}_{13} \in \mathbb{E}(A,C)$. Thus we need only check the conditons (a) and (b) of Definition 2.2.7. Note that if T'_e and T''_e, $e \in E_1$, are the E_2- and E_3-tensor operators for E_1, respectively, then the $E_2 \oplus E_3$-tensor operators for E_1, $T_e, e \in E_1$, are given by $T_e = \psi^{-1}(T'_e \oplus T''_e)$. Thus

$$((F_2 \oplus F_3)T_e^* - T_{S_{E_1}(e)}^* \psi^{-1}(G_2 \oplus G_3)\psi)\psi^{-1} =$$

$$(F_2 T_e'^* - T_{S_{E_1}(e)}'^* G_2) \oplus (F_3 T''_e^* - T_{S_{E_1}(e)}''^* G_3).$$

This identity gives that one of the conditions for $\psi^{-1}(G_2 \oplus G_3)\psi$ to be an $F_2 \oplus F_3$-connection for E_1 is satisfied. The other follows similarly. Note finally that with $\tilde{\phi}_1(a) = \phi_1(a) \otimes_{\phi_2} id$ we have

$$\psi(\phi_1(a) \otimes_{\phi_2 \oplus \phi_3} id[F_1 \otimes_{\phi_2 \oplus \phi_3} id, \psi^{-1}(G_2 \oplus G_3)\psi]\phi_1(a^*) \otimes_{\phi_2 \oplus \phi_3} id)\psi^{-1} =$$

$$\tilde{\phi}_1(a)[F_1 \otimes_{\phi_2} id, G_2]\,\tilde{\phi}_1(a^*) \oplus \phi_1(a) \otimes_{\phi_3} id[F_1 \otimes_{\phi_3} id, G_3]\phi_1(a^*) \otimes_{\phi_3} id,\ a \in A.$$

This identity gives that $\mathcal{F}$ is indeed a Kasparov product of $\mathcal{E}_1$ by $\mathcal{E}_2 \oplus \mathcal{E}_3$.

$\square$

Lemma 2.2.14. *Let* $\mathcal{E}_1^t = (E_1, \phi_1, F_t^1) \in \mathbb{E}(A, B)$, $\mathcal{E}_2^t = (E_2, \phi_2, F_t^2) \in \mathbb{E}(B, C)$, $t \in [0, 1]$. *Assume that* $t \to F_t^1 \in \mathcal{L}_B(E_1)$ *and* $t \to F_t^2 \in \mathcal{L}_C(E_2)$ *are both norm continuous. Then there is a Kasparov product,* $\mathcal{E}_{12}^0$, *of* $\mathcal{E}_1^0$ *by* $\mathcal{E}_2^0$ *which is operator homotopic to a Kasparov product,* $\mathcal{E}_{12}^1$, *of* $\mathcal{E}_1^1$ *by* $\mathcal{E}_2^1$.

Proof. Inspection of the proof of Proposition 2.2.5 reveals that there is a norm continuous path $t \to G_t \in \mathcal{L}_C(E_{12})$, $t \in [0, 1]$, of degree 1 elements such that G_t is an F_t^2 connection for E_1 for each $t \in [0, 1]$. Set $A_1 = \mathcal{K}_B(E_1) \otimes_{\phi_2} id + \mathcal{K}_C(E_{12})$, let A_2 be the C^*-algebra generated by $[G_t, \tilde{\phi}_1(A)], G_t^2 - 1, G_t - G_t^*$ and $[F_t^1 \otimes_{\phi_2} id, G_t]$, $t \in [0, 1]$, and let $\mathcal{F}$ be the closed subspace generated by $\tilde{\phi}_1(A), F_t^1 \otimes_{\phi_2} id$ and G_t, $t \in [0, 1]$. Since everything that varies with t is continuous in norm, the size conditions of Corollary 2.2.3 are satisfied. By repeating the argument from the proof of Theorem 2.2.8 for arbitrary $t \in [0, 1]$ one sees that Corollary 2.2.3 applies to give positive degree 0 elements M, N in $\mathcal{L}_C(E_{12})$ such that $M + N = 1$, $MA_1 \subseteq \mathcal{K}_C(E_{12})$, $NA_2 \subseteq \mathcal{K}_C(E_{12})$ and $[\mathcal{F}, M] \subseteq \mathcal{K}_C(E_{12})$. Set $F_t = M^{\frac{1}{2}}(F_t^1 \otimes_{\phi_2} id) + N^{\frac{1}{2}}G_t$, $t \in [0, 1]$. Then $t \to F_t$ is norm continuous and as in the proof of Theorem 2.2.8 it follows that $\mathcal{E}_{12}^t = (E_{12}, \tilde{\phi}_1, F_t)$ is a Kasparov product of $\mathcal{E}_1^t$ by $\mathcal{E}_2^t$ for each $t \in [0, 1]$. The conclusion follows. $\square$

Theorem 2.2.15. *There is a map*

$$\cdot\ :\ K\hat{K}(A, B) \times K\hat{K}(B, C) \to K\hat{K}(A, C),$$

called the Kasparov product; it is given by $\{\mathcal{E}_1\} \cdot \{\mathcal{E}_2\} = \{\mathcal{E}_{12}\}$, $\mathcal{E}_1 \in \mathbb{E}(A, B)$, $\mathcal{E}_2 \in \mathbb{E}(B, C)$, *where* $\mathcal{E}_{12}$ *is a Kasparov product of* $\mathcal{E}_1$ *by* $\mathcal{E}_2$.

The following relations hold :

(i) $x_1 \cdot (x_2 + x_3) = x_1 \cdot x_2 + x_1 \cdot x_3$, $x_1 \in K\hat{K}(A,B)$, $x_2, x_3 \in K\hat{K}(B,C)$,

(ii) $(x_1 + x_2) \cdot x_3 = x_1 \cdot x_3 + x_2 \cdot x_3$, $x_1, x_2 \in K\hat{K}(A,B)$, $x_3 \in K\hat{K}(B,C)$,

(iii) *when* $f : A_1 \to A$, $g : C \to C_1$ *are homomorphisms of graded C^*-algebras, then*
$$f^*(x) \cdot y = f^*(x \cdot y), \text{ and } x \cdot g_*(y) = g_*(x \cdot y),$$
$$x \in K\hat{K}(A,B), \ y \in K\hat{K}(B,C),$$

(iv) *when* $h : B_1 \to B_2$ *is a homomorphism of graded C^*-algebras, then*

$$h_*(x) \cdot y = x \cdot h^*(y), \quad x \in K\hat{K}(A,B_1), y \in K\hat{K}(B_2,C).$$

Proof. The product is well-defined because of the preceding 6 lemmas. The first two relations, which express the additivity of the product in both variables, follows from Lemma 2.2.12 and Lemma 2.2.13. The relation $f^*(x) \cdot y = f^*(x \cdot y)$ follows from the observation that $f^*(\mathcal{E}_{12})$ is a Kasparov product of $f^*(\mathcal{E}_1)$ by $\mathcal{E}_2$ when $\mathcal{E}_{12}$ is a Kasparov product of $\mathcal{E}_1$ by $\mathcal{E}_2$.

To check the relation $x \cdot g_*(y) = g_*(x \cdot y)$, let $\mathcal{E}_1 = (E_1, \phi_1, F_1) \in \mathbb{E}(A,B)$, $\mathcal{E}_2 = (E_2, \phi_2, F_2) \in \mathbb{E}(B,C)$ and let $\mathcal{E}_{12} = (E_{12}, \tilde{\phi}_1, G)$ be a Kasparov product of $\mathcal{E}_1$ by $\mathcal{E}_2$. Set $\phi = \phi_2 \otimes_g id : B \to \mathcal{L}_{C_1}(E_2 \otimes_g C_1)$. There is an isomorphism

$$\beta : E_{12} \otimes_g C_1 = (E_1 \otimes_{\phi_2} E_2) \otimes_g C_1 \to E_1 \otimes_\phi (E_2 \otimes_g C_1)$$

of graded Hilbert C^*-modules given on simple tensors by

$$\beta((e_1 \otimes_{\phi_2} e_2) \otimes_g d) = e_1 \otimes_\phi (e_2 \otimes_g d), e_1 \in E_1, e_2 \in E_2, d \in C_1.$$

We assert that

$$\mathcal{F} = (E_1 \otimes_\phi (E_2 \otimes_g C_1), \beta(\tilde{\phi}_1 \otimes_g id)\beta^{-1}, \beta(G \otimes_g id)\beta^{-1})$$

is a Kasparov product of $\mathcal{E}_1$ by $g_*(\mathcal{E}_2)$. This will prove the desired relation since β is then an isomorphism of Kasparov $A - C_1$-modules. To check that $\beta(G \otimes_g id)\beta^{-1}$ is an $F_2 \otimes_g id$-connection for E_1, let T_x, $x \in E_1$, be the $E_2 \otimes_g C_1$-tensor operators for E_1 and T'_x, $x \in E_1$, the E_2-tensor operators for E_1. Then

$$\beta^{-1}(T_x(F_2 \otimes_g id) - \beta(G \otimes_g id)\beta^{-1} T_{S_{E_1}(x)}) = (T'_x F_2) \otimes_g id - (GT'_{S_{E_1}(x)}) \otimes_g id$$

which is in $\mathcal{K}_{C_1}(E_2 \otimes_g C_1, (E_1 \otimes_{\phi_2} E_2) \otimes_g C_1)$ because G is an F_2-connection for E_1 and $\mathcal{K}_C(E_1 \otimes_{\phi_2} E_2) \otimes_g id \subseteq \mathcal{K}_{C_1}((E_1 \otimes_{\phi_2} E_2) \otimes_g C_1)$ by Lemma 1.2.8. It follows that

$$T_x(F_2 \otimes_g id) - \beta G \beta^{-1} T_{S_{E_1}(x)} \in \mathcal{K}_{C_1}(E_2 \otimes_g C_1, E_1 \otimes_\phi (E_2 \otimes_g C_1))$$

for all $x \in E_1$. We leave the reader to finish the proof that $\beta(G \otimes_g id)\beta^{-1}$ is an $F_2 \otimes_g id$-connection for E_1. That $\mathcal{F}$ is a Kasparov $A - C_1$-module follows from the fact that $\mathcal{E}_{12}$ is a Kasparov $A - C$-module. That $\mathcal{F}$ is a Kasparov product of $\mathcal{E}_1$ by $g_*(\mathcal{E}_2)$ follows now from the observation that $\beta^{-1}[F_1 \otimes_\phi id, \beta(G \otimes_g id)\beta^{-1}]\beta = [F_1 \otimes_{\phi_2} id, G] \otimes_g id$.

To check the last relation, $h_*(x) \cdot y = x \cdot h^*(y)$, let $\mathcal{E}_1 = (E_1, \phi_1, F_1) \in \mathbb{E}(A, B_1)$, $\mathcal{E}_2 = (E_2, \phi_2, F_2) \in \mathbb{E}(B_2, C)$. There is an isomorphism

$$\psi : (E_1 \otimes_h B_2) \otimes_{\phi_2} E_2 \to E_1 \otimes_{\phi_2 \circ h} E_2$$

of graded Hilbert C-modules given by

$$\psi((e_1 \otimes_h b) \otimes_{\phi_2} e_2) = e_1 \otimes_{\phi_2 \circ h} \phi_2(b)e_2, \ e_1 \in E_1, b \in B_2, e_2 \in E_2.$$

Let $\mathcal{F} = (E_1 \otimes_{\phi_2 \circ h} E_2, \phi_1 \otimes_{\phi_2 \circ h} id, G)$ be a Kasparov product of $\mathcal{E}_1$ by $h^*(\mathcal{E}_2)$. It suffices to check that

$$((E_1 \otimes_h B_2) \otimes_{\phi_2} E_2, (\phi_1 \otimes_h id) \otimes_{\phi_2} id, \psi^{-1}G\psi)$$

is a Kasparov product of $h_*(\mathcal{E}_1)$ by $\mathcal{E}_2$. So let $T_e, e \in E_1$, be the E_2 tensor operators for E_1 and T'_x, $x \in E_1 \otimes_h B_2$, the E_2 tensor operators for $E_1 \otimes_h B_2$. Note that $\psi T'_{e_1 \otimes_h b} = T_{e_1} \phi_2(b)$, $e_1 \in E_1$, $b \in B_2$. Thus

$$\psi(T'_{e_1 \otimes_h b} F_2 - \psi^{-1} G \psi T'_{S_{E_1}(e_1) \otimes_h \beta_{B_2}(b)}) =$$

$$T_{e_1} \phi_2(b) F_2 - G T_{S_{E_1}(e_1)} \phi_2(\beta_{B_2}(b)).$$

Since $(E_2, \phi_2, F_2) \in \mathbb{E}(B_2, C)$, we have in particular that $\phi_2(b)F_2 = F_2 \phi_2(\beta_{B_2}(b))$ mod $\mathcal{K}_C(E_2)$. Using this it is easy to conclude from the above identity that the first condition for $\psi^{-1}G\psi$ to be an F_2 connection for $E_1 \otimes_h B_2$ is satisfied. The other follows from similar considerations. The proof is now completed as most of the preceding lemmas by observing that

$$\psi(\phi_1(a) \otimes_h id)[(F_1 \otimes_h id) \otimes_{\phi_2} id, \psi^{-1}G\psi](\phi_1(a^*) \otimes_h id)\psi^{-1} =$$

$$\phi_1(a) \otimes_{\phi_2 \circ h} id[F_1 \otimes_{\phi_2 \circ h} id, G]\phi_1(a^*) \otimes_{\phi_2 \circ h} id, \ a \in A. \qquad \square$$

The additivity of the Kasparov product in both variables is usually expressed by saying that it is a bilinear pairing.

We proceed now to show that $\mu : K\hat{K}(A,B) \to KK(A,B)$ is an isomorphism. For this we need a "unit" for the product in $K\hat{K}$. To describe it, consider $\mathbb{C}$ as a Hilbert $\mathbb{C}$-module; trivially graded (i.e. graded by the identity operator). Then $u_{\mathbb{C}} = (\mathbb{C}, id, 0) \in \mathbb{E}(\mathbb{C},\mathbb{C})$. Let $u_B = \tau_B(u_{\mathbb{C}}) \in \mathbb{E}(B,B)$, i.e., $u_B = (B, id, 0)$ where B is considered as a Hilbert B-module graded by β_B. Set $1_B = \{u_B\} \in K\hat{K}(B,B)$.

Lemma 2.2.16. $x \cdot 1_B = x$, $x \in K\hat{K}(A,B)$.

Proof.   Let $x = \{\mathcal{E}\}$, $\mathcal{E} = (E, \phi, F) \in \mathbb{E}(A,B)$. Define a map $\psi : E \otimes_{id} B \to E$ by $\psi(e \otimes_{id} b) = eb$, $e \in E$, $b \in B$. Then ψ is an isomorphism of graded Hilbert B-modules. It suffices to check that

$$(E \otimes_{id} B, \phi \otimes_{id} id, \psi^{-1}F\psi)$$

is a Kasparov product of $\mathcal{E}$ by u_B. To see that $\psi^{-1}F\psi$ is a 0-connection for E, let $\{v_i\}$ be an approximate unit for B of degree 0. Then

$$\psi^{-1}F\psi T_{S_E(ev_i)} \to \psi^{-1}F\psi T_{S_E(e)}$$

so to prove the first condition for $\psi^{-1}F\psi$ to be a 0 connection it suffices to check that $\psi^{-1}F\psi T_{S_E(ev_i)} \in \mathcal{K}_B(B, E \otimes_{id} B)$ for fixed i. This follows by observing that

$$\begin{aligned}
\psi^{-1}F\psi T_{S_E(ev_i)}(b) &= FS_E(ev_i) \otimes_{id} b \\
&= FS_E(e) \otimes_{id} v_i b \\
&= \Theta_{x,y}(b), \quad b \in B,
\end{aligned}$$

when $x = FS_E(e) \otimes_{id} v_i^{\frac{1}{2}}$ and $y = v_i^{\frac{1}{2}}$. The second condition follows (as always) similarly. The proof is completed by observing that

$$\begin{aligned}
\psi(\phi(a) \otimes_{id} id\,[F \otimes_{id} id, \psi^{-1}F\psi]\phi(a^*) \otimes_{id} id)\psi^{-1} &= \phi(a)\,[F, F]\phi(a)^* \\
&= 2\phi(a)F^2\phi(a^*) = 2\phi(a)\phi(a^*) \bmod \mathcal{K}_B(E). \qquad \square
\end{aligned}$$

Theorem 2.2.17. *The map* $\mu : K\hat{K}(A,B) \to KK(A,B)$ *is an isomorphism.*

Proof. By Theorem 2.1.23 it suffices to show that when $\mathcal{E} \in \mathbb{E}(A, IB)$, then $\pi_{0*}\{\mathcal{E}\} = \pi_{1*}\{\mathcal{E}\}$ in $K\hat{K}(A, B)$. Let $f_t : C[0, 1] \to \mathbb{C}$ be evaluation at $t \in [0, 1]$. Then clearly $\pi_t^*(u_B) = \tau_B(f_t^*(u_{\mathbb{C}}))$. Using Theorem 2.2.15 and Lemma 2.2.16 we get

$$
\begin{aligned}
\pi_{0*}\{\mathcal{E}\} &= \pi_{0*}\{\mathcal{E}\} \cdot 1_B \\
&= \{\mathcal{E}\} \cdot \pi_0^*(1_B) \\
&= \{\mathcal{E}\} \cdot \{\pi_0^*(u_B)\} \\
&= \{\mathcal{E}\} \cdot \tau_B\{f_0^*(u_{\mathbb{C}})\},
\end{aligned}
$$

and in the same way

$$
\pi_{1*}\{\mathcal{E}\} = \{\mathcal{E}\} \cdot \tau_B\{f_1^*(u_{\mathbb{C}})\}.
$$

Thus it suffices to show that $\{f_0^*(u_{\mathbb{C}})\} = \{f_1^*(u_{\mathbb{C}})\}$ in $K\hat{K}(C[0, 1], \mathbb{C})$. The proof of this proceeds in two steps. The first gives a description of $\{f_t^*(u_{\mathbb{C}})\}$ which is more suitable for the purpose, involving infinite dimensional Hilbert spaces and Fredholm operators. Note that $f_t^*(u_{\mathbb{C}}) = (\mathbb{C}, f_t, 0)$. Let now $\mathcal{H}$ be any infinite dimensional separable Hilbert space, and let $\mathcal{H} \oplus \mathcal{H}$ be graded by the symmetry given in matrix form as $\begin{bmatrix} 1 & 0 \\ 0 & -1 \end{bmatrix}$. Define a graded homomorphism $\tilde{f}_t : C[0, 1] \to \mathcal{B}(\mathcal{H} \oplus \mathcal{H})$ by $\tilde{f}_t(g) = g(t)1$, $g \in C[0, 1]$. (Although $\mathcal{H} \oplus \mathcal{H}$ should be considered as a Hilbert $\mathbb{C}$-module, we write $\mathcal{B}(\mathcal{H} \oplus \mathcal{H})$ for the C^*-algebra $\mathcal{L}_\mathbb{C}(\mathcal{H} \oplus \mathcal{H})$, since it is nothing but the bounded linear operators, $\mathcal{B}(\mathcal{H} \oplus \mathcal{H})$, in this case.) Let $T \in \mathcal{B}(\mathcal{H})$ be a Fredholm operator of index 1 which is unitary mod $\mathcal{K}(\mathcal{H})$.

Assertion A: The triple $\left(\mathcal{H} \oplus \mathcal{H}, \tilde{f}_t, \begin{bmatrix} 0 & T^* \\ T & 0 \end{bmatrix}\right) \in \mathbb{E}(C[0, 1], \mathbb{C})$ represents $\{f_t^*(u_{\mathbb{C}})\}$ in $K\hat{K}(C[0, 1], \mathbb{C})$.

To prove this assertion, observe first that the conditions on T ensure that $\left(\mathcal{H} \oplus \mathcal{H}, \tilde{f}_t, \begin{bmatrix} 0 & T^* \\ T & 0 \end{bmatrix}\right)$ is a Kasparov $C[0, 1] - \mathbb{C}$-module. Furthermore, by appendix A.2, any other Fredholm operator S which is unitary mod $\mathcal{K}(\mathcal{H})$ and has index 1 can be connected to T through a norm continuous path of Fredholm operators (of index 1), all of which are unitary mod $\mathcal{K}(\mathcal{H})$. It follows that $\left(\mathcal{H} \oplus \mathcal{H}, \tilde{f}_t, \begin{bmatrix} 0 & S^* \\ S & 0 \end{bmatrix}\right)$ is operator homotopic to $\left(\mathcal{H} \oplus \mathcal{H}, \tilde{f}_t, \begin{bmatrix} 0 & T^* \\ T & 0 \end{bmatrix}\right)$. Thus, to prove the assertion, it suffices to find one Fredholm operator T of index 1 which is unitary mod $\mathcal{K}(\mathcal{H})$ such that $\left(\mathcal{H} \oplus \mathcal{H}, \tilde{f}_t, \begin{bmatrix} 0 & T^* \\ T & 0 \end{bmatrix}\right)$ represents $\{f_t^*(u_\mathbb{C})\}$ in $K\hat{K}(C[0,1], \mathbb{C})$.

To find T, observe that $\left(\mathcal{H} \oplus \mathcal{H}, \tilde{f}_t, \begin{bmatrix} 0 & 1 \\ 1 & 0 \end{bmatrix}\right)$ is a degenerate Kasparov $C[0,1] - \mathbb{C}$-module. Choose an orthonormal basis $\{\psi_1, \psi_2, \psi_3, \ldots\}$ in $\mathcal{H}$ and define $W_0 : \mathbb{C} \oplus \mathcal{H} \to \mathcal{H}$ by $W_0(1,0) = \psi_1$, $W_0(0, \psi_i) = \psi_{i+1}$, $i \in \mathbb{N}$. Then W_0 is unitary and $W = W_0 \oplus id : \mathbb{C} \oplus \mathcal{H} \oplus \mathcal{H} \to \mathcal{H} \oplus \mathcal{H}$ is an isomorphism of graded Hilbert $\mathbb{C}$-modules. Note that $AdW \circ (f_t \oplus \tilde{f}_t) = \tilde{f}_t$, so that $\left(\mathcal{H} \oplus \mathcal{H}, \tilde{f}_t, W\left(0 \oplus \begin{bmatrix} 0 & 1 \\ 1 & 0 \end{bmatrix}\right) W^*\right)$ represents $\{f_t^*(u_\mathbb{C})\}$ in $K\hat{K}(C[0,1], \mathbb{C})$. Since $W\left(0 \oplus \begin{bmatrix} 0 & 1 \\ 1 & 0 \end{bmatrix}\right) W^*$ is selfadjoint of degree 1, it has the form $\begin{bmatrix} 0 & T^* \\ T & 0 \end{bmatrix}$ for some $T \in \mathcal{B}(\mathcal{H})$. It is easy to check that T is in fact the dual of the unilateral shift, i.e., $T\psi_i = \psi_{i-1}$, $i \geq 2$, $T\psi_1 = 0$. Thus T is unitary mod $\mathcal{K}(\mathcal{H})$ and has index 1. This completes the proof of the assertion.

Now we come to the second and final step in the proof. Let $\mathcal{H} = L^2[0, 2\pi]$ and define $\Phi : C[0, 2\pi] \to \mathcal{B}(L^2[0, 2\pi])$ by $(\Phi(g)\psi)(x) = g(x)\psi(x)$, $x \in [0, 2\pi]$, $g \in C[0, 2\pi]$, $\psi \in L^2[0, 2\pi]$. Consider the orthonormal basis $\{1, \sin x, \cos x, \sin 2x, \cos 2x, \ldots\}$ in $\mathcal{H}$ and define $D \in \mathcal{B}(\mathcal{H})$ by $D(1) = 0$, $D(\sin nx) = \cos nx$, $D(\cos nx) = -\sin nx$, $n \in \mathbb{N}$. Then $D = -D^*$ and $D^2 + 1$ is the projection onto $\mathbb{C}1$ so that, in particular, $D^2 + 1 \in \mathcal{K}(\mathcal{H})$. We assert that $D\Phi(f) - \Phi(f)D \in \mathcal{K}(\mathcal{H})$, $f \in C[0, 2\pi]$. To see this observe that $\Lambda(f) = D\Phi(f) - \Phi(f)D, f \in C[0, 2\pi]$, defines a bounded linear self-adjoint map $\Lambda : C[0, 2\pi] \to \mathcal{B}(\mathcal{H})$ satisfying $\Lambda(fg) = \Lambda(f)\Phi(g) + \Phi(f)\Lambda(g), f, g \in C[0, 2\pi]$. Hence to check the assertion, it suffices to check that $\Lambda(1), \Lambda(\cos nx), \Lambda(\sin nx) \in \mathcal{K}(\mathcal{H})$ for all $n \in \mathbb{N}$ since these functions generate a dense *-algebra in $C[0, 2\pi]$. Since $\Lambda(1) = 0$, it suffices to observe that well known trigonometric identities imply that $\Lambda(\cos nx)$ and $\Lambda(\sin nx)$ map onto subspaces of $\mathcal{H}$ of dimension $\leq 2n + 1$.

Let $\mathcal{S} = \{h \in C[0, 2\pi] : -1 \leq h(x) \leq 1, h(0) = 1, h(2\pi) = -1\}$ and, for each $h \in \mathcal{S}$, let

$$T_h = \Phi(h) - (1 - \Phi(h^2))^{\frac{1}{4}} D (1 - \Phi(h^2))^{\frac{1}{4}}.$$

From the properties of D established above it follows that each T_h is unitary mod $\mathcal{K}(\mathcal{H})$ and $T_h\Phi(f) - \Phi(f)T_h \in \mathcal{K}(\mathcal{H})$ for all $f \in C[0, 2\pi]$. Note that T_h depends continuously on h and that $\mathcal{S}$ is a convex subset of $C[0, 2\pi]$.

Hence each pair, T_{h_1}, T_{h_2}, is connected by a norm continuous path of Fredholm operators (unitary mod $\mathcal{K}(\mathcal{H})$ in fact). It follows from [10], Theorem 5.36, that the Fredholm operators $T_h, h \in \mathcal{S}$, all have the same index. To calculate this common index, let $h = \cos \frac{x}{2}$. Then $T_h(1) = \cos \frac{x}{2}$ and $T_h(\cos nx) = \cos(n - \frac{1}{2})x$, $T_h(\sin nx) = \sin(n - \frac{1}{2})x$, $n \in \mathbb{N}$. Since $\{2\cos \frac{x}{2}, 2\sin \frac{x}{2}, 2\cos \frac{3x}{2}, 2\sin \frac{3x}{2}, \ldots\}$ form an orthonormal basis in $\mathcal{H}$, we see that $\ker T_h = \mathbb{C}(1 - \cos \frac{x}{2})$ and that T_h is surjective. Hence $\text{index}(T_h) = 1$.

Now let

$$P_0 : L^2[0, 2\pi] \rightarrow L^2[0, \frac{\pi}{2}] \subseteq L^2[0, 2\pi]$$

be the projection onto the subspace of functions vanishing outside $[0, \frac{\pi}{2}]$ and

$$Q_0 : L^2[0, 2\pi] \rightarrow L^2[\frac{3\pi}{2}, 2\pi] \subseteq L^2[0, 2\pi]$$

the projection onto the functions vanishing outside $[\frac{3\pi}{2}, 2\pi]$. Note that both P_0 and Q_0 commute with $\Phi(f)$, $f \in C[0, 2\pi]$. Let $h \in \mathcal{S}$ be a function such that $h(x) = -1$, $x \geq \frac{\pi}{2}$, and $g \in \mathcal{S}$ a function such that $g(x) = 1$, $x \leq \frac{3\pi}{2}$. Then $(1 - P_0)(1 - \Phi(h^2))^{\frac{1}{4}} D (1 - \Phi(h^2))^{\frac{1}{4}} = 0$ since $(1 - P_0)(1 - \Phi(h^2)) = 0$. As $i(1 - \Phi(h^2))^{\frac{1}{4}} D (1 - \Phi(h^2))^{\frac{1}{4}}$ is selfadjoint it follows that $1 - P_0$ and hence also P_0 commute with T_h. Similarly, Q_0 commutes with T_g.

Now set $P = \begin{bmatrix} P_0 & 0 \\ 0 & P_0 \end{bmatrix}$, $Q = \begin{bmatrix} Q_0 & 0 \\ 0 & Q_0 \end{bmatrix} \in \mathcal{B}(\mathcal{H} \oplus \mathcal{H})$ and define $\phi : C[0, 2\pi] \rightarrow \mathcal{B}(\mathcal{H} \oplus \mathcal{H})$ by $\phi(f) = \Phi(f) \oplus \Phi(f), f \in C[0, 2\pi]$. It is now straightforward to check that

$$\mathcal{E}_0 = \left(P(\mathcal{H} \oplus \mathcal{H}), P\phi, P \begin{bmatrix} 0 & T_h^* \\ T_h & 0 \end{bmatrix} \right) \in \mathbb{E}(C[0, 2\pi], \mathbb{C}),$$

$$\mathcal{E}_1 = \left(Q(\mathcal{H} \oplus \mathcal{H}), Q\phi, Q \begin{bmatrix} 0 & T_g^* \\ T_g & 0 \end{bmatrix} \right) \in \mathbb{E}(C[0, 2\pi], \mathbb{C}),$$

$$\mathcal{D}_0 = \left((1 - P)(\mathcal{H} \oplus \mathcal{H}), (1 - P)\phi, (1 - P) \begin{bmatrix} 0 & T_h^* \\ T_h & 0 \end{bmatrix} \right) \in \mathbb{D}(C[0, 2\pi], \mathbb{C})$$

and

$$\mathcal{D}_1 = \left((1 - Q)(\mathcal{H} \oplus \mathcal{H}), (1 - Q)\phi, (1 - Q) \begin{bmatrix} 0 & T_g^* \\ T_g & 0 \end{bmatrix} \right) \in \mathbb{D}(C[0, 2\pi], \mathbb{C}).$$

Note that

$$\mathcal{E}_0 = \left(P_0(\mathcal{H}) \oplus P_0(\mathcal{H}), P_0\Phi \oplus P_0\Phi, \begin{bmatrix} 0 & P_0 T_h^* \\ P_0 T_h & 0 \end{bmatrix} \right).$$

Since T_h is unitary mod $\mathcal{K}(\mathcal{H})$ and has index 1, and since $(1 - P_0)T_h = -(1 - P_0)$, it follows that $P_0 T_h$ is unitary mod $\mathcal{K}(P_0\mathcal{H})$ and has index 1. Thus, if we let $\psi : C[0,1] \to C[0, 2\pi]$ be the *-homomorphism given by

$$\psi(g)(x) = \begin{cases} g(0), & x \in [0, \frac{\pi}{2}], \\ g(\frac{x}{\pi} - \frac{1}{2}), & x \in [\frac{\pi}{2}, \frac{3\pi}{2}], \\ g(1), & x \in [\frac{3\pi}{2}, 2\pi] \end{cases} \qquad g \in C[0,1],$$

then Assertion A yields $\psi^*\{\mathcal{E}_0\} = \{f_0^*(u_\mathbb{C})\}$ in $K\hat{K}(C[0,1], \mathbb{C})$. Similarly, $\psi^*\{\mathcal{E}_1\} = \{f_1^*(u_\mathbb{C})\}$. But

$$\mathcal{E}_0 \oplus \mathcal{D}_0 \simeq \left(\mathcal{H} \oplus \mathcal{H}, \phi, \begin{bmatrix} 0 & T_h^* \\ T_h & 0 \end{bmatrix} \right)$$

is operator homotopic to

$$\mathcal{E}_1 \oplus \mathcal{D}_1 \simeq \left(\mathcal{H} \oplus \mathcal{H}, \phi, \begin{bmatrix} 0 & T_g^* \\ T_g & 0 \end{bmatrix} \right)$$

because

$$\left(\mathcal{H} \oplus \mathcal{H}, \phi, \begin{bmatrix} 0 & T_{g_t}^* \\ T_{g_t} & 0 \end{bmatrix} \right),$$

where $g_t = th + (1-t)g, t \in [0,1]$, gives such a homotopy. Thus $\{\mathcal{E}_0\} = \{\mathcal{E}_1\}$ in $K\hat{K}(C[0, 2\pi], \mathbb{C})$ and

$$\{f_0^*(u_\mathbb{C})\} = \psi^*\{\mathcal{E}_0\} = \psi^*\{\mathcal{E}_1\} = \{f_1^*(u_\mathbb{C})\}. \qquad \square$$

Corollary 2.2.18. *There is a map*

$$\cdot : KK(A, B) \times KK(B, C) \to KK(A, C),$$

called the Kasparov product, which is given by $[\mathcal{E}_1] \cdot [\mathcal{E}_2] = [\mathcal{E}_{12}]$, $\mathcal{E}_1 \in \mathbb{E}(A, B), \mathcal{E}_2 \in \mathbb{E}(B, C)$, *where* $\mathcal{E}_{12}$ *is a Kasparov product of* $\mathcal{E}_1$ *by* $\mathcal{E}_2$. *The Kasparov product* $\cdot$ *has all the properties stated in Theorem 2.2.15.*

The essence of Theorem 2.2.17 is that the two apparently very different equivalence relations, $\sim$ and $\approx$, on $\mathbb{E}(A, B)$ actually agree when A is separable.

The Kasparov product is associative in the following sense.

Theorem 2.2.19. *Now let* $x \in KK(A, A_1), y \in KK(A_1, B), z \in KK(B, C)$. *Then* $x \cdot (y \cdot z) = (x \cdot y) \cdot z$ *in* $KK(A, C)$.

The proof of this theorem requires some preparations, but let us first fix the setting for it.

Let $\mathcal{E}_1 = (E_1, \phi_1, F_1) \in \mathbb{E}(A, A_1)$, $\mathcal{E}_2 = (E_2, \phi_2, F_2) \in \mathbb{E}(A_1, B)$, $\mathcal{E}_3 = (E_3, \phi_3, F_3) \in \mathbb{E}(B, C)$, let $\mathcal{E}_{12} = (E_1 \otimes_{\phi_2} E_2, \phi_1 \otimes_{\phi_2} id, F_{12})$ be a Kasparov product of $\mathcal{E}_1$ by $\mathcal{E}_2, \mathcal{E}_{23} = (E_2 \otimes_{\phi_3} E_3, \phi_2 \otimes_{\phi_3} id, F_{23})$ a Kasparov product of $\mathcal{E}_2$ by $\mathcal{E}_3$ and finally $\mathcal{E}_{123} = (E_1 \otimes_{\phi_2 \otimes_{\phi_3} id} (E_2 \otimes_{\phi_3} E_3), \phi_1 \otimes_{\phi_2 \otimes_{\phi_3} id} id, F)$ a Kasparov product of $\mathcal{E}_1$ by $\mathcal{E}_{23}$.

Note that $(E_1, \phi_1, tF_1 + (1-t)\frac{1}{2}(F_1 + F_1^*)) \in \mathbb{E}(A, A_1)$ for all $t \in [0, 1]$. Hence $\mathcal{E}_1$ is operator homotopic to $(E_1, \phi_1, \frac{1}{2}(F_1 + F_1^*))$. Thus for the purpose here (which is to show that $[\mathcal{E}_1] \cdot ([\mathcal{E}_2] \cdot [\mathcal{E}_3]) = ([\mathcal{E}_1] \cdot [\mathcal{E}_2]) \cdot [\mathcal{E}_3])$, we can assume that F_1, F_2 and F_3 are selfadjoint. It follows that we can choose F_{12}, F_{23} and F selfadjoint too.

Set $E = (E_1 \otimes_{\phi_2} E_2) \otimes_{\phi_3} E_3$ and let $\psi : E \to E_1 \otimes_{\phi_2 \otimes_{\phi_3} id} (E_2 \otimes_{\phi_3} E_3)$ be the isomorphism of graded Hilbert C-modules given on simple tensors by $\psi((e_1 \otimes_{\phi_2} e_2) \otimes_{\phi_3} e_3) = e_1 \otimes_{\phi_2 \otimes_{\phi_3} id} (e_2 \otimes_{\phi_3} e_3)$, $e_i \in E_i$, $i = 1, 2, 3$.

Lemma 2.2.20. *Let* $G_3 \in \mathcal{L}_C(E_3), G_2 \in \mathcal{L}_C(E_2 \otimes_{\phi_2} E_3)$ *and* $G_1 \in \mathcal{L}_C(E_1 \otimes_{\phi_2 \otimes_{\phi_3} id}(E_2 \otimes_{\phi_3} E_3))$ *all have degree 0 or 1. If* G_2 *is a* G_3-*connection for* E_2 *and* G_1 *is a* G_2-*connection for* E_1, *then* $\psi^{-1} G_1 \psi \in \mathcal{L}_C(E)$ *is a* G_3-*connection for* $E_1 \otimes_{\phi_2} E_2$.

Proof. Assume first that the common degree is 1.

Let $T_x, x \in E_1 \otimes_{\phi_2} E_2$, be the E_3-tensor operators for $E_1 \otimes_{\phi_2} E_2, T_x'$, $x \in E_1$, the $E_2 \otimes_{\phi_3} E_3$-tensor operators for E_1 and T_x'', $x \in E_2$, the E_3-tensor operators for E_2. Then

$$\psi^{-1} G_1 \psi T_{S_{E_1}(e_1) \otimes_{\phi_2} S_{E_2}(e_2)} = \psi^{-1} G_1 T'_{S_{E_1}(e_1)} T''_{S_{E_2}(e_2)}, e_1 \in E_1, e_2 \in E_2.$$

But

$$G_1 T'_{S_{E_1}(e_1)} = T'_{e_1} G_2 \mod \mathcal{K}_C(E_2 \otimes_{\phi_3} E_3, E_1 \otimes_{\phi_2 \otimes_{\phi_3} id} (E_2 \otimes_{\phi_3} E_3))$$

and

$$G_2 T''_{S_{E_2}(e_2)} = T''_{e_2} G_3 \mod \mathcal{K}_C(E_3, (E_2 \otimes_{\phi_3} E_3)),$$

so we see that

$$\psi^{-1} G_1 \psi T_{S_{E_1}(e_1) \otimes_{\phi_2} S_{E_2}(e_2)} = T_{e_1 \otimes_{\phi_2} e_2} G_3 \mod \mathcal{K}_C(E_3, E).$$

Similarly, the second condition, (2.2.2), for $\psi^{-1} G_1 \psi$ to be a G_3-connection for $E_1 \otimes_{\phi_2} E_2$ is established.

Now let us note that when F_2 and F in Definition 2.2.4 both have degree 0 the condition for F to be an F_2-connection becomes

$$T_x F_2 = F T_x \bmod \mathcal{K}_B(E_2, E_{12})$$

and

$$F_2 T_x^* = T_x^* F \bmod \mathcal{K}_B(E_{12}, E_1), \ x \in E_1.$$

Using this the proof in the remaining case is completely analogue to the preceding argument. $\square$

Lemma 2.2.21. $[\psi(F_{12} \otimes_{\phi_3} id)\psi^{-1}, F]$ *is an* $[F_2 \otimes_{\phi_3} id, F_{23}]$-*connection for* E_1.

Proof. Since the lemma concerns degree 0 operators we have to use the original definition of connections. So fix a homogeneous element $x \in E_1$ and let T_x be the corresponding tensor operator. Set

$$\tilde{T}_x = \begin{bmatrix} 0 & T_x^* \\ T_x & 0 \end{bmatrix} \in \mathcal{L}_C(E_2 \otimes_{\phi_3} E_3 \oplus E_1 \otimes_{\phi_2 \otimes_{\phi_3} id} (E_2 \otimes_{\phi_3} E_3)).$$

We now note that $[F_2 \otimes_{\phi_3} id, F_{23}] \oplus [\psi(F_{12} \otimes_{\phi_3} id)\psi^{-1}, F] = [F_2 \otimes_{\phi_3} id \oplus \psi(F_{12} \otimes_{\phi_3} id)\psi^{-1}, F_{23} \oplus F]$, and by using (1.2.3) we find that

$$(-1)^{deg x}[[F_2 \otimes_{\phi_3} id, F_{23}] \oplus [\psi(F_{12} \otimes_{\phi_3} id)\psi^{-1}, F], \tilde{T}_x] =$$

$$[[F_{23} \oplus F, \tilde{T}_x], F_2 \otimes_{\phi_3} id \oplus \psi(F_{12} \otimes_{\phi_3} id)\psi^{-1}]$$

$$-(-1)^{deg x}[[\tilde{T}_x, F_2 \otimes_{\phi_3} id \oplus \psi(F_{12} \otimes_{\phi_3} id)\psi^{-1}], F_{23} \oplus F] = (*).$$

The first term of $(*)$ is in $\mathcal{K}_C(E_2 \otimes_{\phi_3} E_3 \oplus E_1 \otimes_{\phi_2 \otimes_{\phi_3} id} (E_2 \otimes_{\phi_3} E_3))$ because F is an F_{23}-connection. To handle the second term let T_y', $y \in E_1$, be the E_2 tensor operators for E_1. Then

$$\psi^{-1} T_x = T_x' \otimes_{\phi_3} id$$

(where $T_x' \otimes_{\phi_3} id(e_2 \otimes_{\phi_3} e_3) = T_x'(e_2) \otimes_{\phi_3} e_3$, of course). Then

$$[[\tilde{T}_x, F_2 \otimes_{\phi_3} id \oplus \psi(F_{12} \otimes_{\phi_3} id)\psi^{-1}], F_{23} \oplus F] =$$

$$(id \oplus \psi)[[S_x, F_2 \otimes_{\phi_3} id \oplus F_{12} \otimes_{\phi_3} id], F_{23} \oplus \psi^{-1} F \psi](id \oplus \psi^{-1}),$$

where

$$S_x = \begin{bmatrix} 0 & T_x'^* \otimes_{\phi_3} id \\ T_x' \otimes_{\phi_3} id & 0 \end{bmatrix} \in \mathcal{L}_C(E_2 \otimes_{\phi_3} E_3 \oplus (E_1 \otimes_{\phi_2} E_2) \otimes_{\phi_3} E_3).$$

Define $\phi : E_2 \otimes_{\phi_3} E_3 \oplus E \rightarrow E_2 \otimes_{\phi_3} E_3 \oplus (E_1 \otimes_{\phi_2} E_2) \otimes_{\phi_3} E_3$ as an isomorphism of graded Hilbert C-modules in the natural way. Then

$$\phi[[S_x, F_2 \otimes_{\phi_3} id \oplus F_{12} \otimes_{\phi_3} id], F_{23} \oplus \psi^{-1} F \psi]\phi^{-1} =$$

$$[[\tilde{T}'_x, F_2 \oplus F_{12}] \otimes_{\phi_3} id, \phi(F_{23} \oplus \psi^{-1} F \psi)\phi^{-1}] = (**).$$

Observe that $[\tilde{T}'_x, F_2 \oplus F_{12}] \subseteq \mathcal{K}_B(E_2 \oplus E_1 \otimes_{\phi_2} E_2)$ since F_{12} is an F_2-connection for E_1. By Lemma 2.2.20, $\psi^{-1} F \psi$ is an F_3-connection for $E_1 \otimes_{\phi_2} E_2$. Since F_{23} is an F_3-connection for E_2 it follows readily that $\phi(F_{23} \oplus \psi^{-1} F \psi)\phi^{-1}$ is an F_3-connection for $E_2 \oplus E_1 \otimes_{\phi_2} E_2$. Thus $(**)$ is in $\mathcal{K}_B((E_2 \oplus E_1 \otimes_{\phi_2} E_2) \otimes_{\phi_3} E_3)$ by Lemma 2.2.6 (d). It follows that the second term in $(*)$ is also compact and the proof is complete. $\square$

Corollary 2.2.22. $[F_{12} \otimes_{\phi_3} id, \psi^{-1} F \psi]$ *is a 0-connection for* $E_1 \otimes_{\phi_2} E_2$.

Proof. Now, using the two preceding lemmas it suffices to show that $[F_2 \otimes_{\phi_3} id, F_{23}]$ is a 0-connection for E_2. So let T_x, $x \in E_2$, be the E_3-tensor operators for E_2, and observe that

$$F_{23}(F_2 \otimes_{\phi_3} id)T_{S_{E_2}(e_2)} = -F_{23}T_{S_{E_2}(F_2(e_2))}$$
$$= -T_{F_2(e_2)}F_3 \bmod \mathcal{K}_C(E_3, E_2 \otimes_{\phi_3} E_3)$$

and

$$(F_2 \otimes_{\phi_3} id)F_{23}T_{S_{E_2}(e_2)} = (F_2 \otimes_{\phi_3} id)T_{e_2}F_3 \bmod \mathcal{K}_C(E_3, E_2 \otimes_{\phi_3} E_3).$$

Since $(F_2 \otimes_{\phi_3} id)T_{e_2} = T_{F_2(e_2)}$, we then get the desired conclusion:

$$[F_2 \otimes_{\phi_3} id, F_{23}]T_{S_{E_2}(e_2)} \in \mathcal{K}_C(E_3, E_2 \otimes_{\phi_3} E_3), e_2 \in E_2. \quad \square$$

Proof of Theorem 2.2.19. The strategy of the proof is the following. Consider the Kasparov $A - C$-module $\mathcal{E} = (E, (\phi_1 \otimes_{\phi_2} id) \otimes_{\phi_3} id, \psi^{-1} F \psi)$ which is isomorphic to $\mathcal{E}_{123}$ via ψ. (In the following we will denote $(\phi_1 \otimes_{\phi_2} id) \otimes_{\phi_3} id$ by $\tilde{\tilde{\phi}}_1$ and $(\phi_1(a) \otimes_{\phi_2} id) \otimes_{\phi_3} id$ by $\tilde{\tilde{\phi}}_1(a)$). If $\mathcal{E}$ was a Kasparov product of $\mathcal{E}_{12}$ by $\mathcal{E}_3$ the proof was complete already since then $([\mathcal{E}_1] \cdot [\mathcal{E}_2]) \cdot [\mathcal{E}_3] = [\mathcal{E}] = [\mathcal{E}_1] \cdot ([\mathcal{E}_2] \cdot [\mathcal{E}_3])$. However, we do not know if it is, so we will find $F' \in \mathcal{L}_C(E)$ with the property that $\mathcal{E}' = (E, \tilde{\tilde{\phi}}_1 F')$ is a Kasparov product of $\mathcal{E}_{12}$ by $\mathcal{E}_3$ and at the same time

$$\tilde{\tilde{\phi}}_1(a)[F, F']\tilde{\tilde{\phi}}_1(a)^* \geq 0 \bmod \mathcal{K}_C(E).$$

By Lemma 2.1.18 $\mathcal{E}$ and $\mathcal{E}'$ will then be operator homotopic so that $[\mathcal{E}_1] \cdot ([\mathcal{E}_2] \cdot [\mathcal{E}_3]) = [\mathcal{E}_{123}] = [\mathcal{E}] = [\mathcal{E}'] = ([\mathcal{E}_1] \cdot [\mathcal{E}_2]) \cdot [\mathcal{E}_3]$.

To construct F' we will apply Corollary 2.2.3 to the following triple D_1, D_2 and $\mathcal{F}$. Let

$$D_1 = \mathcal{K}_C(E) + \mathcal{K}_B(E_1 \otimes_{\phi_2} E_2) \otimes_{\phi_3} id + (\mathcal{K}_{A_1}(E_1) \otimes_{\phi_2} id) \otimes_{\phi_3} id,$$

and let D_2 be the C^*-subalgebra of $\mathcal{L}_C(E)$ generated by

$$[(\tilde{\tilde{F}}_1, \psi^{-1}F\psi], \ [\tilde{\tilde{\phi}}_1(A), \psi^{-1}F\psi] \quad \text{and}$$

$$[F_{12} \otimes_{\phi_3} id, \psi^{-1}F\psi]_-$$

(where $\tilde{\tilde{F}}_1 = (F_1 \otimes_{\phi_2} id) \otimes_{\phi_3} id$). (Here, and in the following, x_- will denote the negative part of a selfadjoint element x of a C^*-algebra, i.e., $x_- = \frac{1}{2}((x^2)^{\frac{1}{2}} - x), x_+ = x + x_-$ denotes the positive part of x). Let $\mathcal{F}$ be the closed subspace spanned by $\psi^{-1}F\psi, \tilde{\tilde{F}}_1, F_{12} \otimes_{\phi_3} id$ and $\tilde{\tilde{\phi}}_1(A)$. It is clear that the size conditions in Corollary 2.2.3 are met because the Hilbert C^*-modules under consideration are all countably generated and A separable. We must check that $[\mathcal{F}, D_1] \subseteq D_1$ and that $D_1 D_2 \subseteq \mathcal{K}_C(E)$.

Using Lemma 2.2.6(d) we find that $[\psi^{-1}F\psi, D_1] \subseteq \mathcal{K}_C(E) \subseteq D_1$ because F is an F_{23}-connection for E_1 by assumption and $\psi^{-1}F\psi$ is an F_3-connection for $E_1 \otimes_{\phi_2} E_2$ by Lemma 2.2.20. Similarly $[F_{12} \otimes_{\phi_3} id, D_1] \subseteq D_1$ because F_{12} is an F_2-connection for E_1 and $\mathcal{K}_B(E_1 \otimes_{\phi_2} E_2)$ is an ideal in $\mathcal{L}_B(E_1 \otimes_{\phi_2} E_2)$. $[\tilde{\tilde{F}}_1, D_1] \subseteq D_1$ and $[\tilde{\tilde{\phi}}_1(A), D_1] \subseteq D_1$ for rather trivial reasons. Thus $[\mathcal{F}, D_1] \subseteq D_1$.

Next we check that $D_1 D_2 \subseteq \mathcal{K}_C(E)$. Let $k \in \mathcal{K}_{A_1}(E_1) \otimes_{\phi_2} id$ or $k \in \mathcal{K}_B(E_1 \otimes_{\phi_2} E_2)$. Then (1.2.2) yields

$$[\psi^{-1}F\psi, (\tilde{\tilde{F}}_1)]k \otimes_{\phi_3} id = [\psi^{-1}F\psi, (\tilde{\tilde{F}}_1)k \otimes_{\phi_3} id] + (\tilde{\tilde{F}}_1)[\psi^{-1}F\psi, k \otimes_{\phi_3} id].$$

If $k \in \mathcal{K}_{A_1}(E_1) \otimes_{\phi_2} id$, Lemma 2.2.6 (d) gives that both terms are in $\mathcal{K}_C(E)$ because F is an F_{23}-connection for E_1. If $k \in \mathcal{K}_B(E_1 \otimes_{\phi_2} E_2)$ both terms are in $\mathcal{K}_C(E)$ because $\psi^{-1}F\psi$ is an F_3-connection for $E_1 \otimes_{\phi_2} E_2$. Thus $[\psi^{-1}F\psi, \tilde{\tilde{F}}_1]D_1 \subseteq \mathcal{K}_C(E)$. By exchanging F_1 with $\phi_1(a), a \in A$, in this argument gives that $[\psi^{-1}F\psi, \tilde{\tilde{\phi}}_1(A)]D_1 \subseteq \mathcal{K}_C(E)$. Thus we only have to show that $[F_{12} \otimes_{\phi_3} id, \psi^{-1}F\psi]_- D_1 \subseteq \mathcal{K}_C(E)$. First observe that Corollary 2.2.22 implies that $[F_{12} \otimes_{\phi_3} id, \psi^{-1}F\psi]_-$ is a 0-connection for $E_1 \otimes_{\phi_2} E_2$. Hence

$$\mathcal{K}_B(E_1 \otimes_{\phi_2} E_2) \otimes_{\phi_3} id[F_{12} \otimes_{\phi_3} id, \psi^{-1}F\psi]_- \subseteq \mathcal{K}_C(E)$$

by Lemma 2.2.6 (c). It follows from Lemma 2.2.21 that

$$(*) \qquad T_x[F_2 \otimes_{\phi_3} id, F_{23}]_- = [\psi(F_{12} \otimes_{\phi_3} id)\psi^{-1}, F]_- T_x$$

$$\mod \mathcal{K}_C(E_2 \otimes_{\phi_3} E_3, E_1 \otimes_{\phi_2 \otimes_{\phi_3} id} (E_2 \otimes_{\phi_3} E_3))$$

for every $E_2 \otimes_{\phi_3} E_3$ tensor operator T_x for E_1.

We assert $T_x[F_2 \otimes_{\phi_3} id, F_{23}]_- \in \mathcal{K}_C(E_2 \otimes_{\phi_3} E_3, E_1 \otimes_{\phi_2 \otimes_{\phi_3} id} (E_2 \otimes_{\phi_3} E_3))$. To see this observe first that by (1.2.3)

$$[[F_2 \otimes_{\phi_3} id, F_{23}], \phi_2(a) \otimes_{\phi_3} id] =$$

$$(-1)^{deg(a)}[[F_{23}, \phi_2(a) \otimes_{\phi_3} id], F_2 \otimes_{\phi_3} id]$$

$$+(-1)^{deg(a)}[[\phi_2(a) \otimes_{\phi_3} id, F_2 \otimes_{\phi_3} id], F_{23}],$$

$a \in A_1$. The first term is compact because $\mathcal{E}_{23} \in \mathbb{E}(A_1, C)$ and the last is compact by Lemma 2.2.6(d) since $\mathcal{E}_2 \in \mathbb{E}(A_1, B)$ and F_{23} is an F_3-connection for E_2. Consequently $[F_2 \otimes_{\phi_3} id, F_{23}]$ commutes with $\phi_2(A_1) \otimes_{\phi_3} id \mod \mathcal{K}_C(E_2 \otimes_{\phi_3} E_3)$. But since $\mathcal{E}_{23}$ is a Kasparov product of $\mathcal{E}_2$ by $\mathcal{E}_3$, we know that

$$\phi_2(a) \otimes_{\phi_3} id[F_2 \otimes_{\phi_3} id, F_{23}]\phi_2(a)^* \otimes_{\phi_3} id \geq 0$$
$$\mod \mathcal{K}_C(E_2 \otimes_{\phi_3} E_3)).$$

So it follows now that

$$\phi_2(a) \otimes_{\phi_3} id[F_2 \otimes_{\phi_3} id, F_{23}]\phi_2(a)^* \otimes_{\phi_3} id \in \mathcal{K}_C(E_2 \otimes_{\phi_3} E_3), \ a \in A_1.$$

Our assertion therefore follows from the following equalities involving an approximate unit $\{a_i\}$ for A_1 :

$$T_x[F_2 \otimes_{\phi_3} id, F_{23}]_- T_x^* = \lim_i T_{xa_i}[F_2 \otimes_{\phi_3} id, F_{23}]_- T_{xa_i}^*$$
$$= \lim_i T_x(\phi_2(a_i) \otimes_{\phi_3} id)[F_2 \otimes_{\phi_3} id, F_{23}](\phi_2(a_i) \otimes_{\phi_3} id)T_x^*.$$

Consequently $(*)$ gives that $[\psi(F_{12} \otimes_{\phi_3} id)\psi^{-1}, F]_-$ is a 0-connection for E_1, so

$$\mathcal{K}_{A_1}(E_1) \otimes_{\phi_2 \otimes_{\phi_3} id} id[\psi(F_{12} \otimes_{\phi_3} id)\psi^{-1}, F]_- \subseteq \mathcal{K}_C(E_1 \otimes_{\phi_2 \otimes_{\phi_3} id}(E_2 \otimes_{\phi_3} E_3))$$

by Lemma 2.2.6 (c). It follows that

$$(\mathcal{K}_{A_1}(E_1) \otimes_{\phi_2} id) \otimes_{\phi_3} id[F_{12} \otimes_{\phi_3} id, \psi^{-1}F\psi]_- \subseteq \mathcal{K}_C(E).$$

We have now checked that $D_1 D_2 \subseteq \mathcal{K}_C(E)$.

Corollary 2.2.3 now gives two positive degree 0 elements M, N in $\mathcal{L}_C(E)$ such that $N + M = 1$ and

(a) $MD_1 \subseteq \mathcal{K}_C(E)$,

(b) $ND_2 \subseteq \mathcal{K}_C(E)$ and

(c) $[\mathcal{F}, M] \subseteq \mathcal{K}_C(E)$.

Set $F' = M^{\frac{1}{2}}\tilde{\tilde{F}}_1 + N^{\frac{1}{2}}\psi^{-1}F\psi$. To complete the proof we must check that F' has the properties we want, cf. the beginning of the proof. First we check that $\mathcal{E}' = (E, \tilde{\tilde{\phi}}_1, F') \in \mathbb{E}(A, C)$. Note that

$$[\tilde{\tilde{\phi}}_1(a), F'] = [\tilde{\tilde{\phi}}_1(a), M^{\frac{1}{2}}]\tilde{\tilde{F}}_1 + M^{\frac{1}{2}}[\tilde{\tilde{\phi}}_1(a), \tilde{\tilde{F}}_1]$$
$$+ [\tilde{\tilde{\phi}}_1(a), N^{\frac{1}{2}}]\psi^{-1}F\psi + N^{\frac{1}{2}}[\tilde{\tilde{\phi}}_1(a), \psi^{-1}F\psi], \quad a \in A.$$

The first term is in $\mathcal{K}_C(E)$ because of (c), the second because of (a) and $\mathcal{E}_1 \in \mathbb{E}(A, A_1)$, the third because of (c) and fourth because of (b).

Because of (c) we have

$$F'^2 - 1 = M\tilde{\tilde{F}}_1{}^2 + N\psi^{-1}F^2\psi - 1 \mod \mathcal{K}_C(E).$$

Hence

$$(F'^2 - 1)\tilde{\tilde{\phi}}_1(a) = M((F_1^2 - 1)\phi_1(a)) \otimes_{\phi_2} id \otimes_{\phi_3} id$$
$$+ N\psi^{-1}(F^2 - 1)\psi\tilde{\tilde{\phi}}_1(a) \mod \mathcal{K}_C(E).$$

The first term is in $\mathcal{K}_C(E)$ because of (a) and $\mathcal{E}_1 \in \mathbb{E}(A, A_1)$, the second because $\mathcal{E} \in \mathbb{E}(A, C)$. Thus $(F'^2 - 1)\tilde{\tilde{\phi}}_1(a) \in \mathcal{K}_C(E), a \in A$. And similarly, $(F' - F'^*)\tilde{\tilde{\phi}}_1(a) \in \mathcal{K}_C(E), a \in A$. Thus $\mathcal{E}' \in \mathbb{E}(A, C)$.

Next we check the conditions in Definition 2.2.7 to ensure that $\mathcal{E}'$ is a Kasparov product of $\mathcal{E}_{12}$ by $\mathcal{E}_3$. To show that F' is an F_3-connection for $E_1 \otimes_{\phi_2} E_2$ one proceeds as in the proof of Theorem 2.2.8 (the argument that F is an F_2-connection for E_1) by using that $\psi^{-1}F\psi$ is an F_3-connection for $E_1 \otimes_{\phi_2} E_2$; we leave the details to the reader. To check the positivity condition, observe that

$$[F_{12} \otimes_{\phi_3} id, F'] = [F_{12} \otimes_{\phi_3} id, M^{\frac{1}{2}}]\tilde{\tilde{F}}_1 + M^{\frac{1}{2}}[F_{12} \otimes_{\phi_3} id, \tilde{\tilde{F}}_1]$$
$$+ [F_{12} \otimes_{\phi_3} id, N^{\frac{1}{2}}]\psi^{-1}F\psi + N^{\frac{1}{2}}[F_{12} \otimes_{\phi_3} id, \psi^{-1}F\psi].$$

The first and third terms are in $\mathcal{K}_C(E)$ because of (c). Furthermore, $N^{\frac{1}{2}}[F_{12} \otimes_{\phi_3} id, \psi^{-1}F\psi] = N^{\frac{1}{2}}[F_{12} \otimes_{\phi_3} id, \psi^{-1}F\psi]_+ \bmod \mathcal{K}_C(E)$ since $[F_{12} \otimes_{\phi_3} id, \psi^{-1}F\psi]_- \in D_2$. Since

$$M^{\frac{1}{2}}[F_{12} \otimes_{\phi_3} id, \tilde{\tilde{F}}_1] = M^{\frac{1}{4}}[F_{12} \otimes_{\phi_3} id, \tilde{\tilde{F}}_1]M^{\frac{1}{4}}$$

and

$$N^{\frac{1}{2}}([F_{12} \otimes_{\phi_3} id, \psi^{-1}F\psi]_+ = N^{\frac{1}{4}}[F_{12} \otimes_{\phi_3} id, \psi^{-1}F\psi]_+ N^{\frac{1}{4}} \bmod \mathcal{K}_C(E)$$

by (c), we find that

$$
\begin{aligned}
(\tilde{\tilde{\phi}}_1(a))[F_{12} \otimes_{\phi_3} id, F'](\tilde{\tilde{\phi}}_1(a)^*) &= M^{\frac{1}{4}}(\tilde{\tilde{\phi}}_1(a))[F_{12} \otimes_{\phi_3} id, (\tilde{\tilde{F}}_1)](\tilde{\tilde{\phi}}_1(a)^*)M^{\frac{1}{4}} \\
&\quad + (\tilde{\tilde{\phi}}_1(a))N^{\frac{1}{4}}[F_{12} \otimes_{\phi_3} id, \psi^{-1}F\psi] \\
&\quad + N^{\frac{1}{4}}(\tilde{\tilde{\phi}}_1(a)^*) \bmod \mathcal{K}_C(E), \ a \in A.
\end{aligned}
$$

Since $\mathcal{E}_{12}$ is a Kasparov product of $\mathcal{E}_1$ by $\mathcal{E}_2$, we know that

$$\tilde{\tilde{\phi}}_1(a)[F_{12} \otimes_{\phi_3} id, \tilde{\tilde{F}}_1]\tilde{\tilde{\phi}}_1(a)^*$$

is positive in $\mathcal{L}_C(E)$ mod D_1. Consequently (a) ensures that

$$M^{\frac{1}{4}}\tilde{\tilde{\phi}}_1(a)[F_{12} \otimes_{\phi_3} id, \tilde{\tilde{F}}_1]\tilde{\tilde{\phi}}_1(a)^*M^{\frac{1}{4}}$$

is positive mod $\mathcal{K}_C(E)$. Therefore the positivity condition is satisfied too: $\mathcal{E}'$ is a Kasparov product of $\mathcal{E}_{12}$ by $\mathcal{E}_3$.

Finally, observe that

$$
\begin{aligned}
[\psi^{-1}F\psi, F'] &= [\psi^{-1}F\psi, M^{\frac{1}{2}}]\tilde{\tilde{F}}_1 + M^{\frac{1}{2}}[\psi^{-1}F\psi, \tilde{\tilde{F}}_1] \\
&\quad + [\psi^{-1}F\psi, N^{\frac{1}{2}}]\tilde{\tilde{F}}_1 + N^{\frac{1}{2}}[\psi^{-1}F\psi, \tilde{\tilde{F}}_1]\tilde{\tilde{F}}_1.
\end{aligned}
$$

The first and third terms are in $\mathcal{K}_C(E)$ because of (c). From (c) we also get that

$$\tilde{\tilde{\phi}}_1(a)M^{\frac{1}{2}}[\psi^{-1}F\psi, \tilde{\tilde{F}}_1]\tilde{\tilde{\phi}}_1(a)^* = M^{\frac{1}{4}}\tilde{\tilde{\phi}}_1(a)[\psi^{-1}F\psi, \tilde{\tilde{F}}_1]\tilde{\tilde{\phi}}_1(a)^*M^{\frac{1}{4}} \bmod \mathcal{K}_C(E).$$

Furthermore (c) also gives us a similar expression with N instead of M.

By using that $\mathcal{E}_{123}$ is a Kasparov product of $\mathcal{E}_1$ by $\mathcal{E}_{23}$ one sees that

$$\tilde{\tilde{\phi}}_1(a)[\psi^{-1}F\psi, \tilde{\tilde{F}}_1]\tilde{\tilde{\phi}}_1(a)^*$$

is positive in $\mathcal{L}_C(E)$ mod $\mathcal{K}_C(E)$. Hence, all in all, we get the desired conclusion that

$$\tilde{\phi}_1(a)[\psi^{-1}F\psi, F']\tilde{\phi}_1(a)^*$$

is positive mod $\mathcal{K}_C(E)$, $a \in A$.   □

2.2.23.   *Notes and remarks.*

All the results in this section were obtained by Kasparov in [19]. The proof of Kasparov's technical theorem given here is due to Higson [12] and our exposition follows [28] closely.

Exercise 2.2

E 2.2.1

Let $x \in KK(A, B)$. Prove that $1_A \cdot x = x$.

E 2.2.2

Let $\phi : A \to B$ be a graded *-homomorphism between graded C^*-algebras. Show that $(B, \phi, 0) \in \mathbb{E}(A, B)$. Let $\{\phi\}$ denote the corresponding element of $KK(A, B)$. Show that $\{\phi\} = \phi_*(1_A) = \phi^*(1_B)$.

CHAPTER 3

C^*-Extensions

3.1. The Busby Invariant

Let A and B be (ungraded) C^*-algebras.

Definition 3.1.1. An *extension of A by B* is a short exact sequence

$$(3.1.1) \qquad\qquad 0 \to B \xrightarrow{\ i\ } E \xrightarrow{\ p\ } A \to 0$$

of C^*-algebras, i.e. E is a C^*-algebra, $i : B \to E$ an injective *-homomorphism and $p : E \to A$ a surjective *-homomorphism such that $\operatorname{im} i = \ker p$.

The set of extensions of A by B will be denoted by $\mathcal{E}xt(A, B)$. In the extension (3.1.1) the C^*-algebra E contains $i(B)$ as an ideal in such a way that the quotient $E/i(B)$ is *-isomorphic to A. Conversely, if E is a C^*-algebra containing a copy of B as an ideal such that the quotient is *-isomorphic to A, then one can construct an extension of A by B with E in the middle. Thus the study of extensions of A by B is the study of C^*-algebras that are put together in this way by A and B. However, by studying the whole extension and not only the C^*-algebra E, we keep track of the way B sits inside E and how the quotient is identified with A. When A and B are fixed, the triple (i, E, p) contains all information on the extension.

The first important step in the study of C^*-extensions is to transform an extension into a *-homomorphism between C^*-algebras in such a way that no essential information is lost. To do this we consider the quotient C^*-algebra $\mathcal{Q}(B) = \mathcal{M}(B)/B$ and let $q_B : \mathcal{M}(B) \to \mathcal{Q}(B)$ denote the quotient *-homomorphism. Given an extension as (3.1.1) we can define, for each $e \in E$, a multiplier $e' \in \mathcal{M}(B)$ by

$$(3.1.2) \qquad\qquad i(e'(b)) = ei(b), \ b \in B.$$

Note that $ei(b) \in i(B)$ since $i(B)$ is an ideal in E and that the element $e'(b) \in B$ satisfying (3.1.2) is unique since i is injective. Thus $e' : B \to B$ is at least well-defined as a map, but is actually a multiplier because $i(e'(b)^*c) = i(b^*)e^*i(c) = i(b^*(e^*)'(c))$, $b, c \in B$. In fact it is straightforward to check that the map $e \to e'$ is a *-homomorphism mapping E into $\mathcal{M}(B)$, cf. E 1.1.9.

Definition 3.1.2. Given an extension $(i, E, p) \in \mathcal{E}xt(A, B)$ of A by B we define the Busby invariant of the extension to be the *-homomorphism $\tau : A \to \mathcal{Q}(B) = \mathcal{M}(B)/B$ given by

$$\tau(a) = q_B(e'), \quad p(e) = a, \quad a \in A.$$

This definition requires two remarks. First note that $q_B(e') \in \mathcal{Q}(B)$ does not depend on which element e in the preimage of a under p we choose. To see this let $f \in E$ be another element in this preimage. Then $p(e-f) = 0$ so that $e - f \in i(B)$. Consequently $e' - f' \in B$ so that $q_B(e') = q_B(f')$. Secondly we must check that τ is a *-homomorphism. This is left to the reader.

It is a surprising fact that the Busby invariant contains all essential information on the extension. To make this precise we introduce the following notion of isomorphism of C^*-extension.

Definition 3.1.3. Two extensions (i_1, E_1, p_1), $(i_2, E_2, p_2) \in \mathcal{E}xt(A, B)$ of A by B are *isomorphic* when there is a *-homomorphism $\phi : E_1 \to E_2$ such that the following diagram commutes:

$$
\begin{array}{ccccccccc}
0 & \longrightarrow & B & \overset{i_1}{\longrightarrow} & E_1 & \overset{p_1}{\longrightarrow} & A & \longrightarrow & 0 \\
 & & \| & & {\scriptstyle\phi}\downarrow & & \| & & \\
0 & \longrightarrow & B & \underset{i_2}{\longrightarrow} & E_2 & \underset{p_2}{\longrightarrow} & A & \longrightarrow & 0
\end{array}
$$

Note that such a *-homomorphism ϕ must be a *-isomorphism.

Theorem 3.1.4. *Two extensions of A by B are isomorphic if and only if they have the same Busby invariant. Furthermore, every *-homomorphism $\tau : A \to \mathcal{Q}(B)$ is the Busby invariant of an extension of A by B.*

Proof. Let (i_1, E_1, p_1) and (i_2, E_2, p_2) be two extensions of A by B. Assume first that they are isomorphic, i.e. that there is a *-isomorphism

$\phi : E_1 \rightarrow E_2$ such that

$$0 \longrightarrow B \xrightarrow{\;i_1\;} E_1 \xrightarrow{\;p_1\;} A \longrightarrow 0$$
$$\| \qquad\qquad \phi\downarrow \qquad\qquad \|$$
$$0 \longrightarrow B \xrightarrow[\;i_2\;]{} E_2 \xrightarrow[\;p_2\;]{} A \longrightarrow 0$$

commutes. Let τ_1 be the Busby invariant of the upper extension, τ_2 that of the lower. Let $a \in A$ and choose $e \in E_1$ such that $p_1(e) = a$. Then $p_2(\phi(e)) = a$ and

$$\begin{aligned}
i_2(e'b - \phi(e)'b) &= \phi \circ i_1(e'b) - i_2(\phi(e)'b) \\
&= \phi(ei_1(b)) - \phi(e)i_2(b) \\
&= \phi(e)i_2(b) - \phi(e)i_2(b) = 0
\end{aligned}$$

for all $b \in B$. Thus $e' = \phi(e)'$, so that $\tau_1(a) = q_B(e') = q_B(\phi(e)') = \tau_2(a)$. Hence $\tau_1 = \tau_2$.

Now let $\tau : A \rightarrow Q(B)$ be an arbitrary *-homomorphism. Set

$$E = \{(a, x) \in A \times \mathcal{M}(B) : \tau(a) = q_B(x)\}.$$

Then E is a C^*-algebra and we can define $i : B \rightarrow E$ and $p : E \rightarrow A$ by $i(b) = (0, b)$, $b \in B$, $p(a, x) = a$, $(a, x) \in E$, respectively. Then $\ker p = \operatorname{im} i$, i is injective and p is surjective because q_B is. Hence (i, E, p) is an extension of A by B. A few seconds consideration confirms that τ is the Busby invariant of this extension.

Assume that $\tau_1 = \tau_2 = \tau$. For $e \in E_1$ define $\phi_1(e) = (p_1(e), e') \in A \times \mathcal{M}(B)$. Since $\tau(p_1(e)) = \tau_1(p_1(e)) = q_B(e')$, we see that ϕ_1 defines a *-homomorphism of E_1 into E. It is straightforward to check that the diagram

$$0 \longrightarrow B \xrightarrow{\;i_1\;} E_1 \xrightarrow{\;p_1\;} A \longrightarrow 0$$
$$\| \qquad\qquad \phi_1\downarrow \qquad\qquad \|$$
$$0 \longrightarrow B \xrightarrow[\;i\;]{} E \xrightarrow[\;p\;]{} A \longrightarrow 0$$

commutes, so that in particular ϕ_1 is a *-isomorphism.

Doing the same argument with τ_2 and E_2, replacing τ_1 and E_1 we get all in all the following commutative diagram :

$$0 \longrightarrow B \xrightarrow{\;i_1\;} E_1 \xrightarrow{\;p_1\;} A \longrightarrow 0$$
$$\| \qquad\qquad \phi_1\downarrow \qquad\qquad \|$$
$$0 \longrightarrow B \xrightarrow{\;i\;} E \xrightarrow{\;p\;} A \longrightarrow 0$$
$$\| \qquad\qquad \phi_2\uparrow \qquad\qquad \|$$
$$0 \longrightarrow B \xrightarrow[\;i_2\;]{} E_2 \xrightarrow[\;p_2\;]{} A \longrightarrow 0$$

Thus the two extensions are isomorphic.   □

There is always one extension of A by B, namely the direct sum extension:

$$0 \to B \to B \oplus A \to A \to 0$$

with the obvious *-homomorphisms. An immediate consequence of Theorem 3.1.4 is that when B is unital, this is the only extension (up to isomorphism) of A by B.

In general there are too many isomorphism classes of extensions, even for very reasonable C^*-algebras like $B = \mathcal{K}$, $A = \mathbb{C}$.

3.1.5. *Notes and remarks.*

The material in this section is taken from [3].

Exercise 3.1

E 3.1.1

Let B be abelian and A a C^*-algebra which has no non-zero 1-dimensional representations (like a simple C^*-algebra $\neq \mathbb{C}$).

Show that there is only one extension of A by B (up to isomorphism).

E 3.1.2

(a) Show that if both A and B are abelian, then the C^*-algebra E in an extension of A by B must be abelian.

(b) Let $B = \{f \in C[0,1] : f(0) = f(1) = 0\}, A = \mathbb{C}$. Show that the C^*-algebra E in an extension of A by B must be *-isomorphic to one of the following

$E_1 = B \oplus \mathbb{C}$,

$E_2 = \{f \in C[0,1] : f(0) = 0\}$,

$E_3 = \{f \in C[0,1] : f(1) = 0\}$,

$E_4 = \{f \in C[0,1] : f(0) = f(1)\}$.

(c) Describe all extensions of A by B up to isomorphism.

3.2. The Extension Groups

In order to obtain a more manageable theory of C^*-extensions we introduce a weaker equivalence relation on the set $\mathcal{E}xt(A,B)$.

Definition 3.2.1. Two extensions of A by B, (i_1, E_1, p_1) and $(i_2, E_2, p_2) \in \mathcal{E}xt(A, B)$ are called *unitarily equivalent* when there is a *-homomorphism $\phi : E \to F$ and a unitary $u \in \mathcal{M}(B)$ such that the diagram

$$
\begin{array}{ccccccccc}
0 & \longrightarrow & B & \stackrel{i_1}{\longrightarrow} & E & \stackrel{p_1}{\longrightarrow} & A & \longrightarrow & 0 \\
 & & {\scriptstyle Adu}\downarrow & & {\scriptstyle \phi}\downarrow & & \| & & \\
0 & \longrightarrow & B & \underset{i_2}{\longrightarrow} & F & \underset{p_2}{\longrightarrow} & A & \longrightarrow & 0
\end{array}
$$

commutes.

Note that such a ϕ must be a *-isomorphism, and that unitary equivalence defines an equivalence relation on $\mathcal{E}xt(A, B)$.

Using Theorem 3.1.4 it is easy to prove the following lemma.

Lemma 3.2.2. *Two extensions of A by B, with Busby invariant τ_1 and τ_2, are unitarily equivalent if and only if there is a unitary $u \in \mathcal{M}(B)$ such that $Adq_B(u) \circ \tau_1 = \tau_2$.*

Now and for the rest of this chapter we assume that **all C^*-algebras denoted by $B, B_1, B_2, \ldots$ are stable**, cf. Definition 1.3.1.

By restricting attention to extensions where the kernel B is stable, we obtain the possibility of adding extensions together. To describe the addition we shift the focus from the extensions themselves to the Busby invariant they determine. So let $\mathrm{Hom}\,(A, \mathcal{Q}(B))$ denote the set of *-homomorphisms from A to $\mathcal{Q}(B)$. Two homomorphisms $\psi, \phi \in \mathrm{Hom}\,(A, \mathcal{Q}(B))$ are called *unitarily equivalent* if there is a unitary $u \in \mathcal{M}(B)$ such that $Adq_B(u) \circ \psi = \phi$. We write $\psi \approx \phi$ in this case, and we write $\{\phi\}$ for the equivalence class in $\mathrm{Hom}\,(A, \mathcal{Q}(B))/\approx$ represented by $\phi \in \mathrm{Hom}\,(A, \mathcal{Q}(B))$. Thus Theorem 3.1.4 and Lemma 3.2.2 say that the Busby invariant gives a bijection between the unitary equivalence classes of extensions of A by B and $\mathrm{Hom}\,(A, \mathcal{Q}(B))/\approx$. Instead of adding unitary equivalence classes in $\mathcal{E}xt(A, B)$ we will define an addition in $\mathrm{Hom}\,(A, \mathcal{Q}(B))/\approx$.

To do so we fix an inner isomorphism $\Theta_B : M_2(B) \to B$, cf. Definition 1.3.8 and Lemma 1.3.9. Using the isomorphism $M_2(\mathcal{M}(B)) \simeq \mathcal{M}(M_2(B))$, cf. E 1.2.1 (iv) we obtain the following commutative diagram

$$\begin{array}{ccccccccc}
0 & \longrightarrow & M_2(\mathbb{C}) & \longrightarrow & M_2(\mathcal{M}(B)) & \xrightarrow{q_B \otimes id_{M_2(\mathbb{C})}} & M_2(\mathcal{Q}(B)) & \longrightarrow & 0 \\
 & & {\scriptstyle \Theta_B}\downarrow & & {\scriptstyle \Theta_B}\downarrow & & {\scriptstyle \tilde{\Theta}_B}\downarrow & & \\
0 & \longrightarrow & B & \longrightarrow & \mathcal{M}(B) & \xrightarrow[q_B]{} & \mathcal{Q}(B) & \longrightarrow & 0
\end{array}$$

(3.2.1)

where $\tilde{\Theta}_B$ is the *-isomorphism defined by demanding it to make the diagram commute. Now we can define an addition $+$ in $\mathrm{Hom}\,(A, \mathcal{Q}(B))/\approx$ by

$$\{\psi\} + \{\phi\} = \left\{ \tilde{\Theta}_B \circ \begin{bmatrix} \phi & 0 \\ 0 & \psi \end{bmatrix} \right\}, \quad \psi, \phi \in \mathrm{Hom}\,(A, \mathcal{Q}(B)).$$

To see that the addition is well-defined let $\phi' = Adq_B(u) \circ \phi$ and $\psi' = Adq_B(w) \circ \psi$ for some unitaries $u, w \in \mathcal{M}(B)$. Then

$$\tilde{\Theta}_B \circ \begin{bmatrix} \phi' & 0 \\ 0 & \psi' \end{bmatrix} = \tilde{\Theta}_B \circ Ad\left((q_B \otimes id_{M_2(\mathbb{C})}) \begin{bmatrix} u & 0 \\ 0 & w \end{bmatrix} \right) \circ \begin{bmatrix} \phi & 0 \\ 0 & \psi \end{bmatrix}$$

$$= Ad\left(q_B \Theta_B \begin{bmatrix} u & 0 \\ 0 & w \end{bmatrix} \right) \circ \tilde{\Theta}_B \circ \begin{bmatrix} \phi & 0 \\ 0 & \psi \end{bmatrix}.$$

Hence we see that $\tilde{\Theta}_B \circ \begin{bmatrix} \phi & 0 \\ 0 & \psi \end{bmatrix}$ is unitarily equivalent to $\tilde{\Theta}_B \circ \begin{bmatrix} \phi' & 0 \\ 0 & \psi' \end{bmatrix}$.

Lemma 3.2.3. $\mathrm{Hom}\,(A, \mathcal{Q}(B))/\approx$ *is an abelian semigroup when the addition* $+$ *is defined by*

$$\{\psi\} + \{\phi\} = \left\{ \tilde{\Theta}_B \circ \begin{bmatrix} \phi & 0 \\ 0 & \psi \end{bmatrix} \right\}, \quad \psi, \phi \in \mathrm{Hom}\,(A, \mathcal{Q}(B)).$$

Proof. We give only the proof of the associativity of the addition. (Commutativity is easier.) So let $\phi_1, \phi_2, \phi_3 \in \mathrm{Hom}\,(A, \mathcal{Q}(B))$. We must show that

$$\tilde{\Theta}_B \circ \begin{bmatrix} \phi_1 & 0 \\ 0 & \psi_1 \end{bmatrix} \approx \tilde{\Theta}_B \circ \begin{bmatrix} \psi_2 & 0 \\ 0 & \phi_3 \end{bmatrix},$$

where

$$\psi_1 = \tilde{\Theta}_B \circ \begin{bmatrix} \phi_2 & 0 \\ 0 & \phi_3 \end{bmatrix} \quad \text{and} \quad \psi_2 = \tilde{\Theta}_B \circ \begin{bmatrix} \phi_1 & 0 \\ 0 & \phi_2 \end{bmatrix}.$$

Choose functions $f_i : A \to \mathcal{M}(B)$ such that $\phi_i = q_B \circ f_i$, $i = 1, 2, 3$. Then:

$$\psi_1 = \tilde{\Theta}_B \circ (q_B \otimes id_{M_2(\mathbb{C})}) \circ \begin{bmatrix} f_2 & 0 \\ 0 & f_3 \end{bmatrix}$$

$$= q_B \circ \Theta_B \circ \begin{bmatrix} f_2 & 0 \\ 0 & f_3 \end{bmatrix}$$

$$= q_B(V_1 f_2(\cdot)V_1^*) + q_B(V_2 f_3(\cdot)V_2^*)$$

where $V_1, V_2 \in \mathcal{M}(B)$ are the isometries used to define Θ_B. Similarly,

$$\psi_2 = q_B(V_1 f_1(\cdot)V_1^*) + q_B(V_2 f_2(\cdot)V_2^*).$$

Thus

$$\tilde{\Theta}_B \circ \begin{bmatrix} \phi_1 & 0 \\ 0 & \psi_1 \end{bmatrix} = q_B(AdV_1 \circ f_1 + AdV_2V_1 \circ f_2 + AdV_2V_2 \circ f_3).$$

Set $W_1 = V_1, W_2 = V_2V_1$ and $W_3 = V_2^2$. Then the W_i's satisfy the conditions of Definition 1.3.8 (with $n = 3$), so they define an inner *-isomorphism $\Phi_1 : M_3(B) \to B$ such that

$$\tilde{\Theta}_B \circ \begin{bmatrix} \phi_1 & 0 \\ 0 & \psi_1 \end{bmatrix} = q_B \circ \Phi_1 \circ \begin{bmatrix} f_1 & 0 & 0 \\ 0 & f_2 & 0 \\ 0 & 0 & f_3 \end{bmatrix}.$$

Similarly,

$$\tilde{\Theta}_B \circ \begin{bmatrix} \psi_2 & 0 \\ 0 & \phi_3 \end{bmatrix} = q_B \circ \Phi_2 \circ \begin{bmatrix} f_1 & 0 & 0 \\ 0 & f_2 & 0 \\ 0 & 0 & f_3 \end{bmatrix},$$

for some other inner *-isomorphism $\Phi_2 : M_3(B) \to B$. By Lemma 1.3.9, Φ_1 and Φ_2 differ by conjugation with a unitary in $\mathcal{M}(B)$. It follows that

$$\tilde{\Theta}_B \circ \begin{bmatrix} \phi_1 & 0 \\ 0 & \psi_1 \end{bmatrix} \quad \text{and} \quad \tilde{\Theta}_B \circ \begin{bmatrix} \psi_2 & 0 \\ 0 & \phi_3 \end{bmatrix}$$

are unitarily equivalent, completing the proof of associativity. $\qquad\square$

Definition 3.2.4. An element $\psi \in \mathrm{Hom}\,(A, \mathcal{Q}(B))$ is called *degenerate* when there is a *-homomorphism $\pi : A \to \mathcal{M}(B)$ such that $q_B \circ \pi = \psi$.

It is clear that if $\psi, \phi \in \mathrm{Hom}\,(A, \mathcal{Q}(B))$ are unitarily equivalent and ψ is degenerate, then so is ϕ. We set

$$\mathcal{D} = \{\{\psi\} \in \mathrm{Hom}\,(A, \mathcal{Q}(B))/\approx: \psi \text{ is degenerate}\}.$$

From the commutative diagram (3.2.1) it follows that $\mathcal{D}$ is a subsemigroup of $\mathrm{Hom}\,(A, \mathcal{Q}(B))/\approx$.

Definition 3.2.5. Two elements $\psi, \phi \in \mathrm{Hom}\,(A, \mathcal{Q}(B))$ are called *stably equivalent* when there are degenerate homomorphisms $\lambda_1, \lambda_2 \in \mathrm{Hom}\,(A, \mathcal{Q}(B))$ such that

$$\tilde{\Theta}_B \circ \begin{bmatrix} \psi & 0 \\ 0 & \lambda_1 \end{bmatrix} \approx \tilde{\Theta}_B \circ \begin{bmatrix} \phi & 0 \\ 0 & \lambda_2 \end{bmatrix}.$$

We write $\psi \sim \phi$ in this case.

It follows from Lemma 3.2.3 and the fact that $\mathcal{D}$ is a subsemigroup of $\mathrm{Hom}\,(A, \mathcal{Q}(B))/\approx$ that stable equivalence defines an equivalence relation on $\mathrm{Hom}\,(A, \mathcal{Q}(B))$.

Definition 3.2.6. $Ext(A, B) = \mathrm{Hom}\,(A, \mathcal{Q}(B))/\sim.$

For $\psi \in \mathrm{Hom}\,(A, \mathcal{Q}(B))$ we let $[\psi]$ denote the equivalence class in $Ext(A, B)$ containing ψ. Note that $Ext(A, B)$ is the quotient of the semi-group $\mathrm{Hom}\,(A, \mathcal{Q}(B))/\approx$ by $\mathcal{D}$. Hence $Ext(A, B)$ comes equipped with the structure of an abelian semigroup where the addition $+$ is given by

$$[\psi] + [\phi] = \left[\tilde{\Theta}_B \circ \begin{bmatrix} \psi & 0 \\ 0 & \phi \end{bmatrix}\right], \quad \psi, \phi \in \mathrm{Hom}\,(A, \mathcal{Q}(B)).$$

$Ext(A, B)$ has a neutral element, namely any element $[\psi]$ where ψ is degenerate (like $\psi = 0$). Thus one could hope that $Ext(A, B)$ is an abelian group, but in general it is not. However, it does contain a group, namely the group of invertible elements which we denote by $Ext^{-1}(A, B)$.

There is a neat characterization of the *-homomorphisms in $\mathrm{Hom}\,(A, \mathcal{Q}(B))$ representing invertible elements in $Ext(A, B)$, at least when A is separable. This characterization is based on the notion of completely positive linear maps, cf. [17] pp. 881–885, and on the following theorem due to Kasparov.

Theorem 3.2.7. *Assume that A is separable and let $\phi : A \to \mathcal{M}(B)$ be a completely positive map, $\|\phi\| \leq 1$. Then there is a *-homomorphism $\rho : A \to M_2(\mathcal{M}(B))$ such that*

$$\begin{bmatrix} \phi(a) & 0 \\ 0 & 0 \end{bmatrix} = \begin{bmatrix} 1 & 0 \\ 0 & 0 \end{bmatrix} \rho(a) \begin{bmatrix} 1 & 0 \\ 0 & 0 \end{bmatrix}, \quad a \in A.$$

Proof. First we consider the case where A is unital and $\phi(1) = 1$. Consider the algebraic tensor product $A \otimes_{\mathbb{C}} B$ and define a B-valued "inner product" $< \cdot, \cdot >$ on $A \otimes_{\mathbb{C}} B$ by

$$< \sum_i a_i \otimes_{\mathbb{C}} b_i, \sum_j c_j \otimes_{\mathbb{C}} d_j > = \sum_{i,j} b_i^* \phi(a_i^* c_j) d_j.$$

$A \otimes_{\mathbb{C}} B$ has a right B-module structure such that $(a_1 \otimes_{\mathbb{C}} b_1)b = a_1 \otimes_{\mathbb{C}} b_1 b$, $a_1 \in A$, $b, b_1 \in B$. Then $A \otimes_{\mathbb{C}} B$ satisfies all the conditions on a pre-Hilbert B-module, except (iv), cf. Definition 1.1.1. In particular, condition (iii) follows because ϕ is completely positive. By Lemma 1.1.2 $\mathcal{N} =$

$\{z \in A \otimes_{\mathbb{C}} B : < z, z > = 0\}$ is a vector subspace. Let $q : A \otimes_{\mathbb{C}} B \to A \otimes_{\mathbb{C}} B/\mathcal{N}$ be the quotient map. As in 1.2.3 we can make $A \otimes_{\mathbb{C}} B/\mathcal{N}$ into a pre-Hilbert B-module through the definitions: $q(\Sigma a_i \otimes_{\mathbb{C}} b_i)b = q(\Sigma a_i \otimes_{\mathbb{C}} b_i b)$ and $< q(z), q(z) > = < z, z >$, $z = \Sigma a_i \otimes_{\mathbb{C}} b_i \in A \otimes_{\mathbb{C}} B$, $b \in B$. We let E denote the Hilbert B-module obtained from $A \otimes_{\mathbb{C}} B/\mathcal{N}$ by completion. Note that E is countably generated because B is σ-unital and A-separable. In analogy with the map j considered in 1.2.3, there is a *-homomorphism $\pi : A \to \mathcal{L}_B(E)$ satisfying $\pi(a)q(\Sigma a_i \otimes_{\mathbb{C}} b_i) = q(\Sigma a a_i \otimes_{\mathbb{C}} b_i)$, $a \in A$, $\Sigma a_i \otimes_{\mathbb{C}} b_i \in A \otimes_{\mathbb{C}} B$.

Now define $W : B \to E$ by $Wb = q(1 \otimes_{\mathbb{C}} b)$, $b \in B$. Then $W \in \mathcal{L}_B(B, E)$ with the adjoint given by $W^* q(\Sigma a_i \otimes_{\mathbb{C}} b_i) = \Sigma \phi(a_i)b_i$, $\Sigma a_i \otimes_{\mathbb{C}} b_i \in A \otimes_{\mathbb{C}} B$. To see this it suffices to check that there is a linear map $W^* : E \to B$ satisfying the above equation since W^* then clearly will be the adjoint of W. For this one needs an inequality like

$$\sum_{i,j} b_i^* \phi(a_i)^* \phi(a_j) b_j \leq \sum_{i,j} b_i^* \phi(a_i^* a_j) b_j.$$

This inequality follows straightforwardly from Stinespring's theorem, cf. [17], Exercise 11.5.17 (vi), p. 883.

It is easy to check that $W^* \pi(\cdot) W = \phi(\cdot)$. Since $\phi(1) = 1$ and $\pi(1) = 1$, we have that $W^* W = 1$. Thus $WW^* \in \mathcal{M}(E)$ must be a projection, so that $E' = (1 - WW^*)(E)$ is a countably generated Hilbert B-module. By Kasparov's stabilization theorem, Theorem 1.1.24, and Lemma 1.3.2 there is an isomorphism $S : E' \oplus B \to B$ of Hilbert B-modules. Thus $X = W^* \oplus S : E \oplus B = WW^*(E) \oplus E' \oplus B \to B \oplus B$ is an isomorphism of Hilbert B-modules also. Note that

$$\begin{bmatrix} 1 & 0 \\ 0 & 0 \end{bmatrix} X(e, b) = (W^* WW^* e, 0) = (W^* e, 0), \quad e \in E, \ b \in B,$$

and

$$X^* \begin{bmatrix} 1 & 0 \\ 0 & 0 \end{bmatrix} (b, c) = X^*(b, 0) = (Wb, 0), \ b, c \in B.$$

So if we define $\rho(a) = X(\pi(a) \oplus 0)X^*$, $a \in A$, we obtain a *-homomorphism $A \to \mathcal{L}_B(B \oplus B)$ satisfying

$$\begin{bmatrix} 1 & 0 \\ 0 & 0 \end{bmatrix} \rho(\cdot) \begin{bmatrix} 1 & 0 \\ 0 & 0 \end{bmatrix} = \begin{bmatrix} W^* \pi(\cdot) W & 0 \\ 0 & 0 \end{bmatrix} = \begin{bmatrix} \phi(\cdot) & 0 \\ 0 & 0 \end{bmatrix}.$$

By identifying $\mathcal{L}_B(B \oplus B)$ with $M_2(\mathcal{M}(B))$, cf. E 1.2.1, we have completed the proof when both A and ϕ are unital.

The general case follows from the unital case and the following lemma which guarentees the existence of a completely positive linear map $\hat{\phi}$ from the algebra A_1 obtained by adjoining a unit to A (so that $A_1 = A \oplus \mathbb{C}$ if A is unital already) into $\mathcal{M}(B)$ such that $\hat{\phi}(1) = 1$ and $\hat{\phi}$ extends ϕ.   □

Lemma 3.2.8. *Let* $\phi : A \to D$ *be a completely positive linear map between the* C^*-*algebras* A *and* D, D *unital. Let* A_1 *denote the* C^*-*algebra obtained by adjoining a unit to* A *if* A *has no unit and let* $A_1 = A \oplus \mathbb{C}$ *if* A *has a unit. If* $\|\phi\| \le 1$, *there is a completely posititive linear map* $\hat{\phi} : A_1 \to D$ *such that* $\hat{\phi}(1) = 1$ *and* $\hat{\phi}$ *extends* ϕ.

Proof. Composing ϕ with a faithful unital representation of D we can assume that $D \subseteq \mathcal{B}(\mathcal{H})$ for some Hilbert space $\mathcal{H}$. Since $A_1 = A \oplus \mathbb{C}$ as a Banach space, we can define $\hat{\phi}(a, \lambda) = \phi(a) + \lambda$, $a \in A$, $\lambda \in \mathbb{C}$. Then $\hat{\phi}$ clearly extends ϕ, $\hat{\phi}(1) = 1$ and we must show that $\hat{\phi}$ is completely positive. To do this we need a suitable description of ϕ.

The bidual $\phi^{**} : A^{**} \to \mathcal{B}(\mathcal{H})^{**}$ is a completely positive linear map since ϕ is. Let $j_A : A \to A^{**}$ and $j : \mathcal{B}(\mathcal{H}) \to \mathcal{B}(\mathcal{H})^{**}$ be the canonical inclusions, and let $\Theta : \mathcal{B}(\mathcal{H})^{**} \to \mathcal{B}(\mathcal{H})$ be the dual of the inclusion map $\mathcal{B}(\mathcal{H})_* \subseteq \mathcal{B}(\mathcal{H})^*$. Then $\Theta \circ j$ is the identity map on $\mathcal{B}(\mathcal{H})$, Θ is completely positive and $\Theta \circ \phi^{**} \circ j_A = \Theta \circ j \circ \phi = \phi$. Since A^{**} is unital, we can apply [17], Exercise 11.5.18 to the map $\Theta \circ \phi^{**} : A^{**} \to \mathcal{B}(\mathcal{H})$. It follows that there is a Hilbert space $\mathcal{H}_1$, a *-representation $\tilde{\pi} : A^{**} \to \mathcal{B}(\mathcal{H}_1)$ and a linear operator $T \in \mathcal{B}(\mathcal{H}, \mathcal{H}_1)$ such that $\Theta \circ \phi^{**}(\cdot) = T^* \tilde{\pi}(\cdot) T$. Setting $\pi = \tilde{\pi} \circ j_A$ we obtain a *-representation of A such that $T^* \pi(\cdot) T = \Theta \circ \phi^{**} \circ j_A(\cdot) = \phi$. Since we assume that $\|\phi\| \le 1$, $\|\Theta \circ \phi^{**}\| \le 1$. So after substituting $\tilde{\pi}(1)T$ for T we can assume that $\|T\| \le 1$.

Assuming now that A has no unit, we define $\pi_1 : A_1 \to \mathcal{B}(\mathcal{H}_1) \oplus \mathcal{B}(\mathcal{H})$ by $\pi_1(a, \lambda) = (\pi(a) + \lambda 1, \lambda 1)$, $a \in A$, $\lambda \in \mathbb{C}$, and $T_1 \in \mathcal{B}(\mathcal{H}, \mathcal{H}_1 \oplus \mathcal{H})$ by $T_1 x = (Tx, (1 - T^*T)^{\frac{1}{2}} x)$, $x \in \mathcal{H}$. Then π_1 is a *-representation and $T_1^* \pi_1(a, \lambda) T_1 = T^* \pi(a) T + \lambda = \hat{\phi}(a, \lambda)$, $(a, \lambda) \in A_1$. In particular we find that $\hat{\phi}$ is completely positive.

If A already has a unit, we instead define π_1 by $\pi_1(a, \lambda) = (\pi(a), \lambda 1)$, $a \subset A$, $\lambda \in \mathbb{C}$, and T_1 by $T_1 x = (Tx, x), x \in \mathcal{H}$. Again we find that $T_1^* \pi_1(\cdot) T_1 = \hat{\phi}(\cdot)$ so that $\hat{\phi}$ is completely positive in this case also.   □

With Theorem 3.2.7 at hand it is easy to derive the promised description of the invertible elements in $Ext(A, B)$ when A is separable.

Theorem 3.2.9. *Assume that A is separable. Let $\phi \in \text{Hom}(A, \mathcal{Q}(B))$. Then the following conditions are equivalent :*

(1) *$[\phi]$ is invertible in $\text{Ext}(A, B)$.*

(2) *There is a completely positive map $\psi : A \to \mathcal{M}(B), \|\psi\| \leq 1$, such that $q_B \circ \psi = \phi$.*

(3) *There is a *-homomorphism $\pi : A \to M_2(\mathcal{M}(B))$ such that*

$$\begin{bmatrix} \phi(\cdot) & 0 \\ 0 & 0 \end{bmatrix} = q_B \otimes id_{M_2(\mathbb{C})}(p\pi(\cdot)p), \text{ where } p = \begin{bmatrix} 1 & 0 \\ 0 & 0 \end{bmatrix}.$$

Proof.

(1) $\Rightarrow$ (2) If $[\phi]$ is invertible there is a *-homomorphism $\phi_1 \in \text{Hom}(A, \mathcal{Q}(B))$ such that $\tilde{\Theta}_B \circ \begin{bmatrix} \phi & 0 \\ 0 & \phi_1 \end{bmatrix}$ is degenerate. Consequently there is a *-homomorphism $\pi' : A \to \mathcal{M}(B)$ such that $q_B \circ \pi' = \tilde{\Theta}_B \circ \begin{bmatrix} \phi & 0 \\ 0 & \phi_1 \end{bmatrix}$. Set $\pi = \Theta_B^{-1} \circ \pi'$ and $p = \begin{bmatrix} 1 & 0 \\ 0 & 0 \end{bmatrix} \in M_2(\mathcal{M}(B))$. Then

$$q_B \otimes id_{M_2(\mathbb{C})}(p\pi(\cdot)p) =$$

$$\begin{bmatrix} 1 & 0 \\ 0 & 0 \end{bmatrix} q_B \otimes id_{M_2(\mathbb{C})}(\pi(\cdot)) \begin{bmatrix} 1 & 0 \\ 0 & 0 \end{bmatrix} =$$

$$\begin{bmatrix} 1 & 0 \\ 0 & 0 \end{bmatrix} \tilde{\Theta}_B^{-1} \circ q_B(\pi'(\cdot)) \begin{bmatrix} 1 & 0 \\ 0 & 0 \end{bmatrix} = \begin{bmatrix} \phi(\cdot) & 0 \\ 0 & 0 \end{bmatrix}.$$

Thus if we define $\gamma : M_2(\mathcal{M}(B)) \to \mathcal{M}(B)$ by $\gamma \begin{bmatrix} a & b \\ c & d \end{bmatrix} = a$, with $a, b, c, d \in \mathcal{M}(B)$, and $\psi = \gamma \circ \pi$, we obtain a completely positive linear map ψ such that $q_B \circ \psi = \phi$.

(2) $\Rightarrow$ (3) follows from Theorem 3.2.7.

(3) $\Rightarrow$ (1) We first prove that $p\pi(a) - \pi(a)p \in M_2(B), a \in A$. To see this note that $p\pi(ab)p = p\pi(a)p\pi(b)p \mod M_2(B)$ because ϕ is a *-homomorphism. Inserting $b = a^*$ and using that π is a *-homomorphism, we see that $p\pi(a)p^{\perp}\pi(a)^*p \in M_2(B)$. It follows that $p\pi(a)p^{\perp} \in M_2(B)$. Hence $p\pi(a) - \pi(a)p = p\pi(a)p^{\perp} - p^{\perp}\pi(a)p \in M_2(B), a \in A$.

Define a map $\phi_1 : A \to \mathcal{Q}(B)$ by $\phi_1(\cdot) = q_B \circ \gamma'(\pi(\cdot))$, where $\gamma' : M_2(\mathcal{M}(B)) \to \mathcal{M}(B)$ is given by $\gamma' \begin{bmatrix} a & b \\ c & d \end{bmatrix} = d, a, b, c, d \in \mathcal{M}(B)$. Since

$p^\perp$ commutes with $\pi \mod M_2(B)$ we see that ϕ_1 is a *-homomorphism. Since furthermore

$$\tilde{\Theta}_B \circ \begin{bmatrix} \phi(\cdot) & 0 \\ 0 & \phi_1(\cdot) \end{bmatrix} = \tilde{\Theta}_B \circ q_B \otimes id_{M_2(\mathbb{C})}(p\pi(\cdot)p + p^\perp \pi(\cdot)p^\perp)$$

$$= q_B \circ \Theta_B(p\pi(\cdot) + p^\perp \pi(\cdot))$$

$$= q_B \circ \Theta_B(\pi(\cdot)),$$

we conclude that ϕ_1 represents an inverse for $[\phi]$ in $Ext(A, B)$. $\square$

Corollary 3.2.10. *Assume that A is separable. Every element in $Ext^{-1}(A, B)$ is represented by a *-homomorphism $\psi \in \mathrm{Hom}\,(A, \mathcal{Q}(B))$ given by $\psi(\cdot) = q_B(p\pi(\cdot))$ where $p \in \mathcal{M}(B)$ is a fully complemented projection and $\pi \in \mathrm{Hom}\,(A, \mathcal{M}(B))$.*

Proof. If $\psi_1 \in \mathrm{Hom}\,(A, \mathcal{Q}(B))$ represents an invertible element of $Ext(A, B)$, we have that $\begin{bmatrix} \psi_1(\cdot) & 0 \\ 0 & 0 \end{bmatrix} = q_B \otimes id_{M_2(\mathbb{C})}\left(\begin{bmatrix} 1 & 0 \\ 0 & 0 \end{bmatrix} \pi_1(\cdot)\right)$ for some $\pi_1 \in \mathrm{Hom}\,(A, M_2(\mathcal{M}(B)))$. Set $\psi = \Theta_B \circ \begin{bmatrix} \psi_1 & 0 \\ 0 & 0 \end{bmatrix}$, $p = \Theta_B \begin{bmatrix} 1 & 0 \\ 0 & 0 \end{bmatrix}$ and $\pi = \Theta_B \circ \pi_1$. $\square$

Using Theorem 3.2.9 it can be shown that for a large class of C^*-algebras, $A, B, Ext(A, B)$ is a group. We will not pursue this question here and take instead $Ext^{-1}(A, B)$ as the basic object to study.

Observe that because we took $Ext(A, B)$ to be the quotient of the unitary equivalence classes of extensions by the subsemigroup of degenerate extensions, it is no longer clear what it means that two extensions define the same element in $Ext(A, B)$; they need not be unitary equivalent. As we shall see in the next section, Theorem 2.2.17 solves this problem for $Ext^{-1}(A, B)$.

3.2.11. *Notes and remarks.*

Theorem 3.2.7 is often called "Kasparov's Stinespring Theorem" and it was proved by Kasparov in [18]. Theorem 3.2.9 is due to Arveson [1].

Exercise 3.2

E 3.2.1

Let $\tau \in \text{Hom}(A, \mathcal{Q}(B))$ be the Busby invariant for the extension

$$0 \to B \to E \to A \to 0.$$

(i) Show that τ is degenerate if and only if the extension is *split exact*, i.e. there is *-homomorphism $A \to E$ that is a right inverse for $E \to A$.

(ii) Show that $[\tau]$ is invertible in $Ext(A, B)$ if and only if there is a completely positive contraction $A \to E$ which is a right inverse for $E \to A$. Such an extension is called *semi-split*.

E 3.2.2

To define the addition in $Ext(A, B)$ we used a particuler inner isomorphism $\Theta_B : M_2(B) \to B$ of Hilbert B-modules. Show that any other inner *-isomorphism gives the same composition.

3.3 Connections to KK-theory

For the rest of this chapter **we only consider σ-unital C^*-algebras** and assume that all C^*-algebras denoted $A, A_1, A_2, A_3, \ldots$ are separable. Recall also the standing assumption that all C^*-algebras denoted by $B, B_1, B_2, \ldots$ are assumed to be stable.

Recall that $B_{(1)}$ denotes the C^*-algebra $B \oplus B$ with the odd grading, cf. Example 1.2.11 (b).

Definition 3.3.1. View A as a graded C^*-algebra with trivial grading. Then set $KK^1(A, B) = KK(A, B_{(1)})$.

Our first goal is to show that $Ext^{-1}(A, B) \simeq KK^1(A, B)$. For this purpose we first derive a more manageable description of $KK^1(A, B)$.

A KK^1-cycle for A, B is a pair (v, λ) where $v \in \mathcal{M}(B)$ and $\lambda \in \mathrm{Hom}\,(A, \mathcal{M}(B))$ satisfy

$$(3.3.1) \qquad v\lambda(a) - \lambda(a)v \subset B, \quad a \subset A,$$

$$(3.3.2) \qquad (v^* - v)\lambda(a) \in B, \quad a \in A, \qquad \text{and}$$

$$(3.3.3) \qquad (v^2 - v)\lambda(a) \in B, \quad a \in A.$$

The set of KK^1-cycles for A, B will be denoted by $\mathbb{E}^1(A, B)$. A KK^1-cycle $(v, \lambda) \in \mathbb{E}^1(A, B)$ is *degenerate* when $v\lambda(a) - \lambda(a)v = (v^* - v)\lambda(a) = (v^2 - v)\lambda(a) = 0$, $a \in A$. The set of degenerate KK^1-cycles for A, B will be denoted by $\mathbb{D}^1(A, B)$.

To introduce an equivalence relation on $\mathbb{E}^1(A, B)$ we use the surjections $\pi_t : IB \to B$ obtained by evaluation at $t \in [0, 1]$. Using Corollary 1.1.15 we get strictly continuous *-homomorphisms $\underline{\pi}_t : \mathcal{M}(IB) \to \mathcal{M}(B)$ such that the diagram

$$(3.3.4) \qquad \begin{array}{ccccccccc} 0 & \longrightarrow & IB & \subseteq & \mathcal{M}(IB) & \overset{q_{IB}}{\longrightarrow} & \mathcal{Q}(IB) & \longrightarrow & 0 \\ & & \pi_t \downarrow & & \underline{\pi}_t \downarrow & & \tilde{\pi}_t \downarrow & & \\ 0 & \longrightarrow & B & \subseteq & \mathcal{M}(B) & \underset{q_B}{\longrightarrow} & \mathcal{Q}(B) & \longrightarrow & 0 \end{array}$$

commutes. $\tilde{\pi}_t$ is defined as the *-homomorphism making the diagram commute.

Two KK^1-cycles $(v_1, \lambda_1), (v_2, \lambda_2) \in \mathbb{E}^1(A, B)$ are called *homotopic* when there is a KK^1-cycle $(v, \lambda) \in \mathbb{E}^1(A, IB)$ such that $v_1 = \underline{\pi}_0(v)$, $v_2 = \underline{\pi}_1(v)$, $\underline{\pi}_0 \circ \lambda = \lambda_1$ and $\underline{\pi}_1 \circ \lambda = \lambda_2$.

Lemma 3.3.2. $(v_1, \lambda_1), (v_2, \lambda_2) \in \mathbb{E}^1(A, B)$ *are homotopic if and only if there is a normbounded strictly continuous path* w_t, $t \in [0, 1]$, *in* $\mathcal{M}(B)$ *and a path* ϕ_t, $t \in [0, 1]$, *in* $\mathrm{Hom}\,(A, \mathcal{M}(B))$ *such that*

(i) $(w_t, \phi_t) \in \mathbb{E}^1(A, B)$, $t \in [0, 1]$,

(ii) $t \to \phi_t(a)$ *is strictly continuous for all* $a \in A$,

(iii) $t \to w_t \phi_t(a) - \phi_t(a) w_t$, *and* $t \to (w_t^* - w_t)\phi_t(a)$, $t \to (w_t^2 - w_t)\phi_t(a)$ *are normcontinuous for all* $a \in A$,

(iv) $w_0 = v_1$, $w_1 = v_2$, $\phi_0 = \lambda_1$, *and* $\phi_1 = \lambda_2$.

Proof. If $(v, \lambda) \in \mathbb{E}^1(A, IB)$ gives a homotopy between (v_1, λ_1) and (v_2, λ_2), then $w_t = \underline{\pi}_t(v)$ and $\phi_t = \underline{\pi}_t \circ \lambda$ have all the stated properties.

Conversely, assume that w_t, ϕ_t, $t \in [0, 1]$, are as in the statement of the lemma. Then we can define $v \in \mathcal{M}(IB)$ by

$$(vf)(t) = w_t f(t), \ t \in [0, 1], \ f \in IB,$$

and $\lambda \in \mathrm{Hom}\,(A, \mathcal{M}(IB))$ by

$$(\lambda(a)f)(t) = \phi_t(a)f(t), \ a \in A, \ t \in [0, 1], \ f \in IB.$$

Since $\underline{\pi}_0(v) = w_0 = v_1$, $\underline{\pi}_1(v) = w_1 = v_2$, $\underline{\pi}_0 \circ \lambda = \phi_0 = \lambda_1$ and $\underline{\pi}_1 \circ \lambda = \phi_1 = \lambda_2$, we see that (v_1, λ_1) and (v_2, λ_2) are homotopic. $\square$

It follows immediately from this lemma that homotopy defines an equivalence relation on $\mathbb{E}^1(A, B)$. We denote this equivalence by $\sim$.

In analogy with Lemma 2.1.20 we have

Lemma 3.3.3. *Every* $(v, \lambda) \in \mathbb{D}^1(A, B)$ *is homotopic to* $(0, 0)$.

Proof. By Lemma 1.3.6 it is possible to find a strictly continuous path u_t, $t \in\;]0, 1]$, of isometries in $\mathcal{M}(B)$ such that $u_1 = 1$ and $u_t u_t^* \to 0$ strictly as $t \to 0$. Set $w_t = u_t v u_t^*$, $t \in\;]0, 1]$, $w_0 = 0$, $\phi_t(\cdot) = u_t \lambda(\cdot) u_t^*$, $t \in\;]0, 1]$, $\phi_0 = 0$. Then the pair (w_t, ϕ_t), for $t \in [0, 1]$, satisfy (i), (ii) and (iii) in Lemma 3.3.2. Since $w_1 = v$, $\phi_1 = \lambda$, $w_0 = 0$, $\phi_0 = 0$, we have obtained the desired conclusion. $\square$

Definition 3.3.4. $kK^1(A, B) = \mathbb{E}^1(A, B)/\sim$. The equivalence class in $kK^1(A, B)$ containing $(v, \lambda) \in \mathbb{E}^1(A, B)$ will be denoted by $[v, \lambda]$.

Using the diagram (3.2.1) we can define an addition, $+$, in $kK^1(A, B)$ by

$$(3.3.5) \qquad [v_1, \lambda_1] + [v_2, \lambda_2] = \left[\Theta_B \begin{bmatrix} v_1 & 0 \\ 0 & v_2 \end{bmatrix}, \; \Theta_B \circ \begin{bmatrix} \lambda_1 & 0 \\ 0 & \lambda_2 \end{bmatrix} \right],$$

$$(v_1, \lambda_1), (v_2, \lambda_2) \in \mathbb{E}^1(A, B).$$

Using Lemma 3.3.2 it is easy to see that this composition is well-defined and commutative. Not surprisingly we have

Lemma 3.3.5. $kK^1(A, B)$ *is an abelian group with the zero element represented by any degenerate* $(v, \lambda) \in \mathbb{D}^1(A, B)$.

Proof. We show associativity first. So let $(v_i, \lambda_i) \in \mathbb{E}^1(A, B)$, $i = 1, 2, 3$. Set $w_1 = \Theta_B \begin{bmatrix} v_1 & 0 \\ 0 & v_2 \end{bmatrix}$, $\mu_1 = \Theta_B \circ \begin{bmatrix} \lambda_1 & 0 \\ 0 & \lambda_2 \end{bmatrix}$, $w_2 = \Theta_B \begin{bmatrix} v_2 & 0 \\ 0 & v_3 \end{bmatrix}$ and $\mu_2 = \Theta_B \circ \begin{bmatrix} \lambda_2 & 0 \\ 0 & \lambda_3 \end{bmatrix}$. As in the proof of Lemma 3.2.3 we see that

$$\Theta_B \begin{bmatrix} v_1 & 0 \\ 0 & w_2 \end{bmatrix} = \Phi_1 \begin{bmatrix} v_1 & 0 & 0 \\ 0 & v_2 & 0 \\ 0 & 0 & v_3 \end{bmatrix},$$

$$\Theta_B \circ \begin{bmatrix} \lambda_1 & 0 \\ 0 & \mu_2 \end{bmatrix} = \Phi_1 \circ \begin{bmatrix} \lambda_1 & 0 & 0 \\ 0 & \lambda_2 & 0 \\ 0 & 0 & \lambda_3 \end{bmatrix}$$

and

$$\Theta_B \begin{bmatrix} w_1 & 0 \\ 0 & v_3 \end{bmatrix} = \Phi_2 \begin{bmatrix} v_1 & 0 & 0 \\ 0 & v_2 & 0 \\ 0 & 0 & v_3 \end{bmatrix},$$

$$\Theta_B \circ \begin{bmatrix} \mu_1 & 0 \\ 0 & \lambda_3 \end{bmatrix} = \Phi_2 \circ \begin{bmatrix} \lambda_1 & 0 & 0 \\ 0 & \lambda_2 & 0 \\ 0 & 0 & \lambda_3 \end{bmatrix}$$

for some inner *-isomorphisms $\Phi_1, \Phi_2 : M_3(B) \to B$. By Lemma 1.3.9 these *-isomorphisms differ only by conjugation by a unitary in $\mathcal{M}(B)$. Associativity follows by connecting this unitary to 1 through a strictly continuous path of unitaries, cf. Lemma 1.3.7.

Note that $\Theta_B \begin{bmatrix} m & 0 \\ 0 & 0 \end{bmatrix} = TmT^*$, $m \in \mathcal{M}(B)$, where $T \in \mathcal{M}(B)$ is an isometry. Therefore we can combine Lemma 1.3.7 with Lemma 3.3.2 as above and conclude that $[v_1, \lambda_1] + [0, 0] = \left[\Theta_B \begin{bmatrix} v_1 & 0 \\ 0 & 0 \end{bmatrix}, \Theta_B \circ \begin{bmatrix} \lambda_1 & 0 \\ 0 & 0 \end{bmatrix}\right] = [v_1, \lambda_1]$. Thus $[0, 0]$ is a neutral element in $kK^1(A, B)$.

To show that inverses exist in $kK^1(A, B)$, let $(v, \lambda) \in \mathbb{E}^1(A, B)$ and note that $(1 - v, \lambda) \in \mathbb{E}^1(A, B)$. Let R_t denote the "rotation unitaries" from (1.3.3).

Set $w_t = \begin{bmatrix} v & 0 \\ 0 & 0 \end{bmatrix} + R_t \begin{bmatrix} 1 - v & 0 \\ 0 & 0 \end{bmatrix} R_t^* \in M_2(\mathcal{M}(B))$ and $\phi_t = \begin{bmatrix} \lambda & 0 \\ 0 & \lambda \end{bmatrix} \in \mathrm{Hom}\,(A, M_2(\mathcal{M}(B)))$, $t \in [0, 1]$. Then $\Theta_B(w_t)$ and $\Theta_B \circ \phi_t$, $t \in [0, 1]$, satisfy all relevant conditions of Lemma 3.3.2 and give us a homotopy showing that $[v, \lambda] + [1 - v, \lambda] = \left[\Theta_B \begin{bmatrix} 1 & 0 \\ 0 & 0 \end{bmatrix}, \Theta_B \circ \begin{bmatrix} \lambda & 0 \\ 0 & \lambda \end{bmatrix}\right]$. Since $\left(\Theta_B \begin{bmatrix} 1 & 0 \\ 0 & 0 \end{bmatrix}, \Theta_B \circ \begin{bmatrix} \lambda & 0 \\ 0 & \lambda \end{bmatrix}\right) \in \mathbb{D}^1(A, B)$, Lemma 3.3.3 shows that $[v, \lambda] + [1 - v, \lambda] = [0, 0]$. $\qquad\square$

Now we construct a map $\alpha : kK^1(A, B) \to KK^1(A, B)$ as follows. Let $\phi : \mathcal{M}(B) \oplus \mathcal{M}(B) \to \mathcal{M}(B_{(1)})$ be the isomorphism of Lemma 1.3.3, and let $(v, \lambda) \in \mathbb{E}^1(A, B)$. Define $\phi \circ (\lambda, \lambda) : A \to \mathcal{M}(B_{(1)})$ by $\phi \circ (\lambda, \lambda)(a) = \phi(\lambda(a), \lambda(a))$, $a \in A$. It is now easy to show, using Lemma 1.3.3, that

$$(3.3.6) \qquad \mathcal{E}_{v,\lambda} = (B_{(1)}, \phi \circ (\lambda, \lambda), \phi(2v - 1, 1 - 2v))$$

is a Kasparov $A - B_{(1)}$-module provided B is σ-unital.

Proposition 3.3.6. *There is an isomorphism* $\alpha : kK^1(A, B) \to KK^1(A, B)$ *given by*

$$\alpha[v, \lambda] = [\mathcal{E}_{v,\lambda}], \quad (v, \lambda) \in \mathbb{E}^1(A, B).$$

Proof. We first show that α is well-defined. So let $(v_1, \lambda_1), (v_2, \lambda_2) \in \mathbb{E}^1(A, B)$ be homotopic and let w_t, ϕ_t, for $t \in [0, 1]$, be paths satisfying the conditions in Lemma 3.3.2. Then we can define a *-homomorphism $\pi : A \to \mathcal{M}(IB_{(1)})$ by

$$(\pi(a)f)(t) = \phi \circ (\phi_t, \phi_t)(a)f(t), \quad t \in [0, 1], \quad f \in IB_{(1)}, \quad a \in A,$$

and $V \in \mathcal{M}(IB_{(1)})$ by

$$(Vf)(t) = \phi(2w_t - 1, 1 - 2w_t)f(t), \ f \in IB_{(1)}, \ t \in [0, 1].$$

Using Lemma 1.3.3, it is straightforward to check that $\mathcal{E} = (IB_{(1)}, \pi, V) \in \mathbb{E}(A, IB_{(1)})$. Since $\pi_{0*}(\mathcal{E}) \simeq \mathcal{E}_{v_1, \lambda_1}$ and $\pi_{1*}(\mathcal{E}) \simeq \mathcal{E}_{v_2, \lambda_2}$, we conclude that α is well-defined.

Note that

$$\mathcal{E}_{v_1, \lambda_1} \oplus \mathcal{E}_{v_2, \lambda_2} =$$

$$(B_{(1)} \oplus B_{(1)}, \ \phi \circ (\lambda_1, \lambda_1) \oplus \phi \circ (\lambda_2, \lambda_2), \ \phi(2v_1 - 1, 1 - 2v_1) \oplus \phi(2v_2 - 1, 1 - 2v_2)).$$

Let $V_1, V_2 \in \mathcal{M}(B)$ be the isometries used to define Θ_B. We can use V_1 and V_2 to define an isomorphism $S : B \oplus B \to B$ of Hilbert B-modules by $S(b_1, b_2) = V_1 b_1 + V_2 b_2$, $b_1, b_2 \in B$. Define $T : B_{(1)} \oplus B_{(1)} \to B_{(1)}$ by $T((a,b);(c,d)) = (S(a,c), S(b,d))$, $a, b, c, d \in B$. Then T is an isomorphism of graded Hilbert $B_{(1)}$-modules giving an isomorphism of $\mathcal{E}_{v_1, \lambda_1} \oplus \mathcal{E}_{v_2, \lambda_2}$ with

$$\left(B_{(1)}, \phi \left(\Theta_B \begin{bmatrix} \lambda_1 & 0 \\ 0 & \lambda_2 \end{bmatrix}, \Theta_B \begin{bmatrix} \lambda_1 & 0 \\ 0 & \lambda_2 \end{bmatrix} \right), \phi \left(2\Theta_B \begin{bmatrix} v_1 & 0 \\ 0 & v_2 \end{bmatrix} - 1, 1 - 2\Theta_B \begin{bmatrix} v_1 & 0 \\ 0 & v_2 \end{bmatrix} \right) \right).$$

The latter Kasparov $A - B_{(1)}$-module represents the image of $[v_1, \lambda_1] + [v_2, \lambda_2]$, so we see that α is a homomorphism.

To see that α is surjective, let $\mathcal{E} = (E, \psi, F) \in \mathbb{E}(A, B_{(1)})$. Since $\mathcal{E}_0 = (B_{(1)}, 0, 0)$ is a degenerate Kasparov $A - B_{(1)}$-module, $\mathcal{E} \oplus \mathcal{E}_0$ represents the same element in $KK^1(A, B)$ as $\mathcal{E}$ does. By Kasparov's stabilization theorem for graded $B_{(1)}$-modules, Theorem 1.2.12, and Lemma 1.3.3, we have that $E \oplus B_{(1)}$ is isomorphic to $B_{(1)}$ as graded Hilbert $B_{(1)}$-modules. Thus we can assume from outset that $\mathcal{E} = (B_{(1)}, \psi, F)$. Observe that $(B_{(1)}, \psi, tF + (1-t)\frac{1}{2}(F + F^*)) \in \mathbb{E}(A, B_{(1)})$ for $t \in [0, 1]$, so that we have an operator homotopy connecting $(B_{(1)}, \psi, F)$ to $(B_{(1)}, \psi, \frac{1}{2}(F + F^*))$. Thus we may assume that F is self-adjoint.

By Lemma 1.3.3, $\psi = \phi \circ (\pi, \pi)$ and $F = \phi(H, -H)$ for some $\pi \in \text{Hom}(A, \mathcal{M}(B))$ and some self-adjoint $H \in \mathcal{M}(B)$. Let $g : \mathbb{R} \to \mathbb{R}$ be the function

$$g(t) = \begin{cases} 1, & t \geq 1 \\ t, & t \in [-1, 1] \\ -1, & t \leq -1. \end{cases}$$

Since g is an odd function we can approximate g uniformly on the spectrum of H by odd polynomials taking the value 1 at 1. Doing so we conclude that $g(H)\pi(b) - H\pi(b) \in B$ for all $b \in A$. Therefore

$$(B_{(1)}, \psi, \phi(tH + (1-t)g(H), (t-1)g(H) - tH)) \in \mathbb{E}(A, B_{(1)}), \ t \in [0,1],$$

gives us an operator homotopy that will connect $(B_{(1)}, \psi, F)$ to $(B_{(1)}, \psi, \phi(g(H), -g(H)))$. Thus we can assume that $\|F\| \leq 1$, i.e. that $\mathcal{E} = (B_{(1)}, \phi \circ (\pi, \pi), \phi(H, -H))$ where $H = H^*$, $\|H\| \leq 1$.
Set

$$F_t = \Theta_B \begin{bmatrix} H & it(1-H^2)^{\frac{1}{2}} \\ -it(1-H^2)^{\frac{1}{2}} & -tH \end{bmatrix} \in \mathcal{M}(B), \ t \in [0,1].$$

Define $\pi_1 = \Theta_B \circ \begin{bmatrix} \pi & 0 \\ 0 & 0 \end{bmatrix} \in \text{Hom}(A, \mathcal{M}(B))$. Then

$$(B_{(1)}, \phi \circ (\pi_1, \pi_1), \phi(F_t, -F_t)) \in \mathbb{E}(A, B_{(1)}), \ t \in [0,1],$$

defines an operator homotopy. Note that the isomorphism $T: B_{(1)} \oplus B_{(1)} \to B_{(1)}$ of graded Hilbert $B_{(1)}$-modules used above gives an isomorphism from $\mathcal{E} \oplus (B_{(1)}, 0, 0)$ to $(B_{(1)}, \phi \circ (\pi_1, \pi_1), \phi(F_0, -F_0))$. Since $(B_{(1)}, 0, 0)$ is a degenerate Kasparov $A - B_{(1)}$-module and F_1 is a self-adjoint unitary, the constructed operator homotopy shows that $[\mathcal{E}]$ is represented by a triple $(B_{(1)}, \phi \circ (\pi, \pi), \phi(U, -U)) \in \mathbb{E}(A, B_{(1)})$ where U is a self-adjoint unitary in $\mathcal{M}(B)$. Then $p = \frac{1}{2}(U+1)$ is a projection in $\mathcal{M}(B)$ such that $\pi_1(a)p - p\pi_1(a) \in B$, $a \in A$. It is clear that $(p, \pi_1) \in \mathbb{E}^1(A, B)$, and that $\alpha[p, \pi_1] = [\mathcal{E}]$, so α is surjective.

To see that α is also injective, assume that $[\mathcal{E}_{v,\lambda}] = 0$ for some $(v, \lambda) \in \mathbb{E}^1(A, B)$. Then there is a norm continuous path F_t, $t \in [0,1]$, in $\mathcal{M}(B)$ such that $F_0 = 2v - 1$, $(B_{(1)}, \phi \circ (\lambda, \lambda), \phi(F_t, -F_t)) \in \mathbb{E}^1(A, B_{(1)})$, $t \in [0,1]$, and $(B_{(1)}, \phi \circ (\lambda, \lambda), \phi(F_1, -F_1))$ is a degenerate Kasparov $A - B_{(1)}$-module. Set $w_t = \frac{1}{2}(F_t + 1)$, $\phi_t = \lambda$, $t \in [0,1]$. Then (w_t, ϕ_t), $t \in [0,1]$, meets all relevant conditions of Lemma 3.3.2 and the corresponding homotopy shows us that (v, λ) is homotopic to an element of $\mathbb{D}^1(A, B)$. Hence $[v, \lambda] = 0$ by Lemma 3.3.3. □

Corollary 3.3.7. *Two elements* (v_1, λ_1), $(v_2, \lambda_2) \in \mathbb{E}^1(A, B)$ *are homotopic if and only if there are pairs* (t_1, μ_1), $(t_2, \mu_2) \in \mathbb{D}^1(A, B)$, *a unitary* $u \in \mathcal{M}(B)$ *and a norm continuous path* w_t, $t \in [0,1]$, *such that*

(i) $\left(w_t, \Theta_B \circ \begin{bmatrix} \lambda_1 & 0 \\ 0 & \mu_1 \end{bmatrix} \right) \in \mathbb{E}^1(A, B), \ t \in [0,1],$ *and*

(ii) $w_0 = \Theta_B \begin{bmatrix} v_1 & 0 \\ 0 & t_1 \end{bmatrix}$, $w_1 = u\Theta_B \begin{bmatrix} v_2 & 0 \\ 0 & t_2 \end{bmatrix} u^*$, and

$$\Theta_B \circ \begin{bmatrix} \lambda_1 & 0 \\ 0 & \mu_1 \end{bmatrix} = Adu \circ \Theta_B \circ \begin{bmatrix} \lambda_2 & 0 \\ 0 & \mu_2 \end{bmatrix}.$$

Proof. Assume first that (v_1, λ_1) and (v_2, λ_2) are homotopic. Then $[\mathcal{E}_{v_1,\lambda_1}] = [\mathcal{E}_{v_2,\lambda_2}]$ in $KK^1(A, B)$ by the preceding proposition. Thus there are $\mathcal{F}_1, \mathcal{F}_2 \in \mathbb{D}(A, B_{(1)})$ such that $\mathcal{E}_{v_1,\lambda_1} \oplus \mathcal{F}_1$ is operator homotopic to $\mathcal{E}_{v_2,\lambda_2} \oplus \mathcal{F}_2$. By adding $(B_{(1)}, 0, 0) \in \mathbb{D}(A, B_{(1)})$ to $\mathcal{F}_1$ and $\mathcal{F}_2$, and using Kasparov's stabilization theorem for graded Hilbert C^*-modules, Theorem 1.2.12, and Lemma 1.3.3, we can assume that $\mathcal{F}_i = (B_{(1)}, \psi_i, F_i)$, $i = 1, 2$. By Lemma 1.3.3, $\psi_i = \phi \circ (\mu_i, \mu_i)$ and $F_i = \phi(T_i, -T_i)$ for some $\mu_i \in \text{Hom}(A, \mathcal{M}(B))$, $T_i \in \mathcal{M}(B)$, $i = 1, 2$. Set $t_i = \frac{1}{2}(T_i + 1)$, $i = 1, 2$. Since $\mathcal{F}_i \in \mathbb{D}(A, B_{(1)})$, we have that $(t_i, \mu_i) \in \mathbb{D}^1(A, B)$, $i = 1, 2$. By the proof of the preceding proposition, we have that $\mathcal{E}_{v_i,\lambda_i} \oplus \mathcal{F}_i$ is isomorphic to $\mathcal{E}_{s_i,\nu_i}$, where

$$s_i = \Theta_B \begin{bmatrix} v_i & 0 \\ 0 & t_i \end{bmatrix}, \quad \nu_i = \Theta_B \begin{bmatrix} \lambda_i & 0 \\ 0 & \mu_i \end{bmatrix}, \quad i = 1, 2.$$

Thus there is an operator homotopy $(B_{(1)}, \psi, F_t)$, $t \in [0, 1]$, in $\mathbb{E}(A, B_{(1)})$ such that $(B_{(1)}, \psi, F_0) \simeq \mathcal{E}_{s_1,\nu_1}$ and $(B_{(1)}, \psi, F_1) \simeq \mathcal{E}_{s_2,\nu_2}$. Since a graded Hilbert $B_{(1)}$-module automorphism $B_{(1)} \simeq B_{(1)}$ is given by a unitary in $\mathcal{M}(B_{(1)})$ of degree 0, Lemma 1.3.3 gives us unitaries u_i, $i = 1, 2$, such that $\phi(u_1, u_1)$ implements the first isomorphism, $(B_{(1)}, \psi, F_0) \simeq \mathcal{E}_{s_1,\nu_1}$, and $\phi(u_2, u_2)$ the second, $(B_{(1)}, \psi, F_1) \simeq \mathcal{E}_{s_2,\nu_2}$. Then $(B_{(1)}, Ad\phi(u_1, u_1) \circ \psi, Ad\phi(u_1, u_1) \circ F_t) = (B_{(1)}, \psi', F_t')$, $t \in [0, 1]$, is an operator homotopy in $\mathbb{E}(A, B_{(1)})$ such that $\mathcal{E}_{s_1,\nu_1} = (B_{(1)}, \psi', F_0')$ and $(B_{(1)}, \psi', F_1') \simeq \mathcal{E}_{s_2,\nu_2}$, where the last isomorphism is implemented by $\phi(u_2 u_1^*, u_2 u_1^*)$. Using Lemma 1.3.3 once more we can set $(B_{(1)}, \psi', F_t') = (B_{(1)}, \phi \circ (\nu_1, \nu_1), \phi(2w_t - 1, 1 - 2w_t))$, $t \in [0, 1]$, where w_t, $t \in [0, 1]$, is now a norm continuous path in $\mathcal{M}(B)$ such that $w_0 = s_1$. Setting $u = u_2 u_1^*$, we have achieved what we wanted.

Conversely, assume we have $(t_i, \mu_i) \in \mathbb{E}^1(A, B)$, $i = 1, 2$, a unitary $u \in \mathcal{M}(B)$ and a norm continuous path w_t, $t \in [0, 1]$, in $\mathcal{M}(B)$ with the stated properties. Then clearly $\left[\Theta_B \begin{bmatrix} v_1 & 0 \\ 0 & t_1 \end{bmatrix}, \Theta_B \circ \begin{bmatrix} \lambda_1 & 0 \\ 0 & \mu_1 \end{bmatrix}\right] = \left[u\Theta_B \begin{bmatrix} v_2 & 0 \\ 0 & t_2 \end{bmatrix} u^*, Adu \circ \Theta_B \circ \begin{bmatrix} \lambda_2 & 0 \\ 0 & \mu_2 \end{bmatrix}\right]$ in $kK^1(A, B)$. Connecting u to the identity through a strictly continuous path of unitaries, cf. Lemma

1.3.7, we get

$$
\begin{aligned}
[v_1, \lambda_1] &= [v_1, \lambda_1] + [t_1, \mu_1] \\
&= \left[\Theta_B \begin{bmatrix} v_1 & 0 \\ 0 & t_1 \end{bmatrix}, \Theta_B \circ \begin{bmatrix} \lambda_1 & 0 \\ 0 & \mu_1 \end{bmatrix} \right] \\
&= \left[\Theta_B \begin{bmatrix} v_2 & 0 \\ 0 & t_2 \end{bmatrix}, \Theta_B \circ \begin{bmatrix} \lambda_2 & 0 \\ 0 & \mu_2 \end{bmatrix} \right] \\
&= [v_2, \lambda_2] + [t_2, \mu_2] \\
&= [v_2, \lambda_2]
\end{aligned}
$$

in $kK^1(A, B)$. $\qquad\qquad\qquad\qquad\qquad\qquad\qquad\qquad\qquad\qquad\qquad$ $\square$

Now we can prove that $Ext^{-1}(A, B) \simeq KK^1(A, B)$ by producing an isomorphism $e : Ext^{-1}(A, B) \to kK^1(A, B)$. So let $\psi \in \mathrm{Hom}\,(A, \mathcal{Q}(B))$ such that $[\psi]$ is invertible in $Ext(A, B)$. Then, by Corollary 3.2.10, $\psi(\cdot) = q_B(p_\psi \lambda_\psi(\cdot))$ for a projection $p_\psi \in \mathcal{M}(B)$ and a *-homomorphism $\lambda_\psi \in \mathrm{Hom}\,(A, \mathcal{M}(B))$. Since ψ is a *-homomorphism, we have that $p_\psi \lambda_\psi(a) - \lambda_\psi(a)p_\psi \in B$, $a \in A$, so that $(p_\psi, \lambda_\psi) \in \mathbb{E}^1(A, B)$. This construction gives us at least a homomorphism e.

Lemma 3.3.8. *There is a group homomorphism* $e : Ext^{-1}(A, B) \to kK^1(A, B)$ *such that* $e[\psi] = [p, \lambda]$ *when* $p \in \mathcal{M}(B)$ *is a projection,* $\lambda \in \mathrm{Hom}\,(A, \mathcal{M}(B))$ *and* $\psi(\cdot) = q_B(p\lambda(\cdot)) \in \mathrm{Hom}\,(A, \mathcal{Q}(B))$.

Proof. We first prove that e is well-defined. Let $p_i \in \mathcal{M}(B)$ be projections, $\lambda_i \in \mathrm{Hom}\,(A, \mathcal{M}(B))$ and $\psi_i(\cdot) = q_B(p_i\lambda_i(\cdot)) \in \mathrm{Hom}\,(A, \mathcal{Q}(B))$, $i = 1, 2$, and assume that $[\psi_1] = [\psi_2]$ in $Ext^{-1}(A, B)$. This means that there are *-homomorphisms $\pi_i \in \mathrm{Hom}\,(A, \mathcal{M}(B))$, $i = 1, 2$, such that

$$
Adq_B(T) \circ \tilde{\Theta}_B \circ \begin{bmatrix} \psi_1 & 0 \\ 0 & q_B \circ \pi_1 \end{bmatrix} = \tilde{\Theta}_B \circ \begin{bmatrix} \psi_2 & 0 \\ 0 & q_B \circ \pi_2 \end{bmatrix}
$$

for some unitary $T \in \mathcal{M}(B)$.

Set

$$
p_3 = T\Theta_B \begin{bmatrix} p_1 & 0 \\ 0 & 1 \end{bmatrix} T^*, \quad p_4 = \Theta_B \begin{bmatrix} p_2 & 0 \\ 0 & 1 \end{bmatrix},
$$

$$
\lambda_3 = AdT \circ \Theta_B \circ \begin{bmatrix} \lambda_1 & 0 \\ 0 & \pi_1 \end{bmatrix} \quad \text{and} \quad \lambda_4 = \Theta_B \circ \begin{bmatrix} \lambda_2 & 0 \\ 0 & \pi_2 \end{bmatrix}.
$$

Then

$$(3.3.7) \qquad\qquad q_B(p_3\lambda_3(\cdot)) = q_B(p_4\lambda_4(\cdot)).$$

Furthermore, $[p_1, \lambda_1] = [p_1, \lambda_1] + [1, \pi_1] = [p_3, \lambda_3]$, by the usual procedure af connecting T to 1, and

$$[p_2, \lambda_2] = [p_2, \lambda_2] + [1, \pi_2] = [p_4, \lambda_4].$$

Hence, to conclude that e is well-defined, it suffices to check that $[p_3, \lambda_3] = [p_4, \lambda_4]$ in $kK^1(A, B)$. For this we note that (3.3.7) implies

(3.3.8) $$p_3\lambda_3(a) - p_4\lambda_4(a) \in B, \quad a \in A.$$

Set

$$F_t = (1 + t^2)^{-1} \begin{bmatrix} p_4 & tp_4p_3 \\ tp_3p_4 & t^2p_3 \end{bmatrix} \in M_2(\mathcal{M}(B)), \quad t \in [0, \infty[,$$

and let $F_\infty = \begin{bmatrix} 0 & 0 \\ 0 & p_3 \end{bmatrix}$. Then $F : [0, \infty] \to M_2(\mathcal{M}(B))$ is normcontinuous and $F_0 = \begin{bmatrix} p_4 & 0 \\ 0 & 0 \end{bmatrix}$. Using (3.3.8) and that $p_4\lambda_4(a) - \lambda_4(a)p_4, p_3\lambda_3(a) - \lambda_3(a)p_3 \in B$, $a \in A$, we get $\left(\Theta_B(F_t), \Theta_B \circ \begin{bmatrix} \lambda_4 & 0 \\ 0 & \lambda_3 \end{bmatrix} \right) \in \mathbb{E}^1(A, B)$, $t \in [0, \infty]$. Thus we see that

$$\left[\Theta_B \begin{bmatrix} p_4 & 0 \\ 0 & 0 \end{bmatrix}, \Theta_B \circ \begin{bmatrix} \lambda_4 & 0 \\ 0 & \lambda_3 \end{bmatrix} \right] = \left[\Theta_B \begin{bmatrix} 0 & 0 \\ 0 & p_3 \end{bmatrix}, \Theta_B \circ \begin{bmatrix} \lambda_4 & 0 \\ 0 & \lambda_3 \end{bmatrix} \right]$$

in $kK^1(A, B)$. Since $(0, \lambda_3), (0, \lambda_4) \in \mathbb{D}^1(A, B)$, we see that $[p_4, \lambda_4] = [p_3, \lambda_3]$. Hence e is well-defined.

Since e is a homomorphism by definition of the compositions, the proof is complete. $\square$

To prove that e is injective, we need the following technical lemma.

Lemma 3.3.9. *Let w_t, $t \in [0, 1]$, be a normcontinuous path in $\mathcal{M}(B)$ and $\lambda \in \mathrm{Hom}\, (A, \mathcal{M}(B))$ a *-homomorphism such that $(w_t, \lambda) \in \mathbb{E}^1(A, B)$ for all $t \in [0, 1]$. Then there is a normcontinuous path u_t, $t \in [0, 1]$, of unitaries in $\mathcal{M}(B)$ such that*

(i) $u_t\lambda(a) - \lambda(a)u_t \in B$, *and*

(ii) $(w_t u_t - u_t w_0)\lambda(a) \in B$, $t \in [0, 1]$, $a \in A$.

Proof. Set $F_t = 2w_t - 1$, $t \in [0, 1]$. Then it is easy to see that $(F_t^2 - 1)\lambda(a)$, $(F_t^* - F_t)\lambda(a)$, $F_t\lambda(a) - \lambda(a)F_t \in B$ for all $t \in [0, 1]$, $a \in A$.

Since $t \to F_t$ is normcontinuous, we can find $\delta > 0$ such that $2 - F_s^2 + F_t F_s$ is invertible when $|t - s| \leq \delta$. Choose points $t_0, t_1, t_2, \ldots, t_n$ in $[0,1]$ such that $|t_i - t_{i+1}| \leq \delta$, $i = 0, 1, 2, \ldots, n-1$, $t_0 = 0$ and $t_n = 1$. Set $v_t = \frac{1}{2}(2 - F_0^2 + F_t F_0)$, $t \in [0, t_1]$. Then v_t satisfies (i) and (ii) for $t \in [0, t_1]$ when substituted for u_t. Assume that we have found a continuous path v_t, $t \in [0, t_k]$, $k < n$, of invertibles satisfying (i) and (ii) when replaced for u_t, $t \in [0, t_k]$. Then set $v_t = \frac{1}{2}(2 - F_{t_k}^2 + F_t F_{t_k})v_{t_k}$, $t \in [t_k, t_{k+1}]$. In this way we can construct a continuous path of invertibles in $\mathcal{M}(B)$ with the desired properties, except that it is not unitary. Take $u_t = v_t(v_t^* v_t)^{-\frac{1}{2}}$, for all $t \in [0,1]$. $\square$

Theorem 3.3.10. $e : Ext^{-1}(A, B) \to kK^1(A, B)$ *is an isomorphism.*

Proof. We first prove that e is surjective, so let $(v, \lambda) \in \mathbb{E}^1(A, B)$. Then $(\frac{1}{2}(v + v^*), \lambda) \in \mathbb{E}^1(A, B)$, and $(tv + (1 - t)\frac{1}{2}(v + v^*), \lambda) \in \mathbb{E}^1(A, B)$, $t \in [0, 1]$, provides a homotopy from (v, λ) to $(\frac{1}{2}(v + v^*), \lambda)$. Thus we can assume that $v = v^*$. Now let $h : \mathbb{R} \to \mathbb{R}$ be the function

$$h(t) = \begin{cases} 1, & t \geq 1, \\ t, & t \in [0, 1], \\ 0, & t \leq 0. \end{cases}$$

Then a simple argument shows that $h(v)\lambda(a) = v\lambda(a) \bmod B$, $a \in A$. It follows that $(h(v), \lambda) \in \mathbb{E}^1(A, B)$, and that $(tv + (1 - t)h(v), \lambda) \in \mathbb{E}^1(A, B)$, $t \in [0, 1]$, provides a homotopy connecting (v, λ) to $(h(v), \lambda)$. Thus we can assume that $0 \leq v \leq 1$.

Set $H = 2v - 1$. Then $H\lambda(a) - \lambda(a)H$, $(H^2 - 1)\lambda(a) \in B$, $a \in A$. Set

$$F_t = \Theta_B \begin{bmatrix} H & it(1 - H^2)^{\frac{1}{2}} \\ -it(1 - H^2)^{\frac{1}{2}} & -tH \end{bmatrix} \in \mathcal{M}(B), \ t \in [0, 1].$$

Then F_t depends normcontinuously on t and

$$F_t \Theta_B \circ \begin{bmatrix} \lambda & 0 \\ 0 & 0 \end{bmatrix}(a) - \Theta_B \circ \begin{bmatrix} \lambda & 0 \\ 0 & 0 \end{bmatrix}(a)F_t \in B,$$

$$(F_t^2 - 1)\Theta_B \begin{bmatrix} \lambda & 0 \\ 0 & 0 \end{bmatrix}(a) \in B, \ t \in [0, 1], \ a \in A.$$

Thus

$$\left(\frac{1}{2}(F_t + 1), \Theta_B \begin{bmatrix} \lambda & 0 \\ 0 & 0 \end{bmatrix}\right) \in \mathbb{E}^1(A, B), \ t \in [0, 1],$$

provides us with a homotopy connecting $\left(\Theta_B \begin{bmatrix} v & 0 \\ 0 & 0 \end{bmatrix}, \Theta_B \circ \begin{bmatrix} \lambda & 0 \\ 0 & 0 \end{bmatrix}\right)$ to $\left(\frac{1}{2}(F_1 + 1), \Theta_B \circ \begin{bmatrix} \lambda & 0 \\ 0 & 0 \end{bmatrix}\right)$. Note that $\frac{1}{2}(F_1 + 1) = q$ is a projection. The preceding arguments leads to the identity $[v, \lambda] = [q, \phi]$, where $\phi = \Theta_B \circ \begin{bmatrix} \lambda & 0 \\ 0 & 0 \end{bmatrix}$.

Since $\phi(a)q - q\phi(a) \in B$, $a \in A$, $\psi(\cdot) = q_B(q\phi(\cdot))$ defines an element of $\text{Hom}\,(A, \mathcal{Q}(B))$. Since $a \to q\phi(a)q$ is a completely positive contractive map, it follows from Theorem 3.2.9 that $[\psi] \in Ext^{-1}(A, B)$. Since $e[\psi] = [q, \phi] = [v, \lambda]$, we have shown that e is surjective.

To prove that e is injective, assume that $p \in \mathcal{M}(B)$ is a projection and $\lambda \in \text{Hom}\,(A, \mathcal{M}(B))$ a *-homomorphism such that $p\lambda(a) - \lambda(a)p \in B$, $a \in A$, and $[p, \lambda] = 0$ in $kK^1(A, B)$. We want to conclude that the *-homomorphism $\psi(\cdot) = q_B(p\lambda(\cdot)) \in \text{Hom}\,(A, \mathcal{Q}(B))$ represents 0 in $Ext(A, B)$. We first apply Corollary 3.3.7 to get $(r, \mu) \in \mathbb{D}^1(A, B)$ and a norm continuous path w_t, $t \in [0, 1]$, in $\mathcal{M}(B)$ such that

$$\left(w_t, \Theta_B \circ \begin{bmatrix} \lambda & 0 \\ 0 & \mu \end{bmatrix}\right) \in \mathbb{E}^1(A, B),\ t \in [0, 1],\ w_0 = \Theta_B \begin{bmatrix} p & 0 \\ 0 & r \end{bmatrix}$$

and

$$\left(w_1, \Theta_B \circ \begin{bmatrix} \lambda & 0 \\ 0 & \mu \end{bmatrix}\right) \in \mathbb{D}^1(A, B).$$

Applying next Lemma 3.3.9, we get a unitary $u \in \mathcal{M}(B)$ such that

$$u\Theta_B \circ \begin{bmatrix} \lambda & 0 \\ 0 & \mu \end{bmatrix}(a) - \Theta_B \circ \begin{bmatrix} \lambda & 0 \\ 0 & \mu \end{bmatrix}(a)u \in B,\ a \in A,$$

and

$$\left(w_1 - u\Theta_B \begin{bmatrix} p & 0 \\ 0 & r \end{bmatrix} u^*\right) \Theta_B \circ \begin{bmatrix} \lambda & 0 \\ 0 & \mu \end{bmatrix}(a) \in B,\ a \in A.$$

Thus

$$(3.3.9)\quad Adq_B(u) \circ q_B \left(\Theta_B \begin{bmatrix} p & 0 \\ 0 & r \end{bmatrix}\right) \Theta_B \circ \begin{bmatrix} \lambda & 0 \\ 0 & \mu \end{bmatrix}(\cdot)$$
$$= q_B \left(w_1 \Theta_B \circ \begin{bmatrix} \lambda & 0 \\ 0 & \mu \end{bmatrix}(\cdot)\right).$$

Since $(r, \mu), \left(w_1, \Theta_B \circ \begin{bmatrix} \lambda & 0 \\ 0 & \mu \end{bmatrix}\right) \in \mathbb{D}^1(A, B)$, we can define *-homomorphisms $\phi_1, \phi_2 \in \text{Hom}\,(A, \mathcal{M}(B))$ by

$$\phi_1(a) = r\mu(a) \quad \text{and} \quad \phi_2(a) = w_1 \Theta_B \circ \begin{bmatrix} \lambda & 0 \\ 0 & \mu \end{bmatrix}(a),\ a \in A.$$

Then (3.3.9) says that

$$Adq_B(u) \circ \tilde{\Theta}_B \circ \begin{bmatrix} \psi & 0 \\ 0 & q_B \circ \phi_1 \end{bmatrix} = q_B \circ \phi_2.$$

Thus $[\psi] = 0$ in $Ext(A, B)$, and the proof is complete. $\square$

Corollary 3.3.11. *The map* $\alpha \circ e : Ext^{-1}(A, B) \rightarrow KK^1(A, B)$ *is an isomorphism.*

We have now shown how the theory of C^*-extensions is connected to Kasparov's KK-theory for graded C^*-algebras. Hence the results of Chapter 2 give information on the extension groups. In particular, this makes it possible to see what it means that two extensions of A by B define the same element in $Ext^{-1}(A, B)$.

Definition 3.3.12. An extension $(i, E, p) \in \mathcal{E}xt(A, B)$ is called *semi-split* when there is a completely positive linear contraction $s : B \rightarrow E$ such that $p \circ s = id$.

It is not difficult to see that (i, E, p) represents an invertible element in $Ext^{-1}(A, B)$ if and only if it is semi-split, cf. E3.2.1.

Lemma 3.3.13. *Let* C, C_1 *and* D *be* C^*-*algebras,* $\phi : C \rightarrow C_1$ *a surjective *-homomorphism and* $(i, E, p) \in \mathcal{E}xt(C, D)$ *an extension of* D *by* C. *Then there is an extension* (i_ϕ, E_ϕ, p_ϕ) *of* D *by* C_1, *unique up to isomorphism, such that the diagram*

$$
\begin{array}{ccccccccc}
0 & \longrightarrow & C & \overset{i}{\longrightarrow} & E & \overset{p}{\longrightarrow} & D & \longrightarrow & 0 \\
 & & \phi\downarrow & & \downarrow & & \| & & \\
0 & \longrightarrow & C_1 & \underset{i_\phi}{\longrightarrow} & E_\phi & \underset{p_\phi}{\longrightarrow} & D & \longrightarrow & 0
\end{array}
$$

*commutes for some *-homomorphism* $E \longrightarrow E_\phi$. *The Busby invariant for* (i_ϕ, E_ϕ, p_ϕ) *is* $\tilde{\phi} \circ \psi$, *where* ψ *is the Busby invariant of* (i, E, p) *and* $\tilde{\phi} : \mathcal{Q}(C) \rightarrow \mathcal{Q}(C_1)$ *is the unique *-homomorphism making the diagram*

$$
\begin{array}{ccccccccc}
0 & \longrightarrow & C & \subseteq & \mathcal{M}(C) & \longrightarrow & \mathcal{Q}(C) & \longrightarrow & 0 \\
 & & \phi\downarrow & & \phi\downarrow & & \tilde{\phi}\downarrow & & \\
0 & \longrightarrow & C_1 & \subseteq & \mathcal{M}(C_1) & \longrightarrow & \mathcal{Q}(C_1) & \longrightarrow & 0
\end{array}
$$

commute.

Proof. To prove the existence, we can assume that

$$E = \{(a, x) \in D \times \mathcal{M}(C) : \psi(a) = q_C(x)\},$$

that $i(b) = (0,b)$, $b \in C$, and that $p(a,x) = a$, $(a,x) \in E$, since this extension also has the Busby invariant ψ, cf. Theorem 3.1.4.

Next let us set $E_\phi = \{(a,x) \in D \times \mathcal{M}(C_1) : \tilde{\phi} \circ \psi(a) = q_C(x)\}$, $i_\phi(c) = (0,c)$, $c \in C_1$, and $p_\phi(a,x) = a$, $a \in A$. Define $\lambda : E \to E_\phi$ by $\lambda(a,x) = (a, \underline{\phi}(x))$, $(a,x) \in E$. It is easy to check that with this choice of *-homomorphism $E \to E_\phi$, the diagram commutes. It is also clear from the construction that the Busby invariant of (i_ϕ, E_ϕ, p_ϕ) is $\tilde{\phi} \circ \psi$.

To prove uniqueness, up to isomorphism, we consider an extension (j, F, q) of D by C_1 and a *-homomorphism $\pi : E \to F$ such that

$$
\begin{array}{ccccccccc}
0 & \longrightarrow & C & \xrightarrow{i} & E & \xrightarrow{p} & D & \longrightarrow & 0 \\
& & \phi\downarrow & & \pi\downarrow & & \| & & \\
0 & \longrightarrow & C_1 & \xrightarrow{j} & F & \xrightarrow{q} & D & \longrightarrow & 0
\end{array}
$$

commutes. We must check that the Busby invariant ψ' for (j,F,q) is $\tilde{\phi} \circ \psi$. So fix $a \in D$ and choose $e \in E$ such that $p(e) = a$. Then $q(\pi(e)) = a$, so that $\psi'(a) = q_{C_1}(\pi(e)')$ where $\pi(e)' \in \mathcal{M}(C_1)$ satisfies $j(\pi(e)'c) = \pi(e)j(c)$, $c \in C_1$. For a given $c \in C_1$, take $b \in \phi^{-1}(c)$. Then $\pi(e)j(c) = \pi(e)j(\phi(b)) = \pi(e)\pi(i(b)) = \pi(ei(b)) = \pi(i(e'b)) = j(\phi(e'b)) = j(\underline{\phi}(e')c)$, where $e' \in \mathcal{M}(C)$ satisfies $q_C(e') = \psi(a)$. It follows that $\pi(e)' = \underline{\phi}(e')$, so that $\psi'(a) = q_{C_1}(\pi(e)') = q_{C_1}(\underline{\phi}(e')) = \tilde{\phi}(q_C(e')) = \tilde{\phi} \circ \psi(a)$. $\square$

Theorem 3.3.14. *Let (i_1, E_1, p_1), $(i_2, E_2, p_2) \in \mathcal{E}xt(A,B)$ be two semi-split extensions with Busby invariants ψ_1, $\psi_2 \in \mathrm{Hom}\,(A, \mathcal{Q}(B))$, respectively. Then $[\psi_1] = [\psi_2]$ in $\mathrm{Ext}^{-1}(A,B)$ if and only if there is a semi-split extension $(i, E, p) \in \mathcal{E}xt(A, IB)$ and a commutative diagram*

$$
\begin{array}{ccccccccc}
0 & \longrightarrow & B & \xrightarrow{i_1} & E_1 & \xrightarrow{p_1} & A & \longrightarrow & 0 \\
& & \pi_0\uparrow & & \uparrow & & \| & & \\
0 & \longrightarrow & IB & \xrightarrow{i} & E & \xrightarrow{p} & A & \longrightarrow & 0 \\
& & \pi_1\downarrow & & \downarrow & & \| & & \\
0 & \longrightarrow & B & \xrightarrow{i_2} & E_2 & \xrightarrow{p_2} & A & \longrightarrow & 0
\end{array}
$$

*of C^*-algebras and *-homomorphisms.*

Proof. Assume first that $[\psi_1] = [\psi_2]$ in $\mathrm{Ext}^{-1}(A,B)$. Choose projections $p_i \in \mathcal{M}(B)$ and *-homomorphisms $\lambda_i \in \mathrm{Hom}\,(A, \mathcal{M}(B))$, such that $\psi_i(\cdot) = q_B(p_i\lambda_i(\cdot))$, $i = 1,2$. From Theorem 3.3.10 we conclude that (p_1, λ_1) and (p_2, λ_2) are homotopic in $\mathbb{E}^1(A,B)$. Consequently there is a pair $(v, \lambda) \in \mathbb{E}^1(A, IB)$ such that $\pi_0(v) = p_1$, $\pi_1(v) = p_2$, $\pi_0 \circ \lambda = \lambda_1$ and $\pi_1 \circ \lambda = \lambda_2$. Since $(\frac{1}{2}(v+v^*), \lambda) \in \mathbb{E}^1(A, IB)$ will have the same properties, we can assume that v is self-adjoint. Letting h be the real function from the proof of Theorem 3.3.10, we can then substitute v by $h(v)$ and assume that $0 \leq v \leq 1$. Define $\psi(\cdot) = q_{IB}(v\lambda(\cdot))$. Then ψ is the Busby invariant for an extension (i, E, p) of A by IB which is semi-split because $v\lambda(\cdot)v$ is completely positive and contractive. Since $\pi_0 \circ \psi = \psi_1$ and $\pi_1 \circ \psi = \psi_2$, Lemma 3.3.13 gives us the desired diagram.

Conversely, assume that we have the commuting diagram. Let $\psi \in \text{Hom}(A, \mathcal{Q}(IB))$ be the Busby invariant for the extension $(i, E, p) \in \mathcal{E}xt(A, IB)$. Since this extension is semi-split, there is a projection $f \in \mathcal{M}(IB)$ and a *-homomorphism $\lambda \in \text{Hom}(A, \mathcal{M}(IB))$ such that $\psi(\cdot) = q_{IB}(f\lambda(\cdot))$. By Lemma 3.3.13, $\psi_1(\cdot) = \tilde{\pi}_0 \circ \psi(\cdot) = q_B(\pi_0(f)\pi_0(\lambda(\cdot)))$ and $\psi_2(\cdot) = q_B(\pi_1(f)\pi_1(\lambda(\cdot)))$. Since $(f, \lambda) \in \mathbb{E}^1(A, IB)$ provides a homotopy between $(\pi_0(f), \pi_0 \circ \lambda)$ and $(\pi_1(f), \pi_1 \circ \lambda)$, we see that $e[\psi_1] = [\pi_0(f), \pi_0 \circ \lambda] = [\pi_1(f), \pi_1 \circ \lambda] = e[\psi_2]$. Consequently, $[\psi_1] = [\psi_2]$ in $Ext^{-1}(A, B)$ by Theorem 3.3.10. $\qquad\square$

We remark that although it has been a little obscured by the many details, the preceding theorem should really be considered as a corollary of Theorem 2.2.17.

3.3.15. *Notes and remarks.*

The isomorphism $KK^1(A, B) \simeq Ext^{-1}(A, B)$ goes back to [19]. That this isomorphism together with Theorem 2.2.17 gives Theorem 3.3.14 was observed in [30].

Exercise 3.3

E 3.3.1

Let B be a σ-unital C^*-algebra. Show that there is a natural way to make $Ext^{-1}(\cdot, B)$ into a contravariant functor from separable C^*-algebras to abelian groups. Shows that this functor is homotopy invariant.

Let A be a separable C^*-algebra. Show that there is a natural way to make $Ext^{-1}(A, \cdot)$ into a covariant functor. Show that this functor is homotopy invariant.

CHAPTER 4

The Kasparov Groups
for Ungraded C^*-Algebras

4.1. From Hilbert C^*-Modules to *-Homomorphisms

In this chapter we only consider σ-unital C^*-algebras.

Let A and B be arbitrary (ungraded) C^*-algebras.

Definition 4.1.1. A $KK_h(A, B)$-*cycle* is a pair (ϕ_+, ϕ_-) of *-homomorphisms $\phi_+, \phi_- \in \mathrm{Hom}\,(A, \mathcal{M}(\mathcal{K} \otimes B))$ such that

$$(4.1.1) \qquad \phi_+(a) - \phi_-(a) \in \mathcal{K} \otimes B, \quad a \in A.$$

The set of $KK_h(A, B)$-cycles will be denoted by $\mathbb{F}(A, B)$.

Definition 4.1.2. Two $KK_h(A, B)$-cycles, $(\phi_+, \phi_-), (\psi_+, \psi_-) \in \mathbb{F}(A, B)$ are called *homotopic* when there is a path $(\lambda_+^t, \lambda_-^t) \in \mathbb{F}(A, B)$, $t \in [0, 1]$, such that
  (i) the maps $t \to \lambda_+^t(a)$ and $t \to \lambda_-^t(a)$ from $[0, 1]$ to $\mathcal{M}(\mathcal{K} \otimes B)$ are strictly continuous for all $a \in A$,
 (ii) the map $t \to \lambda_+^t(a) - \lambda_-^t(a)$ from $[0, 1]$ to $\mathcal{K} \otimes B$ is continuous in norm for all $a \in A$, and
(iii) $(\lambda_+^0, \lambda_-^0) = (\phi_+, \phi_-)$, $(\lambda_+^1, \lambda_-^1) = (\psi_+, \psi_-)$.
  We write $(\phi_+, \phi_-) \sim (\psi_+, \psi_-)$ in this case.

It is clear that homotopy defines an equivalence relation in $\mathbb{F}(A, B)$.

Definition 4.1.3. We let $KK_h(A, B)$ denote the homotopy classes of $KK_h(A, B)$-cycles, i.e. $KK_h(A, B) = \mathbb{F}(A, B)/\sim$. The homotopy class in $KK_h(A, B)$ represented by $(\phi_+, \phi_-) \in \mathbb{F}(A, B)$ is denoted by $[\phi_+, \phi_-]$.

Lemma 4.1.4. *Let* $(\phi_+, \phi_-) \in \mathbb{F}(A, B)$. *If* $\phi_+ = \phi_-$, *then* $(\phi_+, \phi_-) \sim (0, 0)$.

Proof. Let v_t, $t \in]0, 1]$, be a path of isometries in $\mathcal{M}(\mathcal{K} \otimes B)$ satisfying the conditions of Lemma 1.3.6. Set $\lambda_+^t = \operatorname{Ad} v_t \circ \phi_+$, $\lambda_-^t = \operatorname{Ad} v_t \circ \phi_-$, $t \in]0, 1]$, and $\lambda_+^0 = \lambda_-^0 = 0$. Then $(\lambda_+^t, \lambda_-^t)$, $t \in [0, 1]$ is a homotopy connecting (ϕ_+, ϕ_-) to $(0, 0)$. $\qquad\square$

Let $\Theta_B : M_2(\mathcal{M}(\mathcal{K} \otimes B)) \to \mathcal{M}(\mathcal{K} \otimes B)$ be an inner *-isomorphism, cf. Definition 1.3.8. Θ_B gives rise to a composition $+$ in $KK_h(A, B)$ given by

$$(4.1.2) \quad [\phi_+, \phi_-] + [\psi_+, \psi_-] = \left[\Theta_B \circ \begin{bmatrix} \phi_+ & 0 \\ 0 & \psi_+ \end{bmatrix}, \ \Theta_B \circ \begin{bmatrix} \phi_- & 0 \\ 0 & \psi_- \end{bmatrix} \right],$$

$(\phi_+, \phi_-), (\psi_+, \psi_-) \in \mathbb{F}(A, B)$.

Proposition 4.1.5. $KK_h(A, B)$ *is an abelian group with* 0-*element represented by* $(0, 0)$ *and with* $-[\phi_+, \phi_-] = [\phi_-, \phi_+]$, $(\phi_+, \phi_-) \in \mathbb{F}(A, B)$.

Proof. That $KK_h(A, B)$ is an abelian semi-group with a zero element represented by $(0, 0)$ is shown as the corresponding statements for $[A, \mathcal{K} \otimes B]$, cf. Lemma 1.3.11 and Lemma 1.3.12.

To show that $[\phi_+, \phi_-] + [\phi_-, \phi_+] = [0, 0]$, let R_t, $t \in [0, 1]$, be the rotation matrices from (1.3.3). Then

$$\lambda_+^t = \Theta_B \circ \begin{bmatrix} \phi_+ & 0 \\ 0 & \phi_- \end{bmatrix}, \quad \lambda_-^t = \Theta_B \circ \operatorname{Ad} R_t \circ \begin{bmatrix} \phi_- & 0 \\ 0 & \phi_+ \end{bmatrix}, \quad t \in [0, 1]$$

defines a homotopy in $\mathbb{F}(A, B)$ showing that

$$[\phi_+, \phi_-] + [\phi_-, \phi_+] = \left[\Theta_B \circ \begin{bmatrix} \phi_+ & 0 \\ 0 & \phi_- \end{bmatrix}, \ \Theta_B \circ \begin{bmatrix} \phi_+ & 0 \\ 0 & \phi_- \end{bmatrix} \right].$$

The conclusion then follows from Lemma 4.1.4. $\qquad\square$

To establish the connection between $KK_h(A, B)$ and the KK-groups of chapter two, we first investigate $KK(A, B)$ in the case where both A and B are ungraded. Fix once and for all a full system of matrix units $\{e_{ij} : i, j \in \mathbb{N}\}$ in $\mathcal{K}$, i.e. the following hold: $e_{ij}^* = e_{ji}$, $e_{ij}e_{kl} = \delta(j, k)e_{il}$ and span $\{e_{ij} : i, j \in \mathbb{N}\}$ is dense in $\mathcal{K}$. By E 1.2.5 and Lemma 1.1.14 there is a *-homomorphism $\Psi_B : \mathcal{M}(\mathcal{K} \otimes B) \to \mathcal{L}_B(H_B)$ such that

$$(4.1.3) \qquad \Psi_B(e_{ij} \otimes bc^*) = \Theta_{\bar{b}_i, \bar{c}_j}, \quad i, j \in \mathbb{N}, \quad b, c \in B$$

Note that (4.1.3) determines Ψ_B uniquely.

Lemma 4.1.6. *Assume that B is trivially graded.   Then $\Phi_B = \Psi_B \otimes id_{M_2(\mathbb{C})} : M_2(\mathcal{M}(\mathcal{K} \otimes B)) \to \mathcal{L}_B(\hat{H}_B) = M_2(\mathcal{L}_B(H_B))$ is a *-isomorphism of graded C^*-algebras when $M_2(\mathcal{M}(\mathcal{K} \otimes B))$ is graded by the inner *-automorphism given by conjugation by $\begin{bmatrix} 1 & 0 \\ 0 & -1 \end{bmatrix}$.*

Proof. $\hat{H}_B = H_B \oplus H_B$ is graded by $id_{H_B} \oplus -id_{H_B} \in \mathcal{L}_B(\hat{H}_B)$. Since $\Phi_B \begin{bmatrix} 1 & 0 \\ 0 & -1 \end{bmatrix} = id_{H_B} \oplus -id_{H_B}$, the proof is complete. $\square$

Let now $(\phi_+, \phi_-) \in \mathbb{F}(A, B)$. It follows from Lemma 4.1.6 that

$$(4.1.4) \qquad \left(\hat{H}_B, \begin{bmatrix} \Psi_B \circ \phi_+ & 0 \\ 0 & \Psi_B \circ \phi_- \end{bmatrix}, \begin{bmatrix} 0 & 1 \\ 1 & 0 \end{bmatrix} \right) \in \mathbb{E}(A, B)$$

when A and B are both trivially graded. For $(\phi_+, \phi_-) \in \mathbb{F}(A, B)$ we let $\mathcal{E}(\phi_+, \phi_-)$ denote the Kasparov A-B-module given by (4.1.4).

For arbitrary C^*-algebras A and B, we let $KK^0(A, B)$ denote the group $KK(A, B)$ obtained by considering A and B as trivially graded C^*-algebras.

Lemma 4.1.7. $\mu : KK_h(A, B) \to KK^0(A, B)$ *defined by* $\mu[\phi_+, \phi_-] = [\mathcal{E}(\phi_+, \phi_-)]$, *for* $(\phi_+, \phi_-) \in \mathbb{F}(A, B)$ *is a homomorphism.*

Proof. Assume that $(\lambda_+^t, \lambda_-^t) \in \mathbb{F}(A, B)$, $t \in [0, 1]$, is a homotopy connecting (ϕ_+, ϕ_-) and (ψ_+, ψ_-). We can define $\lambda_+ : A \to \mathcal{M}(I(\mathcal{K} \otimes B))$ by

$$(\lambda_+(a)f)(t) = \lambda_+^t(a)f(t), \ t \in [0, 1], \ f \in I(\mathcal{K} \otimes B), \ a \in A.$$

$\lambda_- : A \to \mathcal{M}(I(\mathcal{K} \otimes B))$ is defined similarly. Then $(\lambda_+, \lambda_-) \in \mathbb{F}(A, IB)$ (when we identify $\mathcal{K} \otimes IB = I(\mathcal{K} \otimes B)$). We assert that $\mathcal{E}(\lambda_+, \lambda_-)_{\pi_t} \simeq \mathcal{E}(\lambda_+^t, \lambda_-^t)$, $t \in [0, 1]$.

To see this, let us fix $t \in [0, 1]$ and define $\phi_t : H_{IB} \to H_B$ by $\phi_t(b_1, b_2, b_3, \ldots) = (\pi_t(b_1), \pi_t(b_2), \pi_t(b_3), \ldots)$. The map $\phi_t \oplus \phi_t : \hat{H}_{IB} \to \hat{H}_B$ give rise to a map $\chi : (\hat{H}_{IB})_{\pi_t} \to \hat{H}_B$ given, in the notation of 1.2.2, by $\chi(q(x)) = \phi_t \oplus \phi_t(x)$, $x \in \hat{H}_{IB}$. It is straightforward to check that χ is an isomorphism of graded Hilbert B-modules. Since $\chi \cdot \chi^{-1}$ takes $\begin{bmatrix} 0 & 1 \\ 1 & 0 \end{bmatrix} \in \mathcal{L}_{IB}((\hat{H}_{IB})_{\pi_t})$ to $\begin{bmatrix} 0 & 1 \\ 1 & 0 \end{bmatrix} \in \mathcal{L}_B(\hat{H}_B)$, it suffices to check that

(4.1.5) $\chi \circ (\Phi_{IB}(m))_{\pi_t} = (\Phi_B \circ (id_{M_2(\mathbb{C})} \otimes \pi_t)(m)) \circ \chi$

$$m \in M_2(\mathcal{M}(\mathcal{K} \otimes IB)),$$

since this will imply that

$$\chi \circ \left(\begin{bmatrix} \Psi_{IB} \circ \lambda_+(a) & 0 \\ 0 & \Psi_{IB} \circ \lambda_-(a) \end{bmatrix} \right)_{\pi_t} = \begin{bmatrix} \Psi_B \circ \lambda_+^t(a) & 0 \\ 0 & \Psi_B \circ \lambda_-^t(a) \end{bmatrix} \circ \chi$$

for $a \in A$.

To check (4.1.5) it suffices to consider the case where the $\mathcal{M}(\mathcal{K} \otimes IB)$-entries of m is of the form $m = e_{ij} \otimes bc^*$, $i, j \in \mathbb{N}$, $b, c \in IB$, since $M_2(\mathcal{K} \otimes IB)$ is strictly dense in $M_2(\mathcal{M}(\mathcal{K} \otimes IB)) = \mathcal{M}(M_2(\mathcal{K} \otimes IB))$. But for such m (4.1.5) follows by using (4.1.3).

We have shown, in particular, that $\mathcal{E}(\phi_+, \phi_-)$ is homotopic to $\mathcal{E}(\psi_+, \psi_-)$. Thus μ is well-defined. To see that μ is a homomorphism, let $V_1, V_2 \in \mathcal{M}(\mathcal{K} \otimes B)$ be the isometries used to define the inner *-isomorphism $\Theta_B : M_2(\mathcal{M}(\mathcal{K} \otimes B)) \rightarrow \mathcal{M}(\mathcal{K} \otimes B)$.

Set $W_i = \Psi_B(V_i)$, $i = 1, 2$. We can then define an isomorphism $S : \hat{H}_B \oplus \hat{H}_B \rightarrow \hat{H}_B$ of graded Hilbert B-modules by $S((x_1, x_2), (y_1, y_2)) = (W_1 x_1 + W_2 y_1, W_1 x_2 + W_2 y_2)$, $(x_1, x_2), (y_1, y_2) \in \hat{H}_B$.

Using this isomorphism we find

$$\mathcal{E}(\phi_+, \phi_-) \oplus \mathcal{E}(\psi_+, \psi_-) \simeq$$

$$\left(\hat{H}_B, \begin{bmatrix} \Psi_B \circ \Theta_B \begin{bmatrix} \phi_+ & 0 \\ 0 & \psi_+ \end{bmatrix} & 0 \\ 0 & \Psi_B \circ \Theta_B \begin{bmatrix} \psi_- & 0 \\ 0 & \psi_- \end{bmatrix} \end{bmatrix}, \begin{bmatrix} 0 & 1 \\ 1 & 0 \end{bmatrix} \right)$$

$$= \mathcal{E} \left(\Theta_B \circ \begin{bmatrix} \phi_+ & 0 \\ 0 & \psi_+ \end{bmatrix}, \Theta_B \circ \begin{bmatrix} \phi_- & 0 \\ 0 & \psi_- \end{bmatrix} \right).$$

Thus $\mu[\phi_+, \phi_-] + \mu[\psi_+, \psi_-] = \mu([\phi_+, \phi_-] + [\psi_+, \psi_-])$ $\square$

Theorem 4.1.8. $\mu : KK_h(A, B) \rightarrow KK^0(A, B)$ *is an isomorphism.*

Proof. The following proof is a little sketchy in the sense that certain assertions regarding isomorphisms of Kasparov $A - B$-modules are left for the reader to check.

Surjectivity : Let $\mathcal{E} = (E, \phi, F) \in \mathbb{E}(A, B)$ be a Kasparov $A - B$-module. To show that $[\mathcal{E}] \in Im\mu$, we substitute $\mathcal{E}$ by any Kasparov $A - B$-module that represents the same element of $KK^0(A, B)$. By adding the degenerate Kasparov $A - B$-module $\left(\hat{H}_B, 0, \begin{bmatrix} 0 & 1 \\ 1 & 0 \end{bmatrix} \right)$ to $\mathcal{E}$ and then using the graded version of Kasparov's stabilization theorem, Theorem 1.2.12, we can assume that $E = \hat{H}_B$.

Set $F_t = \frac{1}{2}((2-t)F+tF^*)$, $t \in [0,1]$. Then $(\hat{H}_B, \phi, F_t) \in \mathbb{E}(A, B)$, $t \in [0,1]$, is an operator homotopy connecting $\mathcal{E}$ to $(\hat{H}_B, \phi, \frac{1}{2}(F + F^*))$. Thus we can assume that $F = F^*$. Doing so we can consider $h_t(F)$, $t \in [0,1]$, where $h_t : \mathbb{R} \to \mathbb{R}$ is the continuous function

$$h_t(s) = \begin{cases} s & -1 \leq s \leq 1, \\ s - t(s-1) & s \geq 1, \\ s - t(s+1) & s \leq 1, \end{cases} \qquad t \in [0,1]$$

We assert that $(\hat{H}_B, \phi, h_t(F)) \in \mathbb{E}(A, B)$ for all $t \in [0,1]$. To see this we note that h_t is an odd function with $h_t(1) = 1$. Hence it can be approximated, uniformly on the spectrum of F, by odd polynomials P_n taking the value 1 at 1. It is easy to see that $(\hat{H}_B, \phi, P_n(F)) \in \mathbb{E}(A, B)$ for all n. The assertion follows from this.

Since $t \to h_t(F)$ is uniformly continuous, we see that $\mathcal{E}$ is operator homotopic to $(\hat{H}_B, \phi, h_1(F))$. Since $\|h_1(F)\| \leq 1$, this shows that we can assume that $-1 \leq F \leq 1$. Using now that F is of degree 1 and that the grading automorphism of $\mathcal{L}_B(\hat{H}_B) = M_2(\mathcal{L}_B(H_B))$ is conjugation by $\begin{bmatrix} 1 & 0 \\ 0 & -1 \end{bmatrix}$ it follows that

$$F = \begin{bmatrix} 0 & x \\ x^* & 0 \end{bmatrix}$$

for some $x \in \mathcal{L}_B(H_B)$ with $\|x\| \leq 1$. Since $\phi : A \to \mathcal{L}_B(\hat{H}_B)$ maps to elements of degree 0 there are *-homomorphisms $\phi_+, \phi_- : A \to \mathcal{L}_B(H_B)$ such that

$$\phi = \begin{bmatrix} \phi_+ & 0 \\ 0 & \phi_- \end{bmatrix}.$$

Since $(\hat{H}_B, \phi, F) \in \mathbb{E}(A, B)$ we have

$$\begin{aligned} & xx^*\phi_+(a) - \phi_+(a) \in \mathcal{K}_B(H_B) \\ (4.1.6) \quad & x^*x\phi_-(a) - \phi_-(a) \in \mathcal{K}_B(H_B) \\ & \phi_+(a)x - x\phi_-(a) \in \mathcal{K}_B(H_B), \qquad \text{for} \quad a \in A \end{aligned}$$

Choose isometries W_1, $W_2 \in \mathcal{L}_B(H_B)$ such that $W_1 W_1^* + W_2 W_2^* = 1$ and $W_1^* W_2 = 0$, cf. the proof of Lemma 4.1.7. Then we can define an isomorphism $S : \hat{H}_B \oplus \hat{H}_B \to \hat{H}_B$ of graded Hilbert B-modules as in that proof. Let $T : H_B \oplus H_B \to H_B$ be the isomorphism of Hilbert B-modules given by $T(x, y) = W_1 x + W_2 y$, $x, y \in H_B$. By adding the degenerate Kasparov $A - B$-module $(\hat{H}_B, 0, 0)$ to $\mathcal{E}$ and using the isomorphism S, we can assume that

$$\phi = \begin{bmatrix} \psi_+ & 0 \\ 0 & \psi_- \end{bmatrix} \quad \text{and} \quad F = \begin{bmatrix} 0 & v \\ v^* & 0 \end{bmatrix},$$

where $\psi_+(\cdot) = T \begin{bmatrix} \phi_+(\cdot) & 0 \\ 0 & 0 \end{bmatrix} T^*$, $\psi_-(\cdot) = T \begin{bmatrix} \phi_-(\cdot) & 0 \\ 0 & 0 \end{bmatrix} T^*$ and $v = T \begin{bmatrix} x & 0 \\ 0 & 0 \end{bmatrix} T^*$. Set

$$w_t = T \begin{bmatrix} x & t(1 - xx^*)^{\frac{1}{2}} \\ -t(1 - x^*x)^{\frac{1}{2}} & tx^* \end{bmatrix} T^*, \quad t \in [0, 1].$$

Using (4.1.6) it follows that

$$\left(\hat{H}_B, \phi, \begin{bmatrix} 0 & w_t \\ w_t^* & 0 \end{bmatrix} \right) \in \mathbb{E}(A, B), \quad t \in [0, 1].$$

Consequently we have an operator homotopy connecting $(\hat{H}_B, \phi, F)$ to $\left(\hat{H}_B, \phi, \begin{bmatrix} 0 & u \\ u^* & 0 \end{bmatrix} \right)$, where $u \in \mathcal{L}_B(H_B)$ is the unitary

$$u = T \begin{bmatrix} x & (1 - xx^*)^{\frac{1}{2}} \\ -(1 - x^*x)^{\frac{1}{2}} & x^* \end{bmatrix} T^*.$$

Set $x_t = tx + (1 - t)$ and

$$u_t = T \begin{bmatrix} x_t & (1 - x_t x_t^*)^{\frac{1}{2}} \\ -(1 - x_t^* x_t)^{\frac{1}{2}} & x_t^* \end{bmatrix} T^*, \quad t \in [0, 1].$$

Then u_t, $t \in [0, 1]$, is a norm continuous path of unitaries connecting u to 1. Observe that

$$\left(\hat{H}_B, \begin{bmatrix} Adu_t u^* \circ \psi_+ & 0 \\ 0 & \psi_- \end{bmatrix}, \begin{bmatrix} 0 & u_t \\ u_t^* & 0 \end{bmatrix} \right) \in \mathbb{E}(A, B)$$

for all $t \in [0, 1]$.

It follows from E 2.1.3 that $\left(\hat{H}_B, \phi, \begin{bmatrix} 0 & u \\ u^* & 0 \end{bmatrix} \right)$ represents the same element of $KK^0(A, B)$ as

$$\left(\hat{H}_B, \begin{bmatrix} Adu^* \circ \psi_+ & 0 \\ 0 & \psi_- \end{bmatrix}, \begin{bmatrix} 0 & 1 \\ 1 & 0 \end{bmatrix} \right).$$

The latter element of $\mathbb{E}(A, B)$ represents obviously an element in the image of $KK_h(A, B)$ under μ. Thus $[\mathcal{E}] \in \operatorname{im} \mu$.

To prove injectivity of μ, let (ϕ_+, ϕ_-), $(\psi_+, \psi_-) \in \mathbb{F}(A, B)$ and let $\mathcal{F} = (E, \lambda, F) \in \mathbb{E}(A, IB)$ such that $\mathcal{F}_{\pi_0} \simeq \mathcal{E}(\phi_+, \phi_-)$ and $\mathcal{F}_{\pi_1} \simeq \mathcal{E}(\psi_+, \psi_-)$. We want to conclude that $[\phi_+, \phi_-] = [\psi_+, \psi_-]$. To this end note that $\mathcal{E}(0, 0) = \left(\hat{H}_{IB}, 0, \begin{bmatrix} 0 & 1 \\ 1 & 0 \end{bmatrix} \right)$ where $0 \in \operatorname{Hom}(A, IB)$. Set $\mathcal{F}_1 = \mathcal{F} \oplus \mathcal{E}(0, 0)$. Then $(\mathcal{F} \oplus \mathcal{E}(0, 0))_{\pi_0} \simeq \mathcal{F}_{\pi_0} \oplus \mathcal{E}(0, 0)_{\pi_0}$ by Lemma 2.1.11 and Lemma 2.1.13, and $\mathcal{E}(0, 0)_{\pi_0} \simeq \mathcal{E}(0, 0)$ where the last zeros represent $0 \in \operatorname{Hom}(A, B)$. Since $\mathcal{E}(\phi_+, \phi_-) \oplus \mathcal{E}(0, 0) \simeq \mathcal{E} \left(\Theta_B \circ \begin{bmatrix} \phi_+ & 0 \\ 0 & 0 \end{bmatrix}, \Theta_B \circ \begin{bmatrix} \phi_- & 0 \\ 0 & 0 \end{bmatrix} \right)$ by the proof of Lemma 4.1.7 we see that

$$\mathcal{E} \left(\Theta_B \circ \begin{bmatrix} \phi_+ & 0 \\ 0 & 0 \end{bmatrix}, \Theta_B \circ \begin{bmatrix} \phi_- & 0 \\ 0 & 0 \end{bmatrix} \right) \simeq \pi_{0*}(\mathcal{F}_1).$$

Similarly,

$$\mathcal{E} \left(\Theta_B \circ \begin{bmatrix} \psi_+ & 0 \\ 0 & 0 \end{bmatrix}, \Theta_B \circ \begin{bmatrix} \psi_- & 0 \\ 0 & 0 \end{bmatrix} \right) \simeq \pi_{1*}(\mathcal{F}_1).$$

Using Kasparov's stabilization theorem, Theorem 1.2.12, it follows that we can assume that $E = \hat{H}_{IB}$, i.e. we assume that $\mathcal{F} = (\hat{H}_{IB}, \lambda, F)$ and that $\mathcal{F}_{\pi_0} \simeq \mathcal{E}(\phi_+, \phi_-)$, $\mathcal{F}_{\pi_1} \simeq \mathcal{E}(\psi_+, \psi_-)$. In particular, F_{π_0} and F_{π_1} are selfadjoint unitaries. Set $\mathcal{F}^2 = (\hat{H}_{IB}, \lambda, \frac{1}{2}(F + F^*))$. Then $\mathcal{F}_{\pi_0}^2 \simeq \mathcal{E}(\phi_+, \phi_-)$ and $\mathcal{F}_{\pi_1}^2 \simeq \mathcal{E}(\psi_+, \psi_-)$ since $(\frac{1}{2}(F + F^*))_{\pi_i} = \frac{1}{2}(F_{\pi_i} + F_{\pi_i}^*) = F_{\pi_i}$, $i = 0, 1$. Thus we can assume that F is self-adjoint. In the same way we can substitute F by $h_1(F)$ to achieve that $-1 \leq F \leq 1$. Since F has degree 1,

$$F = \begin{bmatrix} 0 & v \\ v^* & 0 \end{bmatrix}$$

for some $v \in \mathcal{L}_{IB}(H_{IB})$ with $\|v\| \leq 1$. Similarly,

$$\lambda = \begin{bmatrix} \lambda_+ & 0 \\ 0 & \lambda_- \end{bmatrix}$$

for some $\lambda_+, \lambda_- \in \mathrm{Hom}\,(A, \mathcal{L}_{IB}(H_{IB}))$. Since

$$\mathcal{F}_{\pi_0} \simeq \left(\hat{H}_B, \begin{bmatrix} \phi_+ & 0 \\ 0 & \phi_- \end{bmatrix}, \begin{bmatrix} 0 & 1 \\ 1 & 0 \end{bmatrix} \right)$$

we see that v_{π_0} is the identity map on $(H_{IB})_{\pi_0} \simeq H_B$. A similar remark applies to v_{π_1}.

Let $U_1, U_2 \in \mathcal{L}_{IB}(H_{IB})$ be isometries with $U_1 U_1^* + U_2 U_2^* = 1$ and $U_1^* U_2 = 0$. Let us now define $T_I : H_{IB} \oplus H_{IB} \rightarrow H_{IB}$ by $T_I(x, y) = U_1 x + U_2 y$ and $S_I : \hat{H}_{IB} \oplus \hat{H}_{IB} \rightarrow \hat{H}_{IB}$ by $S_I((x_1, x_2), (y_1, y_2)) = (T_I(x_1, y_1), T_I(x_2, y_2))$, $x, x_1, x_2, y, y_1, y_2 \in H_{IB}$. Then S_I and T_I are isomorphisms of Hilbert IB-modules; S_I of graded Hilbert IB-modules. Set

$$\lambda_+^1 = T_I \begin{bmatrix} \lambda_+ & 0 \\ 0 & 0 \end{bmatrix} T_I^*, \ \lambda_-^1 = T_I \begin{bmatrix} \lambda_- & 0 \\ 0 & 0 \end{bmatrix} T_I^*$$

and

$$v_1 = T_I \begin{bmatrix} v & (1 - vv^*)^{\frac{1}{2}} \\ -(1 - v^*v)^{\frac{1}{2}} & v^* \end{bmatrix} T_I^*.$$

Then v_1 is a unitary in $\mathcal{L}_{IB}(H_{IB})$. Set

$$\mathcal{F}^3 = \left(\hat{H}_{IB}, \begin{bmatrix} Adv_1^* \circ \lambda_+^1 & 0 \\ 0 & \lambda_-^1 \end{bmatrix}, \begin{bmatrix} 0 & 1 \\ 1 & 0 \end{bmatrix} \right) \in \mathbb{E}(A, IB).$$

It follows that $\mathcal{F}_{\pi_0}^3 \simeq \mathcal{E} \left(\Theta_B \circ \begin{bmatrix} \phi_+ & 0 \\ 0 & 0 \end{bmatrix}, \Theta_B \circ \begin{bmatrix} \phi_- & 0 \\ 0 & 0 \end{bmatrix} \right)$ and that $\mathcal{F}_{\pi_1}^3 \simeq$ $\mathcal{E} \left(\Theta_B \circ \begin{bmatrix} \psi_+ & 0 \\ 0 & 0 \end{bmatrix}, \Theta_B \circ \begin{bmatrix} \psi_- & 0 \\ 0 & 0 \end{bmatrix} \right)$.

All in all it follows that to conclude that $[\phi_+, \phi_-] = [\psi_+, \psi_-]$ we can assume that

$$\mathcal{F} = \left(\hat{H}_{IB}, \begin{bmatrix} \lambda_+' & 0 \\ 0 & \lambda_-' \end{bmatrix}, \begin{bmatrix} 0 & 1 \\ 1 & 0 \end{bmatrix} \right)$$

for some $\lambda_+', \lambda_-' \in \mathrm{Hom}\,(A, \mathcal{L}_{IB}(H_{IB}))$. Then $\lambda_+' = \Psi_{IB} \circ \lambda_+$ and $\lambda_-' = \Psi_{IB} \circ \lambda_-$ for some $\lambda_+, \lambda_- \in \mathrm{Hom}\,(A, \mathcal{M}(I(\mathcal{K} \otimes B)))$. By using (4.1.3) we get that

$$\mathcal{F}_{\pi_0} \simeq \left(\hat{H}_B, \begin{bmatrix} \Psi_B \circ \pi_0 \circ \lambda_+ & 0 \\ 0 & \Psi_B \circ \pi_0 \circ \lambda_- \end{bmatrix}, \begin{bmatrix} 0 & 1 \\ 1 & 0 \end{bmatrix} \right).$$

Thus

$$\mathcal{E}(\phi_+, \phi_-) = \left(\hat{H}_B, \begin{bmatrix} \Psi_B \circ \phi_+ & 0 \\ 0 & \Psi_B \circ \phi_- \end{bmatrix}, \begin{bmatrix} 0 & 1 \\ 1 & 0 \end{bmatrix} \right)$$

$$\simeq \left(\hat{H}_B, \begin{bmatrix} \Psi_B \circ \pi_0 \circ \lambda_+ & 0 \\ 0 & \Psi_B \circ \pi_0 \circ \lambda_- \end{bmatrix}, \begin{bmatrix} 0 & 1 \\ 1 & 0 \end{bmatrix} \right).$$

It follows that there is a unitary $s_0 \in \mathcal{M}(\mathcal{K} \otimes B)$ such that

$$(4.1.7) \qquad \mathrm{Ad}\, s_0 \circ \phi_+ = \underline{\pi}_0 \circ \lambda_+, \quad \mathrm{Ad}\, s_0 \circ \phi_- = \underline{\pi}_0 \circ \lambda_-.$$

Similarly, we find a unitary $s_1 \in \mathcal{M}(\mathcal{K} \otimes B)$ such that

$$(4.1.8) \qquad \mathrm{Ad}\, s_1 \circ \psi_+ = \underline{\pi}_1 \circ \lambda_+, \quad \mathrm{Ad}\, s_1 \circ \psi_- = \underline{\pi}_1 \circ \lambda_-.$$

By using the fact that the unitary group of $\mathcal{M}(\mathcal{K} \otimes B)$ is connected in the strict topology, we see that $[\phi_+, \phi_-] = [\mathrm{Ad}\, s_0 \circ \phi_+, \mathrm{Ad}\, s_0 \circ \phi_-]$ and $[\psi_+, \psi_-] = [\mathrm{Ad}\, s_1 \circ \psi_+, \mathrm{Ad}\, s_1 \circ \psi_-]$. Set $\lambda_+^t = \underline{\pi}_t \circ \lambda_+$ and $\lambda_-^t = \underline{\pi}_t \circ \lambda_-$, $t \in [0,1]$. It is then easy to see that $(\lambda_+^t, \lambda_-^t)$, $t \in [0,1]$, is a homotopy connecting $(\underline{\pi}_0 \circ \lambda_+, \underline{\pi}_0 \circ \lambda_-) \in \mathbb{F}(A,B)$ to $(\underline{\pi}_1 \circ \lambda_+, \underline{\pi}_1 \circ \lambda_-) \in \mathbb{F}(A,B)$. Thus $[\phi_+, \phi_-] = [\psi_+, \psi_-]$ by (4.1.7)-(4.1.8). $\qquad \square$

We now turn to the functoriality properties of KK_h. Let $f : D \to A$ be a *-homomorphism. Then we can define $f^* : KK_h(A,B) \to KK_h(D,B)$ by

$$(4.1.9) \qquad f^*[\phi_+, \phi_-] = [\phi_+ \circ f, \phi_- \circ f], \qquad (\phi_+, \phi_-) \in \mathbb{F}(A,B).$$

It is quite easy to check that f^* is a (well-defined) group homomorphism and that $KK_h(\cdot, B)$ is a contravariant functor from the category of C^*-algebras to the category of abelian groups for any C^*-algebra B, cf. E4.1.1.

It is somewhat more troublesome to make $KK_h(A, \cdot)$ functorial. We first require some preparations.

Let $f : \mathcal{K} \otimes B \to \mathcal{K} \otimes C$ be a quasi-unital *-homomorphism and $\underline{f}$ the unique strictly continuous extension $\underline{f} : \mathcal{M}(\mathcal{K} \otimes B) \to \mathcal{M}(\mathcal{K} \otimes C)$, cf. Corollary 1.1.15. We can then clearly define a map $f_* : KK_h(A,B) \to KK_h(A,C)$ by

$$f_*[\phi_+, \phi_-] = [\underline{f} \circ \phi_+, \underline{f} \circ \phi_-], \quad (\phi_+, \phi_-) \in \mathbb{F}(A,B).$$

To show that f_* is a homomorphism we need a couple of lemmas.

Lemma 4.1.9. *Let $(\phi_+, \phi_-), (\psi_+, \psi_-) \in \mathbb{F}(A,B)$ and assume that $\phi_+(a)\psi_+(a) = \phi_-(a)\psi_-(a) = 0$, $a \in A$.*
Then $[\phi_+, \phi_-] + [\psi_+, \psi_-] = [\phi_+ + \psi_+, \phi_- + \psi_-]$.

Proof. Let R_t be the rotation matrices from (1.3.3). Set

$$\lambda_+^t = \Theta_B \circ \begin{bmatrix} \phi_+ & 0 \\ 0 & 0 \end{bmatrix} + \Theta_B \circ \text{Ad}\, R_t \circ \begin{bmatrix} 0 & 0 \\ 0 & \psi_+ \end{bmatrix},$$

$$\lambda_-^t = \Theta_B \circ \begin{bmatrix} \phi_- & 0 \\ 0 & 0 \end{bmatrix} + \Theta_B \circ \text{Ad}R_t \circ \begin{bmatrix} 0 & 0 \\ 0 & \psi_- \end{bmatrix},$$

$t \in [0,1]$. Then $(\lambda_+^t, \lambda_-^t) \in \mathbb{F}(A,B)$, $t \in [0,1]$, is a homotopy showing that $[\phi_+, \phi_-] + [\psi_+, \psi_-] = [\phi_+ + \psi_+, \phi_- + \psi_-]$. $\qquad\square$

Lemma 4.1.10. *Let $(\phi_+, \phi_-) \in \mathbb{F}(A,B)$ and let $w \in \mathcal{M}(\mathcal{K} \otimes B)$ be a partial isometry with $w^*w \geq \phi_+(1)$ and $w^*w \geq \phi_-(1)$. Then $[w\phi_+(\cdot)w^*, w\phi_-(\cdot)w^*] = [\phi_+, \phi_-]$ in $KK_h(A,B)$.*

Proof. Set

$$S_t = \begin{bmatrix} \cos\frac{\pi}{2}t & -w^*\sin\frac{\pi}{2}t \\ w\sin\frac{\pi}{2}t & ww^*\cos\frac{\pi}{2}t \end{bmatrix}, \ t \in [0,1].$$

Then $S_t^* \begin{bmatrix} \phi_+(\cdot) & 0 \\ 0 & w\phi_-(\cdot)w^* \end{bmatrix} S_t$ is a *-homomorphism for all t. Set

$$\lambda_+^t = \Theta_B \circ \begin{bmatrix} \phi_- & 0 \\ 0 & w\phi_+w^* \end{bmatrix}, \ \lambda_-^t = \Theta_D \left(S_t^* \begin{bmatrix} \phi_+ & 0 \\ 0 & w\phi_-w^* \end{bmatrix} S_t \right), \ t \in [0,1].$$

Then $(\lambda_+^t, \lambda_-^t) \in \mathbb{F}(A,B)$, $t \in [0,1]$, defines a homotopy which shows that $[\phi_-, \phi_+] + [w\phi_+w^*, w\phi_-w^*] = 0$ in $KK_h(A,B)$. The result then follows from Proposition 4.1.5. $\qquad\square$

Lemma 4.1.11. $f_* : KK_h(A,B) \to KK_h(A,C)$ *is a homomorphism.*

Proof. Let $(\phi_+, \phi_-), (\psi_+, \psi_-) \in \mathbb{F}(A,B)$. Let $V_1, V_2 \in \mathcal{M}(\mathcal{K} \otimes B)$ be the two isometries used to define the inner *-isomorphism $\Theta_B :$ $M_2(\mathcal{M}(\mathcal{K} \otimes B)) \to \mathcal{M}(\mathcal{K} \otimes B)$. Then

$f_*([\phi_+, \phi_-] + [\psi_+, \psi_-])$

$= [\text{Ad}\,\underline{f}(V_1) \circ \underline{f} \circ \phi_+ + \text{Ad}\,\underline{f}(V_2) \circ \underline{f} \circ \psi_+, \ \text{Ad}\,\underline{f}(V_1) \circ \underline{f} \circ \phi_- + \text{Ad}\,\underline{f}(V_2) \circ \underline{f} \circ \psi_-]$

(by definition)

$= [\text{Ad}\underline{f}(V_1) \circ \underline{f} \circ \phi_+, \text{Ad}\underline{f}(V_1) \circ \underline{f} \circ \phi_-] + [\text{Ad}\underline{f}(V_2) \circ \underline{f} \circ \psi_+, \text{Ad}\,\underline{f}(V_2) \circ \underline{f} \circ \psi_-]$

(by Lemma 4.1.9)

$= [\underline{f} \circ \phi_+, \underline{f} \circ \phi_-] + [\underline{f} \circ \psi_+, \underline{f} \circ \psi_-]$

(by Lemma 4.1.10)

$= f_*[\phi_+, \phi_-] + f_*[\psi_+, \psi_-]$. $\qquad\square$

Since the composition of quasi-unital *-homomorphisms is again quasi-unital, cf. E 1.3.2, it is quite clear that the usual functoriality equation, $(f_1 \circ f_2)_* = f_{1*} \circ f_{2*}$, holds.

Lemma 4.1.12. *Let $f_1, f_2 : \mathcal{K} \otimes B \to \mathcal{K} \otimes C$ be quasi-unital *-homomorphisms, and assume that f_1 and f_2 are strongly homotopic in* $\mathrm{Hom}_q(\mathcal{K} \otimes B, \mathcal{K} \otimes C)$, *i.e.* $f_1 \tilde{\sim} f_2$.
Then $f_{1*} = f_{2*} : KK_h(A, B) \to KK_h(A, C)$.

Proof. This is an immediate consequence of Lemma 1.3.15. $\qquad\square$

When $f : B \to C$ is a *-homomorphism we let f_q denote any *-homomorphism $f_q : \mathcal{K} \otimes B \to \mathcal{K} \otimes C$ homotopic to $id_{\mathcal{K}} \otimes f$, cf. Theorem 1.3.16, and set $f_* = f_{q*} : KK_h(A, B) \to KK_h(A, C)$. Theorem 1.3.16, Lemma 4.1.12 and Lemma 4.1.11 guarantee that f_* defines a group homomorphism independent of the choice of f_q. Since the composition of quasi-unital *-homomorphisms gives a quasi-unital *-homomorphism by E 1.3.2, it is clear that this definition makes $KK_h(A, \cdot)$ into a covariant functor from the category of σ-unital C^*-algebras to the category of abelian groups. It is then obvious that $KK_h(A, \cdot)$ is a homotopy invariant functor. In the following we investigate the functoriality of KK_h a little further.

Let e_A be the element of $\mathrm{Hom}\,(A, \mathcal{K} \otimes A)$ given by $e_A(a) = e \otimes a$, $a \in A$, where e is a minimal projection in $\mathcal{K}$. Note that another choice of a minimal projection in $\mathcal{K}$ gives rise to a map $A \to \mathcal{K} \otimes A$ which differs from e_A by conjugation by a unitary $U \in \mathcal{M}(\mathcal{K} \otimes A)$. Since the unitary group of $\mathcal{M}(\mathcal{K} \otimes A)$ is connected in the strict topology by Lemma 1.3.7, the two maps will be homotopic. Nonetheless we fix now and for the rest of the book a particular minimal projection $e \in \mathcal{K}$ and define $e_A : A \to \mathcal{K} \otimes A$ for any C^*-algebra A as above.

Lemma 4.1.13. *Let B be an arbitrary C^*-algebra. There is a *-isomorphism $\lambda_B : \mathcal{K} \otimes B \to \mathcal{K} \otimes \mathcal{K} \otimes B$ such that $e_{\mathcal{K} \otimes B}$ is homotopic to λ_B. In fact, $\lambda_B = AdV^* \circ e_{\mathcal{K} \otimes B}$ for some isometry $V \in \mathcal{M}(\mathcal{K} \otimes B)$ with range projection VV^* equal to the relative unit for $e_{\mathcal{K} \otimes B}$.*

Proof. It is clear that $e_{\mathcal{K} \otimes B} : \mathcal{K} \otimes B \to \mathcal{K} \otimes \mathcal{K} \otimes B$ is quasi-unital with a relative unit $e \otimes 1 \otimes 1 \in \mathcal{M}(\mathcal{K} \otimes \mathcal{K} \otimes B)$ given on simple tensors by $e \otimes 1 \otimes 1(k_1 \otimes k_2 \otimes b) = ek_1 \otimes k_2 \otimes b$, $k_1, k_2 \in \mathcal{K}$, $b \in B$. Let $e \otimes 1 \in \mathcal{M}(\mathcal{K} \otimes \mathcal{K}) \simeq \mathcal{B}(l^2)$ be the projection given on simple tensors by

$e \otimes 1(k_1 \otimes k_2) = ek_1 \otimes k_2$. Then $e \otimes 1$ is an infinite projection in $\mathcal{B}(l^2)$, so there is an isometry $W \in \mathcal{M}(\mathcal{K} \otimes \mathcal{K})$ with $e \otimes 1$ as range projection. Let $V = W \otimes 1 \in \mathcal{M}(\mathcal{K} \otimes \mathcal{K} \otimes B)$ be the isometry given on simple tensors by $V(k_1 \otimes k_2 \otimes b) = W(k_1 \otimes k_2) \otimes b$, $k_1, k_2 \in \mathcal{K}$, $b \in B$. Then V is an isometry with $e \otimes 1 \otimes 1$ as range projection. Thus $\lambda_B = \mathrm{Ad}\, V^* \circ e_{\mathcal{K} \otimes B}$: $\mathcal{K} \otimes B \to \mathcal{K} \otimes \mathcal{K} \otimes B$ is a *-isomorphism. It follows from Lemma 1.3.7 that $\lambda_B \sim \mathrm{Ad}\, V \circ \lambda_B = e_{\mathcal{K} \otimes B}$. $\qquad\square$

Lemma 4.1.14. *Let B be a σ-unital C^*-algebra. Then*

$$e_{B*} : KK_h(A, B) \to KK_h(A, \mathcal{K} \otimes B)$$

is an isomorphism.

Proof. We will now define a *-automorphism α of $\mathcal{K} \otimes \mathcal{K} \otimes B$ by $\alpha(k_1 \otimes k_2 \otimes b) = k_2 \otimes k_1 \otimes b$, $k_1, k_2 \in \mathcal{K}$, $b \in B$. Then $id_\mathcal{K} \otimes e_B = \alpha \circ e_{\mathcal{K} \otimes B}$ and $e_{B*} = \alpha_* \circ e_{\mathcal{K} \otimes B*}$. Let λ_B be the *-isomorphism of Lemma 4.1.13. Then $e_{\mathcal{K} \otimes B} \hat{\sim} \lambda_B$ by Theorem 1.3.16. Thus $e_{B*} = \alpha_* \circ \lambda_{B*}$ by Lemma 4.1.12 and e_{B*} has $\lambda_{B*}^{-1} \circ \alpha_*^{-1}$ as inverse. $\qquad\square$

Lemma 4.1.14 says that $KK_h(A, \cdot)$ is a stable functor. The same is true for $KK_h(\cdot, B)$, cf. E 4.1.3.

Proposition 4.1.15. *The isomorphism $\mu : KK_h(A, B) \to KK^0(A, B)$ is natural, i.e. when $g : D \to A$ and $f : B \to C$ are *-homomorphisms, then $\mu \circ g^* = g^* \circ \mu$ and $\mu \circ f_* = f_* \circ \mu$.*

Proof. Only the equality $\mu \circ f_* = f_* \circ \mu$ requires a proof; the other follows immediately from the definitions.

The first step is to reduce to the case where f is quasi-unital. Assume namely that μ is natural with respect to quasi-unital maps and let f be arbitrary. By Theorem 1.3.16 there is a quasi-unital map $g : \mathcal{K} \otimes B \to \mathcal{K} \otimes C$ which is homotopic to $id_\mathcal{K} \otimes f$. Since $(id_\mathcal{K} \otimes f) \circ e_B = e_C \circ f$, we find that

$$
\begin{aligned}
e_{C*} \circ f_* \circ \mu^{-1} &= (id_\mathcal{K} \otimes f)_* \circ e_{B*} \circ \mu^{-1} \\
&= g_* \circ e_{B*} \circ \mu^{-1} \text{(by homotopy invarians)} \\
&= \mu^{-1} \circ g_* \circ e_{B*} \text{ (since } g \text{ and } e_B \text{ are quasi-unital)} \\
&= \mu^{-1} \circ (id_\mathcal{K} \otimes f)_* \circ e_{B*} \text{ (by homotopy invarians)} \\
&= \mu^{-1} \circ e_{C*} \circ f_* \\
&= e_{C*} \circ \mu^{-1} \circ f_* \text{(since } e_C \text{ is quasi-unital).}
\end{aligned}
$$

Thus $\mu^{-1} \circ f_* = f_* \circ \mu^{-1}$ by Lemma 4.1.14.

We assume now that f is quasi-unital with relative unit $p \in \mathcal{M}(C)$. Let $(\phi_+, \phi_-) \in \mathbb{F}(A, B)$. Then $\mu[\phi_+, \phi_-] \in KK^0(A, B)$ is represented by the Kasparov $A - B$-module

$$\left(\hat{H}_B, \begin{bmatrix} \Psi_B \circ \phi_+ & 0 \\ 0 & \Psi_B \circ \phi_- \end{bmatrix}, \begin{bmatrix} 0 & 1 \\ 1 & 0 \end{bmatrix} \right).$$

By the definition of the functoriality of $KK^0(A, \cdot)$, $f_* \circ \mu[\phi_+, \phi_-]$ is then represented by

$$\left(H_B \otimes_f C \oplus H_B \otimes_f C, \begin{bmatrix} \Psi_B \circ \phi_+ \otimes_f id & 0 \\ 0 & \Psi_B \circ \phi_- \otimes_f id \end{bmatrix}, \begin{bmatrix} 0 & 1 \\ 1 & 0 \end{bmatrix} \right),$$

where $H_B \otimes_f C \oplus H_B \otimes_f C$ is graded by $\begin{bmatrix} 1 & 0 \\ 0 & -1 \end{bmatrix}$. There is a linear map $\Phi_1 : H_B \otimes_f C \to H_C$ given on simple tensors by

$$\Phi_1((b_1, b_2, \ldots) \otimes_f c) = (f(b_1)c, f(b_2)c, \ldots), \quad (b_1, b_2, \ldots) \in H_B, \ c \in C.$$

Define $P \in \mathcal{L}_C(H_C)$ by $P(c_1, c_2, \ldots) = (pc_1, pc_2, \ldots)$. Then Φ_1 is an isomorphism from $H_B \otimes_f C$ onto PH_C and by use of (4.1.3) we now find that $\Psi_C((id_\mathcal{K} \otimes f)(x))(\Phi_1(z)) = \Phi_1((\Psi_B(x) \otimes_f id)(z))$, $x \in \mathcal{K} \otimes B$, $z \in H_B \otimes_f C$. Since f is quasi-unital, so is $id_\mathcal{K} \otimes f$, and by using the strict continuity of $\underline{id_{\mathcal{K} \otimes} f}$ we conclude that

$$(4.1.10) \quad \Psi_C((\underline{id_{\mathcal{K} \otimes} f}(m)) \circ \Phi_1 = \Phi_1 \circ (\Psi_B(m) \otimes_f id), \quad m \in \mathcal{M}(\mathcal{K} \otimes B).$$

Let $PH_C \oplus PH_C$ be graded by $\begin{bmatrix} 1 & 0 \\ 0 & -1 \end{bmatrix}$. Then $\Phi = \Phi_1 \oplus \Phi_1 :$ $H_B \otimes_f C \oplus H_B \otimes_f C \to PH_C \oplus PH_C$ is an isomorphism of graded Hilbert C-modules and by (4.1.10) we have that

$$\Phi \circ \begin{bmatrix} \Psi_B \circ \phi_+(\cdot) \otimes_f id & 0 \\ 0 & \Psi_B \circ \phi_-(\cdot) \otimes_f id \end{bmatrix} \circ \Phi^{-1}$$
$$= \begin{bmatrix} \Psi_C \circ \underline{id_{\mathcal{K} \otimes} f} \circ \phi_+ & 0 \\ 0 & \Psi_C \circ \underline{id_{\mathcal{K} \otimes} f} \circ \phi_- \end{bmatrix}.$$

Thus $f_* \circ \mu[\phi_+, \phi_-] \in KK^0(A, C)$ is represented by the Kasparov $A - C$-module

$$\left(PH_C \oplus PH_C, \begin{bmatrix} \Psi_C \circ \underline{id_{\mathcal{K} \otimes} f} \circ \phi_+ & 0 \\ 0 & \Psi_C \circ \underline{id_{\mathcal{K} \otimes} f} \circ \phi_- \end{bmatrix}, \begin{bmatrix} 0 & 1 \\ 1 & 0 \end{bmatrix} \right).$$

$$\left((1-P)H_C \oplus (1-P)H_C,\, 0,\, \begin{bmatrix} 0 & 1 \\ 1 & 0 \end{bmatrix} \right)$$ is a degenerate Kasparov $A-$ C-module and by adding it, we see that $f_* \circ \mu[\phi_+, \phi_-]$ is represented by

$$\left(\hat{H}_C,\, \begin{bmatrix} \Psi_C \circ \underline{id_{\mathcal{K} \otimes} f} \circ \phi_+ & 0 \\ 0 & \Psi_C \circ \underline{id_{\mathcal{K} \otimes} f} \circ \phi_- \end{bmatrix},\, \begin{bmatrix} 0 & 1 \\ 1 & 0 \end{bmatrix} \right).$$

The latter module clearly represents $\mu \circ f_*[\phi_+, \phi_-]$, so the proof is complete.

$\square$

4.1.16   *Notes and remarks.*

The isomorphism $KK_h(A, B) \simeq KK^0(A, B)$ was established by Higson in [12].

Exercise 4.1

E 4.1.1

Show that (4.1.9) defines a group homomorphism $f^* : KK_h(A, B) \rightarrow KK_h(D, B)$ and that this makes $KK_h(\cdot, B)$ into a contravariant functor.

E 4.1.2

Let $\phi : A \rightarrow B$ be a *-homomorphism.

(i) Show that $(e_B \circ \phi, 0) \in \mathbb{F}(A, B)$.

(ii) In the notation of E 2.2.2, show that $\mu[e_B \circ \phi, 0] = \{\phi\} \in KK^0(A, B)$. Conclude that $\mu[e_A, 0] = 1_A$.

(iii) Assume that $\phi, \psi \in \mathrm{Hom}\,(A, B)$ are homotopic. Show that $[e_B \circ \phi, 0] = [e_B \circ \psi, 0]$.

E 4.1.3

The purpose of this exercise is to show that $KK_h(\cdot, B)$ is a stable and homotopy invariant functor.

(i) Show that $KK_h(\cdot, B)$ is a homotopy invariant functor.

(ii) Show that there is a *-homomorphism

$$\Phi : \mathcal{K} \otimes \mathcal{M}(\mathcal{K} \otimes B) \rightarrow \mathcal{M}(\mathcal{K} \otimes \mathcal{K} \otimes B)$$

such that $\Phi(k \otimes m)(k_1 \otimes k_2 \otimes b) = kk_1 \otimes m(k_2 \otimes b)$, $k, k_1, k_2 \in \mathcal{K}$, $m \in \mathcal{M}(\mathcal{K} \otimes B)$.

Let $\lambda_B : \mathcal{K} \otimes B \to \mathcal{K} \otimes \mathcal{K} \otimes B$ be the *-isomorphism from Lemma 4.1.13.

(iii) Show that there is a map $G : KK_h(A, B) \to KK_h(\mathcal{K} \otimes A, B)$ given by
$G[\phi_+, \phi_-] = [\lambda_B^{-1} \circ \Phi \circ (id_{\mathcal{K}} \otimes \phi_+), \lambda_B^{-1} \circ \Phi \circ (id_{\mathcal{K}} \otimes \phi_-)]$, $(\phi_+, \phi_-) \in \mathbb{F}(A, B)$.

(iv) Show that G is the inverse of $e_A^* : KK_h(\mathcal{K} \otimes A, B) \to KK_h(A, B)$.

E 4.1.4

Note that for any C^*-algebras A and B, $[A, \mathcal{K} \otimes B]$ and $[\mathcal{K} \otimes A, \mathcal{K} \otimes B]$ are both abelian semi-groups with a zero element, cf. Lemma 1.3.12. Adopt the notation of Lemma 4.1.13. Define $E : [\mathcal{K} \otimes A, \mathcal{K} \otimes B] \to [A, \mathcal{K} \otimes B]$ by $E[\phi] = [\phi \circ e_A]$, $\phi \in \mathrm{Hom}\,(\mathcal{K} \otimes A, \mathcal{K} \otimes B)$ and $F : [A, \mathcal{K} \otimes B] \to [\mathcal{K} \otimes A, \mathcal{K} \otimes B]$ by $F[\phi] = [\lambda_B^{-1} \circ (id_{\mathcal{K}} \otimes \phi)]$, $\phi \in \mathrm{Hom}\,(A, \mathcal{K} \otimes B)$. Show that E is an isomorphism of abelian semi-groups with inverse F.

E 4.1.5

Let $(\phi_+, \phi_-), (\psi_+, \psi_-) \in \mathbb{F}(A, B)$. Show that $(\phi_+, \phi_-) \sim (\psi_+, \psi_-)$ if and only if there is a $(\lambda_+, \lambda_-) \in \mathbb{F}(A, IB)$ such that $\underline{\pi}_0 \circ \lambda_\pm = \phi_\pm$ and $\underline{\pi}_1 \circ \lambda_\pm = \psi_\pm$.

4.2. Split-Exactness of KK_h

The main purpose of this section is to prove that KK_h is split exact in both variables. The basic tool in the proof of this fact is the Kasparov product. Since $\mu : KK_h(A, B) \to KK^0(A, B)$ is a natural map, we transfer the Kasparov product of Chapter 2 to KK_h.

Note that $(e_A, 0) \in \mathbb{F}(A, A)$. We set $1_A = [e_A, 0] \in KK_h(A, A)$ and note that $\mu(1_A) = 1_A$ by E 4.1.2.

Theorem 4.2.1. *Let A, A_1 be separable C^*-algebras and B, C, and C_1 σ-unital C^*-algebras. There is a bilinear pairing*

$$\cdot \ : KK_h(A, B) \times KK_h(B, C) \to KK_h(A, C),$$

called the Kasparov product, with the following properties :

(a) $1_A \cdot x = x \cdot 1_B = x, \ \ x \in KK_h(A, B)$.

(b) *When $f : A_1 \to A, g : C \to C_1$ are $*$-homomorphisms, then*

$$f^*(x) \cdot y = f^*(x \cdot y), \ \ x \cdot g_*(y) = g_*(x \cdot y),$$

$x \in KK_h(A, B), \ y \in KK_h(B, C)$.

(c) *When $h : B \to C_1$ is a $*$-homomorphism, then*

$$h_*(x) \cdot y = x \cdot h^*(y),$$

$x \in KK_h(A, B), \ y \in KK_h(C_1, C)$.

(d) $x_1 \cdot (x_2 \cdot x_3) = (x_1 \cdot x_2) \cdot x_3$,

$x_1 \in KK_h(A, A_1), \ x_2 \in KK_h(A_1, B), \ x_3 \in KK_h(B, C)$.

Proof. In view of Theorem 4.1.8 and Proposition 4.1.15, Theorem 4.2.1 is a translation of Theorem 2.2.15 and Theorem 2.2.19, including Lemma 2.1.16 and the result of E 2.2.1. $\qquad\qquad\square$

Now fix a C^*-extension

$$0 \ \longrightarrow \ J \ \xrightarrow{\ j\ } \ E \ \xrightarrow{\ p\ } \ B \ \longrightarrow \ 0.$$

Then the following short sequence is also exact :

$$0 \longrightarrow \mathcal{K} \otimes J \overset{id_{\mathcal{K}} \otimes j}{\longrightarrow} \mathcal{K} \otimes E \overset{id_{\mathcal{K}} \otimes p}{\longrightarrow} \mathcal{K} \otimes B \longrightarrow 0.$$

By E1.1.9 there is a unique map $r_J : \mathcal{M}(\mathcal{K} \otimes E) \to \mathcal{M}(\mathcal{K} \otimes J)$ such that $(id_{\mathcal{K}} \otimes j)(r_J(m)x) = m(id_{\mathcal{K}} \otimes j)(x)$, $m \in \mathcal{M}(\mathcal{K} \otimes E)$, $x \in \mathcal{K} \otimes J$.

Lemma 4.2.2. *Assume that J is σ-unital. There is a strictly continuous path $\{V_t : t \in [0,1]\}$ of isometries in $\mathcal{M}(\mathcal{K} \otimes J)$ and an isometry $T \in \mathcal{M}(\mathcal{K} \otimes E)$ such that $r_J(\mathcal{K} \otimes E)V_t = \mathcal{K} \otimes J$, $t \in [0,1[$, and $V_1 = r_J(T)$.*

Proof. Let $w_1, w_2, w_3, \ldots$ be a sequence of isometries in $\mathcal{M}(\mathcal{K}) \simeq \mathcal{B}(l^2)$ such that $w_i^* w_j = 0$, $i \neq j$. Set $W_n = w_n \otimes 1 \in \mathcal{M}(\mathcal{K} \otimes E)$. Since J is σ-unital so is $\mathcal{K} \otimes J$, and there is a countable approximate unit $\{a_n\}_{n=1}^\infty$ for $\mathcal{K} \otimes J$. Let $\delta_0 < \delta_1 < \delta_2 < \cdots$ be an increasing sequence in $[0,1]$ with $\delta_0 = 0$ and $\delta_n \uparrow 1$. For each $n = 1, 2, 3, \ldots$ we can construct continuous and increasing functions $b_n : [0,1[\to \mathcal{K} \otimes J$ such that

(i) $b_n(\delta_k) = a_{n+k}$, $k = 0, 1, 2, 3 \ldots$, and

(ii) $a_{n+k} \leq b_n(t) \leq a_{n+k+1}$, $t \in [\delta_k, \delta_{k+1}]$.

The actual construction of the b_n's does not present any problem and is left to the reader. Then $\{b_n(t)\}_{n \in \mathbb{N}}$ is an approximate unit in $\mathcal{K} \otimes J$ for all $t \in [0,1[$. Set $s_n(t) = [b_n(t) - b_{n-1}(t)]^{\frac{1}{2}}$, $n \geq 2$, and $s_1(t) = b_1(t)^{\frac{1}{2}}$ for $t \in [0,1[$. It is easy to see that $\lim\limits_{t \to 1} s_n(t) = 0$, $n \geq 2$, and that $\lim\limits_{t \to 1} s_1(t) = 1$ in the strict topology of $\mathcal{M}(\mathcal{K} \otimes J)$. Set

$$V_t^k = \sum_{n=1}^k r_J(W_n) s_n(t), \quad k = 1, 2, 3, \ldots, \ t \in [0,1[.$$

Then

(4.2.1)
$$V_t^{k*} V_t^k = \sum_{n=1}^k s_n(t)^2 = b_k(t),$$

so $\|V_t^k\| \leq 1$, for all t and k. Let $x \in \mathcal{K} \otimes J$. Then

(4.2.2) $\| (V_t^k - V_t^m)x\|^2 = \left\| x^* \left(\sum_{n=m+1}^k s_n(t)^2 \right) x \right\| = \|x^*(b_k(t) - b_m(t))x\|$

for all $t \in [0,1[$ when $k \geq m$. It follows that $\{V_t^k x : k = 1, 2, \ldots\}$ converges in $\mathcal{K} \otimes J$ for all $t \in [0,1[$.

When $p \in \mathcal{K}$ is a projection, $pw_n = 0$ except for a finite number of n's. Since $\mathcal{K}$ has an approximate unit of projections, there is a set $M \subseteq \mathcal{K} \otimes E$ which spans a dense subspace in $\mathcal{K} \otimes E$ with the following property :

(4.2.3)   For each $y \in M$, there is an $N_y \in \mathbb{N}$ such that

$$r_J(y)V_t^k = \sum_{n=1}^{N_y} r_J(yW_n)s_n(t), \ k \geq N_y, \ t \in [0,1[\, .$$

In particular, $\lim_k r_J(y)V_t^k \in \mathcal{K} \otimes J$ for all $t \in [0,1[$, $y \in M$. Since M spans a dense subspace of $\mathcal{K} \otimes E$ and the norm of the V_t^k's are uniformly bounded by 1, it follows that $\lim_k xV_t^k \in \mathcal{K} \otimes J$, for all $t \in [0,1[$ and all $x \in r_J(\mathcal{K} \otimes E)$. Since $\mathcal{K} \otimes J \subseteq r_J(\mathcal{K} \otimes E)$, it follows in particular that $\{V_t^k : k = 1, 2, \ldots\}$ converges in the strict topology to some element $V_t \in \mathcal{M}(\mathcal{K} \otimes J)$ for every $t \in [0,1[$. Note that $r_J(\mathcal{K} \otimes E)V_t \subseteq \mathcal{K} \otimes J$, $t \in [0,1[$.

Since $\lim_{n \to \infty} b_n(t) = 1$, we conclude from (4.2.1) that V_t is an isometry for all $t \in [0,1[$. Thus $\mathcal{K} \otimes J = (\mathcal{K} \otimes J)V_t^*V_t \subseteq (\mathcal{K} \otimes J)V_t \subseteq r_J(\mathcal{K} \otimes E)V_t$, $t \in [0,1[$. Letting $k \to \infty$ in (4.2.2) and using that $b_m(t) \geq b_m(0) = a_m$ we get the inequality

(4.2.4) $\|(V_t - V_t^m)x\| \leq \|x^*(1 - a_m)x\|, \ x \in \mathcal{K} \otimes J, \ m \in \mathbb{N}, \ t \in [0,1[.$

Thus the convergence of $\{V_t^k x : k = 1, 2, 3, \ldots\}$ is uniform in $t \in [0,1[$ for all $x \in \mathcal{K} \otimes J$. It follows from (4.2.3) that so is the convergence of $\{xV_t^k : k = 1, 2, \ldots\}$ for all $x \in \mathcal{K} \otimes J$. In particular, this shows that the map $t \to V_t$ is continuous in the strict topology.

To complete the proof it suffices to show that $\lim_{t \to 1} V_t = r_J(W_1)$ in the strict topology. So let $x \in \mathcal{K} \otimes J$. By the uniform convergence of $\{V_t^k x : k = 1, 2, 3 \ldots\}$ it follows that, $\lim_{t \to 1} V_t x = \lim_m \lim_{t \to 1} V_t^m x$. Using the convergence properties of the s_n's, it follows that $V_t x \to r_J(W_1)$. Similarly, $xV_t \to xr_J(W_1)$. $\square$

Now fix an arbitrary C^*-algebra A.

Lemma 4.2.3. *Let* $(\phi_+, \phi_-) \in \mathbb{F}(A, E)$ *and assume that we have* $\phi_+(a) - \phi_-(a) \in id_{\mathcal{K}} \otimes j(\mathcal{K} \otimes J) = \mathcal{K} \otimes j(J)$ *for all* $a \in A$.
    Then $(r_J \circ \phi_+, r_J \circ \phi_-) \in \mathbb{F}(A, J)$ *and* $j_*[r_J \circ \phi_+, r_J \circ \phi_-] = [\phi_+, \phi_-]$.

Proof. Since $r_J \circ (id_K \otimes j) = id_{K \otimes J}$, it is clear from our assumptions that $(r_J \circ \phi_+, r_J \circ \phi_-) \in \mathbb{F}(A, J)$.

The remaining assertion is proved in 3 steps.

STEP I : Assume that J is an essential ideal in E. Then $K \otimes J$ is an essential ideal in $K \otimes E$ and $r_J : M(K \otimes E) \to M(K \otimes J)$ is injective by E 1.1.9. Let $\{V_t : t \in [0,1]\}$ and T be as in Lemma 4.2.2. Set $V = V_0$. Then $r_J(K \otimes E)V = K \otimes J$ so $VV^* r_J(K \otimes E) = V(K \otimes J) \subseteq K \otimes J \subseteq r_J(K \otimes E)$. Since r_J is injective, we can define $P \in M(K \otimes E)$ by $r_J(Py) = VV^* r_J(y)$, $y \in K \otimes E$. Then P is a projection and $r_J(P) = VV^*$. Note that $r_J[id_K \otimes j(V(K \otimes J)V^*)(K \otimes E)] = V(K \otimes J)V^* r_J(K \otimes E) = V(K \otimes J) = r_J(P(K \otimes E))$. It follows that $(id_K \otimes j) \circ \operatorname{Ad} V : K \otimes J \to K \otimes E$ is quasi-unital with relative unit P. Since $(id_K \otimes j) \circ \operatorname{Ad} V \sim id_K \otimes j$ by Lemma 1.3.7, we have

$$j_*[r_J \circ \phi_+, r_J \circ \phi_-] = [(id_K \otimes j) \circ \operatorname{Ad} V \circ r_J \circ \phi_+, (id_K \otimes j) \circ \operatorname{Ad} V \circ r_J \circ \phi_-].$$

Note that $\operatorname{Ad} V$ leaves $r_J(M(K \otimes E))$ globally invariant. In fact $V r_J(m)V^* = r_J(n)$, where $n \in M(K \otimes E)$ is given by

$$nz = id_K \otimes j(V r_J(m)V^* r_J(z)), \; z \in K \otimes E.$$

We assert that $r_J^{-1} \circ \operatorname{Ad} V = (id_K \otimes j) \circ \operatorname{Ad} V$ on $r_J(M(K \otimes E))$. This is equivalent to $\operatorname{Ad} V = r_J \circ (id_K \otimes j) \circ \operatorname{Ad} V$ in $\operatorname{Hom}(M(K \otimes J))$. It is clear that this equality holds on $K \otimes J$. Since both *-homomorphisms in question are strictly continuous, it holds throughout. It follows that

$$j_*[r_J \circ \phi_+, r_J \circ \phi_-] = [r_J^{-1} \circ \operatorname{Ad} V \circ r_J \circ \phi_+, r_J^{-1} \circ \operatorname{Ad} V \circ r_J \circ \phi_-].$$

Note that $\operatorname{Ad} V_t$ leaves $r_J(M(K \otimes E))$ globally invariant for all $t \in [0,1]$. Thus $\lambda_+^t = r_J^{-1} \circ \operatorname{Ad} V_t \circ r_J \circ \phi_+, \lambda_-^t = r_J^{-1} \circ \operatorname{Ad} V_t \circ r_J \circ \phi_-$, $t \in [0,1]$, define a homotopy in $\mathbb{F}(A, E)$, showing that we can substitute $r_J(T)$ for V in the above expression for $j_*[r_J \circ \phi_+, r_J \circ \phi_-]$. Since $r_J^{-1} \circ \operatorname{Ad} r_J(T) \circ r_J = \operatorname{Ad} T$, we see that

$$j_*[r_J \circ \phi_+, r_J \circ \phi_-] = [\operatorname{Ad} T \circ \phi_+, \operatorname{Ad} T \circ \phi_-].$$

Thus $j_*[r_J \circ \phi_+, r_J \circ \phi_-] = [\phi_+, \phi_-]$ by an easy application of Lemma 1.3.7.

STEP II : Assume that $E = J \oplus B$ and that $j(x) = (x, 0)$, $x \in J$. In this case $M(K \otimes E) \simeq M(K \otimes J) \oplus M(K \otimes B)$ under a *-isomorphism making $id_K \otimes j$ quasi-unital with $id_K \otimes j$ the natural map

$$M(K \otimes J) \to M(K \otimes J) \oplus M(K \otimes B)$$

and

$$r_J : \mathcal{M}(\mathcal{K} \otimes J) \oplus \mathcal{M}(\mathcal{K} \otimes B) \to \mathcal{M}(\mathcal{K} \otimes J)$$

the natural projection. Thus $j_* [r_J \circ \phi_+, r_J \circ \phi_-] = [\phi_+^1, \phi_-^1]$ where $\phi_\pm^1$ is the component of $\phi_\pm$ under the decomposition

$$\mathcal{M}(\mathcal{K} \otimes E) \simeq \mathcal{M}(\mathcal{K} \otimes J) \oplus \mathcal{M}(\mathcal{K} \otimes B).$$

By the assumption on (ϕ_+, ϕ_-) the other components, $\phi_\pm^2$, of $\phi_\pm$ (taking values in $0 \oplus \mathcal{M}(\mathcal{K} \otimes B)$) agree. Thus Lemma 4.1.4 and Lemma 4.1.9 in combination gives that $[\phi_+^1, \phi_-^1] = [\phi_+, \phi_-]$.

STEP III : The general case. Set $J^\perp = \{x \in E : xj(J) = j(J)x = 0\}$. Then $J^\perp$ is an ideal in E and $j(J) + J^\perp$ is an essential ideal in E. Note that $j(J) + J^\perp \simeq J \oplus J^\perp$. Let $\psi : J \oplus J^\perp \to j(J) + J^\perp$ be a *-isomorphism such that $\psi \circ j_0 = j_1$ where $j_0 : J \to J \oplus J^\perp$ is the natural inclusion and $j_1 : J \to j(J) + J^\perp$ is given by $j_1(x) = j(x)$, $x \in J$. Let $j_2 : j(J) + J^\perp \to E$ be the inclusion map. Let $r_1 : \mathcal{M}(\mathcal{K} \otimes (j(J) + J^\perp)) \to \mathcal{M}(\mathcal{K} \otimes J)$ and $r_2 : \mathcal{M}(\mathcal{K} \otimes E) \to \mathcal{M}(\mathcal{K} \otimes (j(J) + J^\perp))$ be the *-homomorphisms corresponding to j_1 and j_2, respectively. Then $r_J = r_1 \circ r_2$ and $j = j_2 \circ j_1$. Note that $r_1 = r_3 \circ (\underline{id_\mathcal{K} \otimes \psi})^{-1}$ where $\underline{id_\mathcal{K} \otimes \psi} : \mathcal{M}(\mathcal{K} \otimes (J \oplus J^\perp)) \to \mathcal{M}(\mathcal{K} \otimes (j(J) + J^\perp))$ is the strictly continuous extension of $id_\mathcal{K} \otimes \psi$ and $r_3 : \mathcal{M}(\mathcal{K} \otimes (J \oplus J^\perp)) \to \mathcal{M}(\mathcal{K} \otimes J)$ is the map corresponding to $j_0 : J \to J \oplus J^\perp$. Thus

$$j_* [r_J \circ \phi_+, r_J \circ \phi_-]$$

$$= j_{2*} \circ \psi_* \circ j_{0*} [r_3 \circ (\underline{id_\mathcal{K} \otimes \psi})^{-1} \circ r_2 \circ \phi_+, r_3 \circ (\underline{id_\mathcal{K} \otimes \psi})^{-1} \circ r_2 \circ \phi_-]$$

$$= j_{2*} \circ \psi_* [(\underline{id_\mathcal{K} \otimes \psi})^{-1} \circ r_2 \circ \phi_+, (\underline{id_\mathcal{K} \otimes \psi})^{-1} \circ r_2 \circ \phi_-] \quad \text{(by Step II)}$$

$$= j_{2*} [r_2 \circ \phi_+, r_2 \circ \phi_-] \quad \text{(by definition of } \psi_*)$$

$$= [\phi_+, \phi_-] \quad \text{(by Step I).} \qquad \qquad \square$$

Lemma 4.2.4. *Let $f, g : A \to B$ be *-homomorphisms between σ-unital C^*-algebras A and B. Then $(e_B \circ f, e_B \circ g) \in \mathbb{F}(A, B)$ and furthermore $[e_B \circ f, e_B \circ g] = f_*(1_A) - g_*(1_A)$ in $KK_h(A, B)$.*

Proof. It is obvious that $(e_B \circ f, e_B \circ g) \in \mathbb{F}(A, B)$ since both *-homomorphisms takes values in $\mathcal{K} \otimes B$.

Assume first that f and g are quasi-unital. Then $id_\mathcal{K} \otimes f$ and $id_\mathcal{K} \otimes g$

are quasi-unital, so

$$f_*(1_A) - g_*(1_A) = f_*[e_A, 0] - g_*[e_A, 0]$$
$$= [(id_\mathcal{K} \otimes f) \circ e_A, 0] + [0, (id_\mathcal{K} \otimes g) \circ e_A]$$
$$= [e_B \circ f, 0] + [0, e_B \circ g]$$
$$= [e_B \circ f, e_B \circ g].$$

where the last equality follows from Lemma 4.1.9.

Now let f, g be arbitrary. Let $\lambda, \mu : \mathcal{K} \otimes A \to \mathcal{K} \otimes B$ be quasi-unital *-homomorphisms homotopic to $id_\mathcal{K} \otimes f$ and $id_\mathcal{K} \otimes g$, respectively. λ and μ exist by Theorem 1.3.16. Using that the lemma holds for quasi-unital f and g, that $e_B \circ f = (id_\mathcal{K} \otimes f) \circ e_A$ and that $KK_h(A, \cdot)$ is homotopy invariant, we find

$$e_{B*} \circ f_*(1_A) - e_{B*} \circ g_*(1_A) = \lambda_* \circ e_{A*}(1_A) - \mu_* \circ e_{A*}(1_A)$$
$$= [e_{\mathcal{K} \otimes B} \circ \lambda \circ e_A, e_{\mathcal{K} \otimes B} \circ \mu \circ e_A]$$
$$= (*).$$

Since $e_{\mathcal{K} \otimes B} \sim id_\mathcal{K} \otimes e_B$, $\lambda \sim id_\mathcal{K} \otimes f$ and $\mu \sim id_\mathcal{K} \otimes g$, it follows that

$$(*) = e_{B*}[(id_\mathcal{K} \otimes f) \circ e_A, (id_\mathcal{K} \otimes g) \circ e_A]$$
$$= e_{B*}[e_B \circ f, e_B \circ g].$$

Thus the lemma follows from the fact that e_{B*} is injective, Lemma 4.1.14.
$\square$

Proposition 4.2.5. *Let*

$$0 \longrightarrow J \xrightarrow{j} E \xrightarrow{p} B \longrightarrow 0$$

*be a split exact sequence of σ-unital C^*-algebras with splitting *-homomorphism $s : B \to E$. Let A be a separable C^*-algebra.*

Then the map $j_* \oplus s_* : KK_h(A, J) \oplus KK_h(A, B) \to KK_h(A, E)$ is an isomorphism.

Proof. Since we have that $x - s \circ p(x) \in j(J)$, $x \in E$, it follows that $(r_J \circ e_E, r_J \circ e_E \circ s \circ p) \in \mathbb{F}(E, J)$. Set $\pi = [r_J \circ e_E, r_J \circ e_E \circ s \circ p] \in KK_h(E, J)$. Then $j_*(\pi) = [e_E, e_E \circ s \circ p] = 1_E - s_* \circ p_*(1_E)$ by Lemmas 4.2.3 and 4.2.4. It is easy to check that $r_J \circ e_E \circ j = e_J$ so we get that $j^*(\pi) = [r_J \circ e_E \circ j, r_J \circ e_E \circ s \circ p \circ j] = [r_J \circ e_E \circ j, 0] = 1_J$.

Let $\phi_\pi : KK_h(A, E) \to KK_h(A, J)$ be the homomorphism obtained by taking the Kasparov product with π, cf. Theorem 4.2.1. Then $\phi_\pi \circ j_*(x) = j_*(x) \cdot \pi = x \cdot j^*(\pi) = x \cdot 1_J = x$, $x \in KK_h(A, J)$. Thus j_* is injective. Since $p_*(j_*(x) + s_*(y)) = y$, $x \in KK_h(A, J)$, $y \in KK_h(A, B)$, it follows that $j_* \oplus s_*$ is injective.

To see that $j_* \oplus s_*$ is also surjective note that

$$
\begin{aligned}
j_* \circ \phi_\pi(x) &= j_*(x \cdot \pi) \\
&= x \cdot j_*(\pi) \\
&= x \cdot (1_E - s_* \circ p_*(1_E)) \\
&= x - s_* \circ p_*(x), \quad x \in KK_h(A, E).
\end{aligned}
$$

Thus $x = j_* \circ \phi_\pi(x) + s_* \circ p_*(x)$, $x \in KK_h(A, E)$. In particular, $j_* \oplus s_*$ is surjective. $\qquad\square$

Proposition 4.2.5 says that $KK_h(A, \cdot)$ is a split exact functor for separable A, cf. Appendix C. The same is true in the first variable :

Proposition 4.2.6. *Let*

$$
0 \longrightarrow J \overset{j}{\longrightarrow} E \overset{p}{\longrightarrow} B \longrightarrow 0
$$

be a split exact sequence of separable C^*-algebras with splitting $*$-homomorphism $s : B \to E$. Let A be a σ-unital C^*-algebra.

Then $j^* \oplus s^* : KK_h(E, A) \to KK_h(J, A) \oplus KK_h(B, A)$ is an isomorphism.

Proof. Let $\pi \in KK_h(E, J)$ be the same element as in the previous proof and let $\psi_\pi : KK_h(J, A) \to KK_h(E, A)$ be the homomorphism obtained by taking the Kasparov product with π (on the left this time), cf. Theorem 4.2.1. Then $\psi_\pi \circ j^*(x) = \pi \cdot j^*(x) = j_*(\pi) \cdot x = (1_E - s_* \circ p_*(1_E)) \cdot x = x - p^* \circ s^*(x)$, $x \in KK_h(E, A)$. In particular, the last equality shows that $j^* \oplus s^*$ is injective.

To see that $j^* \oplus s^*$ is also surjective, note that

$$
\begin{aligned}
j^* \circ \psi_\pi(x) &= j^*(\pi \cdot x) \\
&= j^*(\pi) \cdot x \\
&= 1_J \cdot x = x, \quad x \in KK_h(J, A),
\end{aligned}
$$

and that $s^*(\pi) = [r_J \circ e_E \circ s, r_J \circ e_E \circ s] = 0$. Thus if $x \in KK_h(J, A)$ and $y \in KK_h(B, A)$, we find that $j^* \oplus s^*(\psi_\pi(x) + p^*(y)) = (x, y)$. $\qquad\square$

Given a homotopy invariant, stable and split exact covariant functor F from the full sub-category of σ-unital C^*-algebras to abelian groups, there is a canonical way to produce, for any $KK_h(A, B)$-cycle $\Phi = (\phi_+, \phi_-) \in \mathbb{F}(A, B)$, a homomorphism $\Phi_F : F(A) \to F(B)$, see E 4.2.1. With Proposition 4.2.5 at hand we know that $KK_h(A, \cdot)$ is such a functor when A is separable.

Proposition 4.2.7. *Let $\Phi = (\phi_+, \phi_-) \in \mathbb{F}(A, B)$ and assume that A is separable and B is σ-unital. Then the map*

$$\Phi_{KK_h(A, \cdot)} : KK_h(A, A) \to KK_h(A, B)$$

takes 1_A to $[\phi_+, \phi_-]$, i.e. $\Phi_{KK_h(A, \cdot)}(1_A) = [\phi_+, \phi_-] \in KK_h(A, B)$.

Proof. We will now use the notation from E 4.2.1. Note that $\hat{\phi}_{+*}(1_A) - \hat{\phi}_{-*}(1_A) = [e_{A_*} \circ \hat{\phi}_+, e_{A_*} \circ \hat{\phi}_-]$ by Lemma 4.2.4. On the other hand $j_*^{-1}[e_{A_*} \circ \hat{\phi}_+, e_{A_*} \circ \hat{\phi}_-] = [r_J \circ e_{A_*} \circ \hat{\phi}_+, r_J \circ e_{A_*} \circ \hat{\phi}_-]$ by Lemma 4.2.3. It is straightforward to check that $r_J \circ e_{A_*} \circ \hat{\phi}_\pm = \underline{e}_{\mathcal{K} \otimes B} \circ \phi_\pm$. Since $e_{\mathcal{K} \otimes B} \sim id_{\mathcal{K}} \otimes e_B$, it follows by combining Theorem 1.3.16 with Lemma 1.3.15 that

$$[r_J \circ e_{A_*} \circ \hat{\phi}_+, r_J \circ e_{A_*} \circ \hat{\phi}_-] = [\underline{e}_{\mathcal{K} \otimes B} \circ \phi_+, \underline{e}_{\mathcal{K} \otimes B} \circ \phi_-]$$
$$= [(id_{\mathcal{K}} \otimes e_B) \circ \phi_+, (id_{\mathcal{K}} \otimes e_B) \circ \phi_-].$$
$$= e_{B*}[\phi_+, \phi_-]$$

Thus $\hat{\phi}_{+*}(1_A) - \hat{\phi}_{-*}(1_A) = j_* \circ e_{B*}[\phi_+, \phi_-]$. By the construction of $\Phi_{KK_h(A, \cdot)}$, this finishes the proof. $\square$

When G is a contravariant functor from the category of separable C^*-algebras to abelian groups which is homotopy invariant, stable and split-exact, we have, for each $KK_h(A, B)$-cycle $\Phi = (\phi_+, \phi_-) \in \mathbb{F}(A, B)$ $(A, B$ separable), a group homomorphism $\Phi_G : G(B) \to G(A)$ constructed in a way analoque to the covariant case, cf. E 4.2.2. In particular, $KK_h(\cdot, C)$ is such a functor when C is σ-unital.

Proposition 4.2.8. *Let $\Phi = (\phi_+, \phi_-) \in \mathbb{F}(A, B)$ and assume that both A and B are separable. Then the map*

$$\Phi_{KK_h(\cdot, B)} : KK_h(B, B) \to KK_h(A, B)$$

takes 1_B to $[\phi_+, \phi_-]$, i.e. $\Phi_{KK_h(\cdot, B)}(1_B) = [\phi_+, \phi_-] \in KK_h(A, B)$.

Proof. Now consider the C^*-extension of E 4.2.1. By construction $\Phi_{KK_h}(\cdot, B) = (\hat{\phi}^*_+ - \hat{\phi}^*_-) \circ \psi_\pi \circ e_B^{*-1}$ where $\psi_\pi : KK_h(\mathcal{K} \otimes B, B) \to KK_h(A_\Phi, B)$ is the right inverse of j^* given by left Kasparov multiplication with an element $\pi \in KK_h(A_\Phi, \mathcal{K} \otimes B)$, cf. the proof of Proposition 4.2.6. Let $x \in KK_h(A, A)$. Using Theorem 4.2.1 we get

$$
\begin{aligned}
x \cdot \Phi_{KK_h}(\cdot, B)(1_B) &= (\hat{\phi}_{+*} - \hat{\phi}_{-*})(x) \cdot (\pi \cdot e_B^{*-1}(1_B)) \\
&= (\hat{\phi}_{+*} - \hat{\phi}_{-*})(x) \cdot \pi) \cdot e_B^{*-1}(1_B) \\
&= e_{B^*}^{-1}((\hat{\phi}_{+*} - \hat{\phi}_{-*})(x) \cdot \pi) \cdot 1_B \\
&= \Phi_{KK_h}(A, \cdot)(x).
\end{aligned}
$$

Thus if we insert $x = 1_A$, we find $\Phi_{KK_h}(\cdot, B)(1_B) = \Phi_{KK_h}(A, \cdot)(1_A)$. The conclusion therefore follows from Proposition 4.2.7.   □

Theorem 4.2.9. *Assume that $\#$ is a bilinear pairing*

$$
\# : KK_h(A, B) \times KK_h(B, C) \to KK_h(A, C)
$$

for all separable C^*-algebras A, B, C. Assume that $\#$ satisfies the following two conditions for separable C^*-algebras A, B, C, D :

(i) $x \# 1_B = x$, $x \in KK_h(A, B)$,

(ii) $x \# f^*(y) = f_*(x) \# y$, *when* $x \in KK_h(A, B)$, $y \in KK_h(D, C)$, *and* $f \in \operatorname{Hom}(B, D)$.

Then $\#$ agrees with the Kasparov product.

Proof. Assume first that we can prove the following identity for all separable C^*-algebras A, B, C :

$$
(4.2.3) \quad x \# \Phi_{KK_h}(\cdot, C)(1_C) = \Phi_{KK_h}(A, \cdot)(x), \quad x \in KK_h(A, B),
$$

$$
\Phi = (\phi_+, \phi_-) \in \mathbb{F}(B, C).
$$

Let $\Psi = (\psi_+, \psi_-) \in \mathbb{F}(A, B)$. Then

$$
\begin{aligned}
[\psi_+, \psi_-] \# [\phi_+, \phi_-] &= [\psi_+, \psi_-] \# \Phi_{KK_h}(\cdot, C)(1_C) \\
&= \Phi_{KK_h}(A, \cdot)([\psi_+, \psi_-]) \\
&= \Phi_{KK_h}(A, \cdot) \circ \Psi_{KK_h}(A, \cdot)(1_A)
\end{aligned}
$$

by Proposition 4.2.8, (4.2.3) and Proposition 4.2.7. Since $\cdot$ also satisfies (i) and (ii) by Theorem 4.2.1, we must also have $[\psi_+, \psi_-] \cdot [\phi_+, \phi_-] = \Phi_{KK_h}(A, \cdot) \circ \Psi_{KK_h}(A, \cdot)(1_A)$. Thus $[\psi_+, \psi_-] \# [\phi_+, \phi_-] = [\psi_+, \psi_-] \cdot [\phi_+, \phi_-]$.

So it suffices to prove (4.2.3). For this purpose recall that $\Phi_{KK_h(\cdot,C)} = (\hat{\phi}^*_+ - \hat{\phi}^*_-) \circ r \circ e_C^{*-1}$, where $r : KK_h(\mathcal{K} \otimes C, C) \to KK_h(B_\Phi, C)$ is a right inverse for $j^* : KK_h(B_\Phi, C) \to KK_h(\mathcal{K} \otimes C, C)$, cf. E 4.2.1 and E 4.2.2. Thus by using (ii) and additivity in both variables we get

$$x \# \Phi_{KK_h(\cdot,C)}(1_C) = (\hat{\phi}_{+*} - \hat{\phi}_{-*})(x) \# r \circ e_C^{*-1}(1_C) = (*).$$

By split-exactness of $KK_h(A, \cdot), (\hat{\phi}_{+*} - \hat{\phi}_{-*})(x) = j_* \circ t \circ (\hat{\phi}_{+*} - \hat{\phi}_{-*})(x)$, where $t : KK_h(A, B_\Phi) \to KK_h(A, \mathcal{K} \otimes B)$ is a right inverse for j_*. Thus we can use (ii) again to get

$$(*) = t \circ (\hat{\phi}_{+*} - \hat{\phi}_{-*})(x) \# e_C^{*-1}(1_C) = e_{C*}^{-1} \circ t \circ (\hat{\phi}_{+*} - \hat{\phi}_{-*})(x) \# 1_C.$$

Since the lefthand side in this "product" is exactly $\Phi_{KK_h(A,\cdot)}(x)$, (4.2.3) follows now from (i).   □

4.2.10 *Notes and remarks.*

The exposition in this section follows a piece of the track laid down in [12].

Exercise 4.2

E 4.2.1

Let F be a homotopy invariant, stable and split exact covariant functor from the full sub-category of σ-unital C^*-algebras to abelian groups and let $\Phi = (\phi_+, \phi_-) \in \mathbb{F}(A, B)$ be a $KK_h(A, B)$-cycle where A and B are σ-unital C^*-algebras. Set

$$A_\Phi = \{(a, x) \in A \oplus M(\mathcal{K} \otimes B) : x - \phi_+(a) \in \mathcal{K} \otimes B\}.$$

(i) Show that A_Φ is a C^*-subalgebra of $A \oplus M(\mathcal{K} \otimes B)$.

(ii) Let $h_1 \in A$ and $h_2 \in \mathcal{K} \otimes B$ be strictly positive elements of A and $\mathcal{K} \otimes B$, respectively. Show that A_Φ is σ-unital by proving that the element $(h_1, \phi_+(h_1) + h_2)$ is strictly positive in A_Φ. Show that A_Φ is separable if A and B are.

(iii) Define $\hat{\phi}_{\pm} : A \to A_\Phi$ by $\hat{\phi}_{\pm}(a) = (a, \phi_{\pm}(a))$, $a \in A$, $j : \mathcal{K} \otimes B \to A_\Phi$ by $j(x) = (0, x)$, $x \in \mathcal{K} \otimes B$ and $p : A_\Phi \to A$ by $p(a, x) = a$ for $(a, x) \in A_\Phi$. Show that the sequence

$$0 \longrightarrow \mathcal{K} \otimes B \overset{j}{\longrightarrow} A_\Phi \overset{p}{\longrightarrow} A \longrightarrow 0$$

is exact and that both $\hat{\phi}_+$ and $\hat{\phi}_-$ are splitting maps for p.

(iv) Show that there are unique homomorphisms $\Phi_F : F(A) \to F(B)$ such that $j_* \circ e_{B*} \circ \Phi_F = \hat{\phi}_{+*} - \hat{\phi}_{-*}$.

(v) Let H be another covariant functor from the category of σ-unital C^*-algebras to abelian groups which is also stable, homotopy invariant and split-exact. Let $T : F \to H$ be a natural transformation of functors. Let $\Phi = (\phi_+, \phi_-) \in \mathbb{F}(A, B)$. Show that $T_B \circ \Phi_F = \Phi_H \circ T_A$.

E 4.2.2

Adopt the notation from the previous exercise, but now let G be a contravariant functor from the full subcategory of separable C^*-algebras to abelian groups. Assume that G is homotopy invariant, stable and split-exact. Let $\Psi = (\psi_+, \psi_-) \in \mathbb{F}(A, B)$.

Show that there is a unique homomorphism $\Psi_G : G(B) \to G(A)$ such that $\Psi_G \circ e_B^* \circ j^* = \hat{\phi}_+^* - \hat{\phi}_-^*$.

4.3 The Kasparov Product as a Generalization of Composition of *-Homomorphisms

In Section 4.2 we transferred the Kasparov product as introduced in Section 2.2 to KK_h by using the results of Section 4.1. The product was then used to establish split exactness of the KK_h-functors, and we proved a quite strong uniquenes result for the product in Theorem 4.2.9. However, no recipe for a calculation of $[\phi_+, \phi_-] \cdot [\psi_+, \psi_-]$, $(\phi_+, \phi_-) \in \mathbb{F}(A, B)$, $(\psi_+, \psi_-) \in \mathbb{F}(B, C)$, was given.

In this section we give a description of the Kasparov product which shows that it can be considered as an extension of the composition of *-homomorphisms.

Let A and B be arbitrary C^*-algebras.

Definition 4.3.1. A $kK_h(A, B) - cycle$ (ϕ_+, ϕ_-) is a pair of strictly continuous *-homomorphisms $\phi_+, \phi_- : \mathcal{M}(\mathcal{K} \otimes A) \to \mathcal{M}(\mathcal{K} \otimes B)$ with the property that $\phi_+(x) - \phi_-(x) \in \mathcal{K} \otimes B$ for all $x \in \mathcal{K} \otimes A$.
We let $\mathbb{G}(A,B)$ denote the set of $kK_h(A, B)$-cycles.

Definition 4.3.2. (ϕ_+, ϕ_-), $(\psi_+, \psi_-) \in \mathbb{G}(A, B)$ are called *homotopic* when there is a path $(\lambda_+^t, \lambda_-^t) \in \mathbb{G}(A, B)$, $t \in [0, 1]$, such that

(i) $t \to \lambda_+^t(m)$ and $t \to \lambda_-^t(m)$ are strictly continuous maps from $[0, 1]$ to $\mathcal{M}(\mathcal{K} \otimes B)$ for all $m \in \mathcal{M}(\mathcal{K} \otimes A)$,

(ii) $t \to \lambda_+^t(x) - \lambda_-^t(x)$ is a norm continuous map from $[0, 1]$ to $\mathcal{K} \otimes B$ for all $x \in \mathcal{K} \otimes A$, and

(iii) $(\lambda_+^0, \lambda_-^0) = (\phi_+, \phi_-)$, $(\lambda_+^1, \lambda_-^1) = (\psi_+, \psi_-)$.

We write $(\phi_+, \phi_-) \sim (\psi_+, \psi_-)$ in this case and call $(\lambda_+^t, \lambda_-^t)$, $t \in [0, 1]$, a homotopy connecting (ϕ_+, ϕ_-) to (ψ_+, ψ_-).

It is clear that homotopy defines an equivalence relation on $\mathbb{G}(A, B)$ and we let $kK_h(A, B)$ denote the set of homotopy classes in $\mathbb{G}(A, B)$, i.e. $kK_h(A, B) = \mathbb{G}(A, B)/ \sim$. For $(\phi_+, \phi_-) \in \mathbb{G}(A, B)$ we let $[\phi_+, \phi_-]$ denote the corresponding element in $kK_h(A, B)$.

In order to make $kK_h(A, B)$ into a group we use an inner *-isomorphism $\Theta_B : M_2(\mathcal{M}(\mathcal{K} \otimes B)) \to \mathcal{M}(\mathcal{K} \otimes B)$ and proceed as in Section 4.1.

We define

$$[\phi_+, \phi_-] + [\psi_+, \psi_-] = \left[\Theta_B \circ \begin{bmatrix} \phi_+ & 0 \\ 0 & \psi_+ \end{bmatrix}, \Theta_B \circ \begin{bmatrix} \phi_- & 0 \\ 0 & \psi_- \end{bmatrix}\right],$$

for $(\phi_+, \phi_-), (\psi_+, \psi_-) \in \mathbb{G}(A, B)$. It is quite easy to see that $+$ is a well-defined composition in $kK_h(A, B)$ and is independent of the particular choice of the inner *-isomorphism Θ_B. By essentially the same arguments that proved Lemma 4.1.4 and Proposition 4.1.5 we get the following two results.

Lemma 4.3.3. *Let $(\phi_+, \phi_-) \in \mathbb{G}(A, B)$ and assume that $\phi_+(x) = \phi_-(x)$, $x \in \mathcal{K} \otimes A$. Then $(\phi_+, \phi_-) \sim (0, 0)$.*   □

Proposition 4.3.4. $kK_h(A, B)$ *is an abelian group with neutral element $[0, 0]$. The inverse of $[\phi_+, \phi_-]$ is $[\phi_-, \phi_+]$, $(\phi_+, \phi_-) \in \mathbb{G}(A, B)$.*   □

Let $f : \mathcal{K} \otimes B \to \mathcal{K} \otimes C$ be a quasi-unital *-homomorphism and $\underline{f}$ the unique strictly continuous extension $\underline{f} : \mathcal{M}(\mathcal{K} \otimes B) \to \mathcal{M}(\mathcal{K} \otimes C)$, cf. Corollary 1.1.15. It is then clear that we can define maps

$$f_* : kK_h(A, B) \to kK_h(A, C) \quad \text{and} \quad f^* : kK_h(C, A) \to kK_h(B, A)$$

by

$$f_*[\phi_+, \phi_-] = [\underline{f} \circ \phi_+, \underline{f} \circ \phi_-], \quad (\phi_+, \phi_-) \in \mathbb{G}(A, B), \quad \text{and}$$
$$f^*[\phi_+, \phi_-] = [\phi_+ \circ \underline{f}, \phi_- \circ \underline{f}], \quad (\phi_+, \phi_-) \in \mathbb{G}(C, A), \quad \text{respectively.}$$

To show that f^* and f_* are homomorphisms we need a couple of lemmas which have the same proofs as Lemma 4.1.9 and 4.1.10, respectively.

Lemma 4.3.5. *Let $(\phi_+, \phi_-), (\psi_+, \psi_-) \in \mathbb{G}(A, B)$ and assume that $\phi_+(m)\psi_+(m) = \phi_-(m)\psi_-(m) = 0$, $m \in \mathcal{M}(\mathcal{K} \otimes A)$. Then $[\phi_+, \phi_-] + [\psi_+, \psi_-] = [\phi_+ + \psi_+, \phi_- + \psi_-]$ in $kK_h(A, B)$.*   □

Lemma 4.3.6. *Let $(\phi_+, \phi_-) \in \mathbb{G}(A, B)$ and let $w \in \mathcal{M}(\mathcal{K} \otimes B)$ be a partial isometry with $w^*w \geq \phi_+(1)$ and $w^*w \geq \phi_-(1)$. Then $[w\phi_+(\cdot)w^*, w\phi_-(\cdot)w^*] = [\phi_+, \phi_-]$ in $kK_h(A, B)$.*   □

Lemma 4.3.7. *The following two maps are both homomorphisms:*

$$f_* : kK_h(A, B) \to kK_h(A, C)$$

$$f^* : kK_h(C, A) \to kK_h(B, A).$$

Proof. Let $(\phi_+, \phi_-), (\psi_+, \psi_-) \in \mathbb{G}(C, A)$. Then

$f^*([\phi_+, \phi_-] + [\psi_+, \psi_-])$

$\quad = [\mathrm{Ad}\, V_1 \circ \phi_+ \circ \underline{f} + \mathrm{Ad}\, V_2 \circ \psi_+ \circ \underline{f}, \ \mathrm{Ad}\, V_1 \circ \phi_- \circ \underline{f} + \mathrm{Ad}\, V_2 \circ \psi_- \circ \underline{f}]$

$\quad\quad \text{(by definition)}$

$\quad = [\mathrm{Ad}\, V_1 \circ \phi_+ \circ \underline{f}, \mathrm{Ad}\, V_1 \circ \phi_- \circ \underline{f}] + [\mathrm{Ad}\, V_2 \circ \psi_+ \circ \underline{f}, \mathrm{Ad}\, V_2 \circ \psi_- \circ \underline{f}]$

$\quad\quad \text{(by Lemma 4.3.5)}$

$\quad = [\phi_+ \circ \underline{f}, \phi_- \circ \underline{f}] + [\psi_+ \circ \underline{f}, \psi_- \circ \underline{f}]$

$\quad\quad \text{(by Lemma 4.3.6)}$

$\quad = f^*[\phi_+, \phi_-] + f^*[\psi_+, \psi_-].$

The proof for f_* is the same as the corresponding argument in Section 4.1, namely the proof of Lemma 4.1.11. $\square$

Since the composition of quasi-unital *-homomorphisms is again quasi-unital by E 1.3.2, it is clear that the usual functoriality equations, $(f_1 \circ f_2)_* = f_{1*} \circ f_{2*}$ and $(f_1 \circ f_2)^* = f_2^* \circ f_1^*$, hold.

Lemma 4.3.8. *Let $f_1, f_2 : \mathcal{K} \otimes B \to \mathcal{K} \otimes C$ be quasi-unital *-homomorphisms. Assume that $f_1 \sim f_2$, i.e. f_1 and f_2 are strongly homotopic in* $\mathrm{Hom}_q(\mathcal{K} \otimes B, \mathcal{K} \otimes C)$.
  Then

$$f_{1*} = f_{2*} : kK_h(A, B) \to kK_h(A, C)$$

and

$$f_1^* = f_2^* : kK_h(C, A) \to kK_h(B, A).$$

Proof. This is an immediate consequence of Lemma 1.3.15. $\square$

We can now very easily make kK_h functorial. If $f : B \to C$ is a *-homomorphism, we let f_q denote any quasi-unital *-homomorphism

$\mathcal{K} \otimes B \to \mathcal{K} \otimes C$ homotopic to $id_\mathcal{K} \otimes f$, cf. Theorem 1.3.16, and set

$$f^* = f_q^* : kK_h(C, A) \to kK_h(B, A),$$
$$f_* = f_{q*} : kK_h(A, B) \to kK_h(A, C).$$

Theorem 1.3.16, Lemma 4.3.8 and Lemma 4.3.7 guarentee that f^* and f_* define group homomorphisms independent of the choice of f_q. Since the composition of quasi-unital *-homomorphisms gives a quasi-unital *-homomorphism by E 1.3.2, it is clear that these definitions make kK_h into a contravariant functor in the first variable and a covariant one in the second.

We now turn to the construction of the Kasparov product in this setting. So let $(\phi_+, \phi_-) \in \mathbb{G}(B, C)$ and $(\psi_+, \psi_-) \in \mathbb{G}(A, B)$. We first introduce the following two subspaces of $M_2(\mathcal{M}(\mathcal{K} \otimes C))$:

$$S_1 = \left\{ \begin{bmatrix} \phi_+(\psi_+(a) - \psi_-(a)) & \phi_+(\psi_+(b) - \psi_-(b)) \\ \phi_+(\psi_+(c) - \psi_-(c)) & \phi_+(\psi_+(d) - \psi_-(d)) \end{bmatrix} : a, b, c, d \in \mathcal{K} \otimes A \right\}$$

and

$$S_2 = \left\{ \begin{bmatrix} \phi_+(\psi_+(a)) & 0 \\ 0 & \phi_-(\psi_+(a)) \end{bmatrix} : a \in \mathcal{K} \otimes A \right\}.$$

Definition 4.3.9. A *composition operator* for (ϕ_+, ϕ_-) and (ψ_+, ψ_-) is a partial isometry $V \in M_2(\mathcal{M}(\mathcal{K} \otimes C))$ with the following properties :

(i) $VV^* = V^*V = \begin{bmatrix} \phi_+(1) & 0 \\ 0 & \phi_-(1) \end{bmatrix}$,

(ii) $V + V^* \geq 0$,

(iii) $VZ - \begin{bmatrix} 0 & -1 \\ 1 & 0 \end{bmatrix} Z \in M_2(\mathcal{K} \otimes C), \ Z \in S_1$,

(iv) $V^*Z - \begin{bmatrix} 0 & 1 \\ -1 & 0 \end{bmatrix} Z \in M_2(\mathcal{K} \otimes C), \ Z \in S_1$,

(v) $VZ - ZV \in M_2(\mathcal{K} \otimes C), \ Z \in S_2$.

Lemma 4.3.10. *Assume that A is a separable C^*-algebra and let $(\phi_+, \phi_-) \in \mathbb{G}(B, C)$, $(\psi_+, \psi_-) \in \mathbb{G}(A, B)$. Then there exists a composition operator for (ϕ_+, ϕ_-) and (ψ_+, ψ_-).*

Proof. We apply Kasparov's technical theorem, Theorem 2.2.1. Set $E_1 = \phi_+(\mathcal{K} \otimes B)$, $\mathcal{F} = \phi_+ \circ \psi_+(\mathcal{K} \otimes A) + \mathbb{C}\phi_+(1)$. Then $\mathcal{F}$ is separable since A is and E_1 is σ-unital since B is. Let

$$X = \psi_+(\mathcal{K} \otimes A) + \psi_-(\mathcal{K} \otimes A) + \mathbb{C}1 \subseteq \mathcal{M}(\mathcal{K} \otimes B)$$

and let E_2 be the C^*-subalgebra of $\mathcal{M}(\mathcal{K} \otimes C)$ generated by the following set $\{\phi_+(x) - \phi_-(x) : x \in X\}$. Then E_2 is separable because A is, $\mathcal{F}E_1 \subseteq E_1$ and $E_1 E_2 \subseteq \mathcal{K} \otimes C$. Thus Kasparov's technical theorem provides us with $M, N \in \mathcal{M}(\mathcal{K} \otimes C)$ so that $M, N \geq 0$, $N + M = 1$, $ME_1 \subseteq \mathcal{K} \otimes C$, $NE_2 \subseteq \mathcal{K} \otimes C$ and $[N, \mathcal{F}] \subseteq \mathcal{K} \otimes C$.

Let $p : \mathcal{M}(\mathcal{K} \otimes C) \to \mathcal{M}(\mathcal{K} \otimes C)/\mathcal{K} \otimes C$ denote the quotient map and set

$$U = p \otimes id_{M_2(\mathbb{C})} \begin{bmatrix} \phi_+(1)M^{\frac{1}{2}} & -\phi_+(1)N^{\frac{1}{2}} \\ \phi_+(1)N^{\frac{1}{2}} & \phi_-(1)M^{\frac{1}{2}} \end{bmatrix} \in M_2(\mathcal{M}(\mathcal{K} \otimes C)/\mathcal{K} \otimes C)).$$

Since N commutes with $\phi_+(1) \in \mathcal{F}$ mod $\mathcal{K} \otimes C$ and $N\phi_+(1) = N\phi_-(1) \bmod \mathcal{K} \otimes C$, we see that N and M commute with $\phi_-(1)$ as well as with $\phi_+(1) \bmod \mathcal{K} \otimes C$. It follows that $U + U^* \geq 0$ and that

$$UU^* = U^*U = p \otimes id_{M_2(\mathbb{C})} \begin{bmatrix} \phi_+(1) & 0 \\ 0 & \phi_-(1) \end{bmatrix}.$$

Furthermore, the conditions on M and N imply that

$$UZ = \begin{bmatrix} 0 & -1 \\ 1 & 0 \end{bmatrix} Z \quad \text{and} \quad U^*Z = \begin{bmatrix} 0 & 1 \\ -1 & 0 \end{bmatrix} Z, \qquad Z \in p \otimes id_{M_2(\mathbb{C})}(S_1),$$

and that

$$UZ = ZU, \ Z \in p \otimes id_{M_2(\mathbb{C})}(S_2).$$

Since $U + U^* \geq 0$, we can use spectral theory to find $V \in M_2(\mathcal{M}(\mathcal{K} \otimes C))$ such that

$$V + V^* \geq 0, \ VV^* = V^*V = \begin{bmatrix} \phi_+(1) & 0 \\ 0 & \phi_-(1) \end{bmatrix} \quad \text{and} \quad p \otimes id_{M_2(\mathbb{C})}(V) = U.$$

Then V will be a composition operator for (ϕ_+, ϕ_-) and (ψ_+, ψ_-). $\square$

Lemma 4.3.11. *Let $(\phi_+, \phi_-) \in \mathbb{G}(B, C)$, $(\psi_+, \psi_-) \in \mathbb{G}(A, B)$. Assume that V is a composition operator for (ϕ_+, ϕ_-) and (ψ_+, ψ_-). Then*
(4.3.1)

$$\left(\Theta_C \circ \begin{bmatrix} \phi_+ \circ \psi_+ & 0 \\ 0 & \phi_- \circ \psi_- \end{bmatrix}, \Theta_C \circ \mathrm{Ad}\, V \circ \begin{bmatrix} \phi_+ \circ \psi_- & 0 \\ 0 & \phi_- \circ \psi_+ \end{bmatrix} \right) \in \mathbb{G}(A, C).$$

If W is another composition operator for (ϕ_+, ϕ_-) and (ψ_+, ψ_-), then
(4.3.2)

$$\left(\Theta_C \circ \begin{bmatrix} \phi_+ \circ \psi_+ & 0 \\ 0 & \phi_- \circ \psi_- \end{bmatrix}, \Theta_C \circ \mathrm{Ad}\, V \circ \begin{bmatrix} \phi_+ \circ \psi_- & 0 \\ 0 & \phi_- \circ \psi_+ \end{bmatrix} \right)$$

is homotopic to

(4.3.3)
$$\left(\Theta_C \circ \begin{bmatrix} \phi_+ \circ \psi_+ & 0 \\ 0 & \phi_- \circ \psi_- \end{bmatrix}, \ \Theta_C \circ \mathrm{Ad}\,W \circ \begin{bmatrix} \phi_+ \circ \psi_- & 0 \\ 0 & \phi_- \circ \psi_+ \end{bmatrix} \right).$$

Proof. First we remark that $\mathrm{Ad}\,V \circ \begin{bmatrix} \phi_+ \circ \psi_+ & 0 \\ 0 & \phi_- \circ \psi_- \end{bmatrix}$ is a $*$-homomorphism because

$$V^* V = \begin{bmatrix} \phi_+(1) & 0 \\ 0 & \phi_-(1) \end{bmatrix}.$$

From the properties of V one gets that the following equalities hold mod $M_2(\mathcal{K} \otimes C)$ for all $a \in \mathcal{K} \otimes A$:

$$\mathrm{Ad}\,V \begin{bmatrix} \phi_+ \circ \psi_-(a) & 0 \\ 0 & \phi_- \circ \psi_+(a) \end{bmatrix} =$$

$$\mathrm{Ad}\,V \begin{bmatrix} \phi_+ \circ \psi_-(a) - \phi_+ \circ \psi_+(a) & 0 \\ 0 & 0 \end{bmatrix} + \mathrm{Ad}\,V \begin{bmatrix} \phi_+ \circ \psi_+(a) & 0 \\ 0 & \phi_- \circ \psi_+(a) \end{bmatrix} =$$

$$\mathrm{Ad} \begin{bmatrix} 0 & -1 \\ 1 & 0 \end{bmatrix} \begin{bmatrix} \phi_+ \circ \psi_-(a) - \phi_+ \circ \psi_+(a) & 0 \\ 0 & 0 \end{bmatrix} + \begin{bmatrix} \phi_+ \circ \psi_+(a) & 0 \\ 0 & \phi_- \circ \psi_+(a) \end{bmatrix} =$$

$$\begin{bmatrix} \phi_+ \circ \psi_+(a) & 0 \\ 0 & \phi_- \circ \psi_+(a) + \phi_+ \circ \psi_-(a) - \phi_+ \circ \psi_+(a) \end{bmatrix}.$$

Since $\phi_+(\psi_+(a) - \psi_-(a)) = \phi_-(\psi_+(a) - \psi_-(a)) \mod \mathcal{K} \otimes C$ it follows that the last expression equals

$$\begin{bmatrix} \phi_+ \circ \psi_+(a) & 0 \\ 0 & \phi_- \circ \psi_-(a) \end{bmatrix}$$

mod $M_2(\mathcal{K} \otimes C)$. This proves (4.3.1), without using that V has property (ii) of Definition 4.3.9.

If W is another composition operator for (ϕ_+, ϕ_-) and (ψ_+, ψ_-) then V^t and W^t are defined for all $t \in [0,1]$ because $V + V^* \geq 0$ and $W + W^* \geq 0$. When we take the exponentials relative to the C^*-algebra

$$\begin{bmatrix} \phi_+(1) & 0 \\ 0 & \phi_-(1) \end{bmatrix} M_2(\mathcal{M}(\mathcal{K} \otimes C)) \begin{bmatrix} \phi_+(1) & 0 \\ 0 & \phi_-(1) \end{bmatrix}$$

we obtain that $V^t W^{(1-t)}$ is a composition operator for (ϕ_+, ϕ_-) and (ψ_+, ψ_-) for all $t \in [0,1]$, except that the positivity condition (ii) of Definition 4.3.9 may fail. Nonetheless we can substitute V in the expresssion of

(4.3.1) by $V^t W^{(1-t)}$ and this will give us a $kK_h(A, C)$-cycle for all $t \in [0,1]$, by the first part of this proof. It is clear that we obtain a homotopy from (4.3.3) to (4.3.2). □

To show that the existence of composition operators leads to a map $kK_h(A, B) \times kK_h(B, C) \to kK_h(A, C)$ we need the following lemmas.

Lemma 4.3.12. *Assume that A is separable, $(\phi_+, \phi_-) \in \mathbb{G}(B, C)$ and let $(\lambda_+^t, \lambda_-^t) \in \mathbb{G}(A, B)$, $t \in [0,1]$, be a homotopy. Then there is a strictly continuous path V_t, $t \in [0,1]$, in $M_2(\mathcal{M}(\mathcal{K} \otimes C)) = \mathcal{M}(M_2(\mathcal{K} \otimes C))$ such that for every $t \in [0,1]$, V_t is a composition oprator for (ϕ_+, ϕ_-) and $(\lambda_+^t, \lambda_-^t)$.*

Proof. For every $m \in \mathcal{M}(\mathcal{K} \otimes B)$ there is a unique element $\bar{\phi}_+(m) \in \mathcal{M}(I(\mathcal{K} \otimes C))$ so that $\pi_t(\bar{\phi}_+(m)) = \phi_+(m)$, $t \in [0,1]$. In the same way we define $\psi_{+,+}(m) \in \mathcal{M}(I(\mathcal{K} \otimes C))$ by $\pi_t(\psi_{+,+}(m)) = \phi_+ \circ \lambda_+^t(m)$, $t \in [0,1]$. Similarly we define $\psi_{+,-}(m)$, $\psi_{-,+}(m)$ and $\psi_{-,-}(m)$ by $\pi_t(\psi_{+,-}(m)) = \phi_+ \circ \lambda_-^t(m)$, $\pi_t(\psi_{-,+}(m)) = \phi_- \circ \lambda_+^t(m)$ and $\pi_t(\psi_{-,-}(m)) = \phi_- \circ \lambda_-^t(m)$, $t \in [0,1]$, respectively. Set $E_1 = \bar{\phi}_+(\mathcal{K} \otimes B)$, $\mathcal{F} = \psi_{+,+}(\mathcal{K} \otimes A) + \mathbb{C}\bar{\phi}_+(1)$ and let E_2 be the C^*-subalgebra of $\mathcal{M}(I(\mathcal{K} \otimes C))$ generated by $\psi_{+,+}(a) - \psi_{-,+}(a)$, $\psi_{+,-}(a) - \psi_{-,-}(a)$, $a \in \mathcal{K} \otimes A$, and $\bar{\phi}_+(1) - \bar{\phi}_-(1)$ (where $\pi_t(\bar{\phi}_-(m)) = \phi_-(m)$, $t \in [0,1]$, $m \in \mathcal{M}(\mathcal{K} \otimes B)$, of course). Then $\mathcal{F}E_1 \subseteq E_1$, $E_1 E_2 \subseteq I(\mathcal{K} \otimes C)$ and Kasparov's technical theorem, Theorem 2.2.1, applies. Proceeding as in the proof of Lemma 4.3.10, we get $V \in M_2(\mathcal{M}(I(\mathcal{K} \otimes C)))$ such that:

(i) $VV^* = V^*V = \begin{bmatrix} \bar{\phi}_+(1) & 0 \\ 0 & \bar{\phi}_-(1) \end{bmatrix}$,

(ii) $V + V^* \geq 0$,

(iii) $VZ - \begin{bmatrix} 0 & -1 \\ 1 & 0 \end{bmatrix} Z$, $V^*Z - \begin{bmatrix} 0 & 1 \\ -1 & 0 \end{bmatrix} Z \in M_2(I(\mathcal{K} \otimes C))$, for all Z of the form
$$\begin{bmatrix} \psi_{+,+}(a) - \psi_{+,-}(a) & \psi_{+,+}(b) - \psi_{+,-}(b) \\ \psi_{+,+}(c) - \psi_{+,-}(c) & \psi_{+,+}(d) - \psi_{+,-}(d) \end{bmatrix}, \quad a, b, c, d \in \mathcal{K} \otimes A, \text{ and}$$

(iv) $VZ - ZV \in M_2(I(\mathcal{K} \otimes C))$, for Z of the form
$$Z = \begin{bmatrix} \psi_{+,+}(a) & 0 \\ 0 & \psi_{-,+}(a) \end{bmatrix}, \quad a \in \mathcal{K} \otimes A.$$

Then $V_t = \pi_t \otimes id_{M_2(\mathbb{C})}(V)$, $t \in [0,1]$, will have the properties required in the lemma. □

Lemma 4.3.13. *Assume that A is separable. Let $(\lambda_+^t, \lambda_-^t) \in \mathbb{G}(B, C)$ $t \in [0, 1]$, be a homotopy and $(\psi_+, \psi_-) \in \mathbb{G}(A, B)$. Then there is a strictly continuous path V_t, $t \in [0, 1]$, in $M_2(\mathcal{M}(\mathcal{K} \otimes C)) = \mathcal{M}(M_2(\mathcal{K} \otimes C))$ such that for each t, V_t is a composition operator for $(\lambda_+^t, \lambda_-^t)$ and (ψ_+, ψ_-).*

Proof. Define $\lambda_\pm : \mathcal{M}(\mathcal{K} \otimes B) \to \mathcal{M}(I(\mathcal{K} \otimes C))$ to be the *-homomorphisms satisfying $\pi_t(\lambda_\pm(m)) = \lambda_\pm^t(m)$, $t \in [0, 1]$, $m \in \mathcal{M}(\mathcal{K} \otimes B)$. Upon identifying $I(\mathcal{K} \otimes C)$ with $\mathcal{K} \otimes IC$ in the natural way we obtain $(\lambda_+, \lambda_-) \in \mathbb{G}(B, IC)$. Let V be a composition operator for (λ_+, λ_-) and (ψ_+, ψ_-), cf. Lemma 4.3.10. Then $V_t = \pi_t \otimes id_{M_2(\mathbb{C})}(V)$, $t \in [0, 1]$, will be a path with the right properties. $\square$

It follows from the last two lemmas that the element of $kK_h(A, C)$ represented by (4.3.1) only depends on the images of (ϕ_+, ϕ_-) in $kK_h(B, C)$ and (ψ_+, ψ_-) in $kK_h(A, B)$. To state the properties of the product obtained this way, we introduce the notation $1_A \in kK_h(A, A)$ for the element represented by $(id_{\mathcal{M}(\mathcal{K} \otimes A)}, 0) \in \mathbb{G}(A, A)$.

Theorem 4.3.14. *Assume that A and A_1 are separable, B, C and D are σ-unital C^*-algebras.*

There exists a bilinear pairing

$$\cdot : kK_h(A, B) \times kK_h(B, C) \to kK_h(A, C)$$

with the following properties:

(i) *If $f : A_1 \to A$ is a *-homomorphism , then*

$$f^*(x \cdot y) = f^*(x) \cdot y, \quad x \in kK_h(A, B), \ y \in kK_h(B, C).$$

(ii) *If $g : B \to C$ is a *-homomorphism, then*

$$g_*(x) \cdot z = x \cdot g^*(z), \quad x \in kK_h(A, B), \ z \in kK_h(C, D).$$

(iii) *If $h : C \to D$ is a *-homomorphism, then*

$$h_*(x \cdot y) = x \cdot h_*(y), \quad x \in kK_h(A, B), \ y \in kK_h(B, C).$$

(iv) $1_A \cdot x = x \cdot 1_B = x$, $x \in kK_h(A, B)$.

Proof. For $(\psi_+, \psi_-) \in \mathbb{G}(A, B)$, $(\phi_+, \phi_-) \in \mathbb{G}(B, C)$, the product $[\psi_+, \psi_-] \cdot [\phi_+, \phi_-]$ is defined as the element of $kK_h(A, C)$ represented by (4.3.1) for some composition operator V. To simplify notation, we write $(\phi_+, \phi_-)V(\psi_+, \psi_-)$ for the element of $\mathbb{G}(A, C)$ given by (4.3.1) and $[(\phi_+, \phi_-)V(\psi_+, \psi_-)]$ for the corresponding element of $kK_h(A, C)$.

We first prove additivity of $\cdot$ in the second variable. Let $(\psi_+, \psi_-) \in \mathbb{G}(A, B)$, (ϕ_+^1, ϕ_-^1), $(\phi_+^2, \phi_-^2) \in \mathbb{G}(B, C)$. Let W_i be a composition operator for (ϕ_+^i, ϕ_-^i) and (ψ_+, ψ_-), $i = 1, 2$. When V_1, V_2 are the isometries in $\mathcal{M}(\mathcal{K} \otimes C)$ used to define Θ_C, it is easy to see that

$$
W = \begin{bmatrix} V_1 & 0 \\ 0 & V_1 \end{bmatrix} W_1 \begin{bmatrix} V_1 & 0 \\ 0 & V_1 \end{bmatrix}^* + \begin{bmatrix} V_2 & 0 \\ 0 & V_2 \end{bmatrix} W_2 \begin{bmatrix} V_2 & 0 \\ 0 & V_2 \end{bmatrix}^*
$$

is a composition operator for $\left(\Theta_C \circ \begin{bmatrix} \phi_+^1 & 0 \\ 0 & \phi_+^2 \end{bmatrix}, \Theta_C \circ \begin{bmatrix} \phi_-^1 & 0 \\ 0 & \phi_-^2 \end{bmatrix} \right)$ and (ψ_+, ψ_-). Set

$$
S_i = \begin{bmatrix} V_i & 0 \\ 0 & V_i \end{bmatrix} W_i \begin{bmatrix} V_i & 0 \\ 0 & V_i \end{bmatrix}^*, \quad i = 1, 2.
$$

Then Lemma 4.3.5 yields

$$
\left[\left(\Theta_C \circ \begin{bmatrix} \phi_+^1 & 0 \\ 0 & \phi_+^2 \end{bmatrix}, \Theta_C \circ \begin{bmatrix} \phi_-^1 & 0 \\ 0 & \phi_-^2 \end{bmatrix} \right) W(\psi_+, \psi_-) \right] =
$$

$$
[(\mathrm{Ad}\, V_1 \circ \phi_+^1,\ \mathrm{Ad}\, V_1 \circ \phi_-^1) S_1(\psi_+, \psi_-)] + [(\mathrm{Ad}\, V_2 \circ \phi_+^2,\ \mathrm{Ad}\, V_2 \circ \phi_-^2) S_2(\psi_+, \psi_-)].
$$

By applying Lemma 4.3.6 we get the desired conclusion; i.e.

$$
[\psi_+, \psi_-] \cdot ([\phi_+^1, \phi_-^1] + [\phi_+^2, \phi_-^2]) = [\psi_+, \psi_-] \cdot [\phi_+^1, \phi_-^1] + [\psi_+, \psi_-] \cdot [\phi_+^2, \phi_-^2].
$$

To prove additivity in the first variable, let (ψ_+^1, ψ_-^1), $(\psi_+^2, \psi_-^2) \in \mathbb{G}(A, B)$ and $(\phi_+, \phi_-) \in \mathbb{G}(B, C)$. Let W_i be a composition operator for (ϕ_+, ϕ_-) and (ψ_+^i, ψ_-^i), $i = 1, 2$, and now let V_1, V_2 be the isometries in $\mathcal{M}(\mathcal{K} \otimes B)$ used to define Θ_B. Set

$$
S_i = \begin{bmatrix} \phi_+(V_i) & 0 \\ 0 & \phi_-(V_i) \end{bmatrix} W_i \begin{bmatrix} \phi_+(V_i) & 0 \\ 0 & \phi_-(V_i) \end{bmatrix}^*, \quad i = 1, 2.
$$

Then $W = S_1 + S_2$ is a composition operator for (ϕ_+, ϕ_-) and

$$
\left(\Theta_B \circ \begin{bmatrix} \psi_+^1 & 0 \\ 0 & \psi_+^2 \end{bmatrix}, \Theta_B \circ \begin{bmatrix} \psi_-^1 & 0 \\ 0 & \psi_-^2 \end{bmatrix} \right).
$$

Using Lemma 4.3.5 again we find that

$$
\left[(\phi_+,\phi_-)W\left(\Theta_B\circ\begin{bmatrix}\psi_+^1 & 0\\ 0 & \psi_+^2\end{bmatrix},\ \Theta_B\circ\begin{bmatrix}\psi_-^1 & 0\\ 0 & \psi_-^2\end{bmatrix}\right)\right]=
$$

$$
[(\mathrm{Ad}\,\phi_+(V_1)\circ\phi_+,\ \mathrm{Ad}\,\phi_-(V_1)\circ\phi_-)S_1(\psi_+^1,\psi_-^1)]
$$

$$
+\,[(\mathrm{Ad}\,\phi_+(V_2)\circ\phi_+,\ \mathrm{Ad}\,\phi_-(V_2)\circ\phi_-)S_2(\psi_+^2,\psi_-^2)]\,.
$$

Since, by Lemma 4.3.6, these terms equal $[(\phi_+,\phi_-)W_i(\psi_+^i,\psi_-^i)]$, $i=1,2$, respectively, we get

$$
([\psi_+^1,\psi_-^1]+[\psi_+^2,\psi_-^2])\cdot[\phi_+,\phi_-]=[\psi_+^1,\psi_-^1]\cdot[\phi_+,\phi_-]+[\psi_+^2,\psi_-^2]\cdot[\phi_+,\phi_-],
$$

proving additivity in the first variable.

Let us now prove relation (iii) of the theorem. By definition of the functoriality we can assume that $h:\mathcal{K}\otimes C\to\mathcal{K}\otimes D$ is quasi-unital and then prove that $h_*(x\cdot y)=x\cdot h_*(y)$. By Lemma 1.3.19 we can also assume that $1-\underline{h}(1)=WW^*$ for some isometry $W\in\mathcal{M}(\mathcal{K}\otimes D)$. Let $U_1,U_2\in\mathcal{M}(\mathcal{K}\otimes D)$ be two isometries with $U_1U_1^*+U_2U_2^*=1$ (so that $U_1^*U_2=U_2^*U_1=0$ automatically) and let $V_1,V_2\in\mathcal{M}(\mathcal{K}\otimes C)$ be isometries satisfying the same equality. Set $S_i=\underline{h}(V_i)+WU_iW^*$, $i=1,2$. Then $S_1,S_2\in\mathcal{M}(\mathcal{K}\otimes D)$ are isometries such that $S_1S_1^*+S_2S_2^*=1$. We use these isometries to define Θ_D. Let $V\in M_2(\mathcal{M}(\mathcal{K}\otimes C))$ be a composition operator for (ϕ_+,ϕ_-) and (ψ_+,ψ_-), where $x=[\psi_+,\psi_-]$ and $y=[\phi_+,\phi_-]$. Then $\underline{h}\otimes id_{M_2(\mathbb{C})}(V)=\tilde{V}$ is a composition operator for $(\underline{h}\circ\phi_+,\underline{h}\circ\phi_-)$ and (ψ_+,ψ_-). Thus

$$
x\cdot h_*(y)=
$$

$$
\left[\Theta_D\circ\begin{bmatrix}\underline{h}\circ\phi_+\circ\psi_+ & 0\\ 0 & \underline{h}\circ\phi_-\circ\psi_-\end{bmatrix},\ \Theta_D\circ\mathrm{Ad}\,\tilde{V}\circ\begin{bmatrix}\underline{h}\circ\phi_+\circ\psi_- & 0\\ 0 & \underline{h}\circ\phi_-\circ\psi_+\end{bmatrix}\right].
$$

By our careful choice of Θ_D we have $\Theta_D\circ(\underline{h}\otimes id_{M_2(\mathbb{C})})=\underline{h}\circ\Theta_C$ when Θ_C is defined from the V_i's. Therefore the above expression equals $h_*(x\cdot y)$.

To prove relation (ii) we can assume as above that $g:\mathcal{K}\otimes B\to\mathcal{K}\otimes C$ is quasi-unital and then prove that $g_*(x)\cdot z=x\cdot g^*(z)$. Choose $(\psi_+,\psi_-)\in\mathcal{G}(A,B)$ and $(\lambda_+,\lambda_-)\in\mathcal{G}(C,D)$ such that $x=[\psi_+,\psi_-]$ and $y=[\lambda_+,\lambda_-]$. Let S be a composition operator for $(\lambda_+\circ\underline{g},\ \lambda_-\circ\underline{g})$ and (ψ_+,ψ_-). Then

$$
W=S+\begin{bmatrix}\lambda_+(1-\underline{g}(1)) & 0\\ 0 & \lambda_-(1-\underline{g}(1))\end{bmatrix}
$$

is a composition operator for (λ_+, λ_-) and $(\underline{g} \circ \psi_+, \underline{g} \circ \psi_-)$. Thus

$$g_*(x) \cdot z =$$

$$\left[\Theta_D \circ \begin{bmatrix} \lambda_+ \circ \underline{g} \circ \psi_+ & 0 \\ 0 & \lambda_- \circ \underline{g} \circ \psi_- \end{bmatrix}, \Theta_D \circ \operatorname{Ad} W \circ \begin{bmatrix} \lambda_+ \circ \underline{g} \circ \psi_- & 0 \\ 0 & \lambda_- \circ \underline{g} \circ \psi_+ \end{bmatrix} \right] =$$

$$\left[\Theta_D \circ \begin{bmatrix} \lambda_+ \circ \underline{g} \circ \psi_+ & 0 \\ 0 & \lambda_- \circ \underline{g} \circ \psi_- \end{bmatrix}, \Theta_D \circ \operatorname{Ad} S \circ \begin{bmatrix} \lambda_+ \circ \underline{g} \circ \psi_- & 0 \\ 0 & \lambda_- \circ \underline{g} \circ \psi_+ \end{bmatrix} \right] =$$

$$x \cdot g^*(z).$$

To prove relation (iv) let $(\psi_+, \psi_-) \in G(A, B)$. Then $\begin{bmatrix} 1 & 0 \\ 0 & 0 \end{bmatrix}$ is a composition operator for $(id_{\mathcal{M}(\mathcal{K} \otimes B)}, 0)$ and (ψ_+, ψ_-). The equality $[\psi_+, \psi_-] \cdot 1_B = [\psi_+, \psi_-]$ follows immediately from this. Let V be a composition operator for (ψ_+, ψ_-) and $(id_{\mathcal{M}(\mathcal{K} \otimes A)}, 0)$. Then V is unitary in

$$\begin{bmatrix} \psi_+(1) & 0 \\ 0 & \psi_-(1) \end{bmatrix} M_2(\mathcal{M}(\mathcal{K} \otimes B)) \begin{bmatrix} \psi_+(1) & 0 \\ 0 & \psi_-(1) \end{bmatrix}$$

and since $V + V^* \geq 0$ we can define V^t, $t \in [0, 1]$, where the exponential is taken in the algebra above. Set $W_t = \begin{bmatrix} 0 & -1 \\ 1 & 0 \end{bmatrix}^{1-t} V^t$, $t \in [0, 1]$. Then $W_t^* W_t = \begin{bmatrix} \psi_+(1) & 0 \\ 0 & \psi_-(1) \end{bmatrix}$ and W_t satisfies all conditions on a composition operator for (ψ_+, ψ_-) and $(id_{\mathcal{M}(\mathcal{K} \otimes A)}, 0)$ except possibly (ii) and part of (i) in Definition 4.3.9. Set

$$\lambda_+^t = \Theta_B \circ \begin{bmatrix} \psi_+ & 0 \\ 0 & 0 \end{bmatrix}$$

and

$$\lambda_-^t = \Theta_B \circ \operatorname{Ad} W_t \circ \begin{bmatrix} 0 & 0 \\ 0 & \psi_- \end{bmatrix}, \quad t \in [0, 1].$$

Then $(\lambda_+^t, \lambda_-^t)$, $t \in [0, 1]$, is a homotopy which justifies the second equality in the following calculation:

$$1_A \cdot [\psi_+, \psi_-] = \left[\Theta_B \circ \begin{bmatrix} \psi_+ & 0 \\ 0 & 0 \end{bmatrix}, \Theta_B \circ \operatorname{Ad} V \circ \begin{bmatrix} 0 & 0 \\ 0 & \psi_- \end{bmatrix} \right]$$

$$= \left[\Theta_B \circ \begin{bmatrix} \psi_+ & 0 \\ 0 & 0 \end{bmatrix}, \Theta_B \circ \begin{bmatrix} \psi_- & 0 \\ 0 & 0 \end{bmatrix} \right]$$

$$= [\psi_+, \psi_-].$$

Since relation (i) is trivial, the proof is complete.   □

By use of Theorem 4.3.14 we can now show that $kK_h(A, B) \simeq KK_h(A, B)$ when A is separable.

It is obvious from the definitions that there is a group homomorphism $\Lambda : kK_h(A, B) \to KK_h(A, B)$ given

$$\Lambda[\phi_+, \phi_-] = [\phi_+ \circ e_A, \phi_- \circ e_A], \quad (\phi_+, \phi_-) \in \mathbb{G}(A, B).$$

We will show that Λ is an isomorphism when either A is unital or A is separable. For the last case we need the fact that Λ is a natural transformation of functors.

Lemma 4.3.15. *Let $f : D \to A$ be a *-homomorphism. Then*

$$\Lambda \circ f^* = f^* \circ \Lambda : kK_h(A, B) \to KK_h(D, B) \quad and$$

$$\Lambda \circ f_* = f_* \circ \Lambda : kK_h(B, D) \to KK_h(B, A).$$

Proof. Let $(\phi_+, \phi_-) \in \mathbb{G}(A, B)$. Then $f^*[\phi_+, \phi_-] = [\phi_+ \circ \underline{f_q}, \phi_- \circ \underline{f_q}]$ where $f_q : \mathcal{K} \otimes D \to \mathcal{K} \otimes A$ is a quasi-unital *-homomorphism homotopic to $id_{\mathcal{K}} \otimes f$. Hence $\Lambda \circ f^*[\phi_+, \phi_-] = [\phi_+ \circ f_q \circ e_D, \phi_- \circ f_q \circ e_D]$. Since $f_q \circ e_D$ is homotopic to $e_A \circ f$ it follows that $[\phi_+ \circ f_q \circ e_D, \phi_- \circ f_q \circ e_D] = [\phi_+ \circ e_A \circ f, \phi_- \circ e_A \circ f] = f^* \circ \Lambda[\phi_+, \phi_-]$.

That $\Lambda \circ f_* = f_* \circ \Lambda$ follows immediately from the definitions.   □

Note that $e_{\mathcal{K} \otimes B}$ is quasi-unital with relative unit $e \otimes 1_{\mathcal{M}(\mathcal{K} \otimes B)} \in \mathcal{M}(\mathcal{K} \otimes \mathcal{K} \otimes B)$ given by $e \otimes 1_{\mathcal{M}(\mathcal{K} \otimes B)}(k_1 \otimes k_2 \otimes b) = ek_1 \otimes k_2 \otimes b$, $k_1, k_2 \in \mathcal{K}$, $b \in B$. The unique strictly continuous extension $\underline{e}_{\mathcal{K} \otimes B}$ is given by $\underline{e}_{\mathcal{K} \otimes B}(m) = e \otimes m$, $m \in \mathcal{M}(\mathcal{K} \otimes B)$. The unique strictly continuous extension $\underline{\lambda}_B$ of the *-isomorphism $\lambda_B : \mathcal{K} \otimes B \to \mathcal{K} \otimes \mathcal{K} \otimes B$ given by Lemma 4.1.13 has the form $V^* \underline{e}_{\mathcal{K} \otimes B}(\cdot)V$ for some isometry $V \in \mathcal{M}(\mathcal{K} \otimes \mathcal{K} \otimes B)$ with $VV^* = e \otimes 1_{\mathcal{M}(\mathcal{K} \otimes B)}$. Connecting V to 1 through a strictly continuous path of isometries, we get the following

Lemma 4.3.16. *There is a path μ_t, $t \in [0, 1]$, in $\mathrm{Hom}\,(\mathcal{M}(\mathcal{K} \otimes B))$ such that*

(i) *each* $\mu_t : M(\mathcal{K} \otimes B) \to M(\mathcal{K} \otimes B)$ *is strictly continuous and maps* $\mathcal{K} \otimes B$ *into* $\mathcal{K} \otimes B$,

(ii) $t \to \mu_t(m)$ *is strictly continuous for all* $m \in M(\mathcal{K} \otimes B)$,

(iii) $t \to \mu_t(x)$ *is norm continuous from* $[0, 1]$ *into* $\mathcal{K} \otimes B$ *for all* $x \in \mathcal{K} \otimes B$, *and*

(iv) $\mu_0 = \underline{\lambda}_B^{-1} \circ \underline{e}_{\mathcal{K} \otimes B}, \ \mu_1 = id_{M(\mathcal{K} \otimes B)}.$ $\qquad\qquad\square$

Lemma 4.3.17. *Assume that* A *is a unital* C^*-*algebra.* *Then* $\Lambda : kK_h(A, B) \to KK_h(A, B)$ *is an isomorphism.*

Proof. The proof is modelled on the argument needed in E 4.1.4. We shall construct an inverse for Λ. Given a *-homomorphism $\phi : A \to M(\mathcal{K} \otimes B)$ we define $\tilde{\phi} : \mathcal{K} \otimes A \to M(\mathcal{K} \otimes \mathcal{K} \otimes B)$ by $\tilde{\phi}(k \otimes a)(k_1 \otimes k_2 \otimes a) = kk_1 \otimes \phi(a)(k_2 \otimes b)$, $k, k_1, k_2 \in \mathcal{K}$, $b \in B$. We have

$$\overline{\tilde{\phi}(\mathcal{K} \otimes A)(\mathcal{K} \otimes \mathcal{K} \otimes B)} = (1 \otimes \phi(1))\mathcal{K} \otimes \mathcal{K} \otimes B,$$

so by Corollary 1.1.15 there is a unique strictly continuous extension $\tilde{\tilde{\phi}} : M(\mathcal{K} \otimes A) \to M(\mathcal{K} \otimes \mathcal{K} \otimes B)$. We can define a map $\Phi : KK_h(A, B) \to kK_h(A, B)$ by

$$\Phi[\phi_+, \phi_-] = [\underline{\lambda}_B^{-1} \circ \tilde{\tilde{\phi}}_+, \underline{\lambda}_B^{-1} \circ \tilde{\tilde{\phi}}_-], \quad (\phi_+, \phi_-) \in \mathbb{F}(A, B).$$

To see that Φ is well-defined, assume that $(\phi_+, \phi_-) \sim (\psi_+, \psi_-)$ in $\mathbb{F}(A, B)$. Then there is $(\lambda_+, \lambda_-) \in \mathbb{F}(A, IB)$ such that $\underline{\pi}_0 \circ \lambda_\pm = \phi_\pm$ and $\underline{\pi}_1 \circ \lambda_\pm = \psi_\pm$, cf. E 4.1.5. But then $\underline{\pi}_0 \circ \tilde{\lambda}_\pm = \tilde{\tilde{\phi}}_\pm$ and $\underline{\pi}_1 \circ \tilde{\lambda}_\pm = \tilde{\tilde{\psi}}_\pm$ where π_t is now the evaluation map $I(\mathcal{K} \otimes \mathcal{K} \otimes B) \to \mathcal{K} \otimes \mathcal{K} \otimes B$. It follows readily from this that $(\underline{\lambda}_B^{-1} \circ \tilde{\tilde{\phi}}_+, \underline{\lambda}_B^{-1} \circ \tilde{\tilde{\phi}}_-) \sim (\underline{\lambda}_B^{-1} \circ \tilde{\tilde{\psi}}_+, \underline{\lambda}_B^{-1} \circ \tilde{\tilde{\psi}}_-)$ in $\mathbb{G}(A, B)$.

The proof is now completed by showing that $\Lambda\Phi = id$ on $KK_h(A, B)$ and $\Phi\Lambda = id$ on $kK_h(A, B)$. Let first $(\phi_+, \phi_-) \in \mathbb{F}(A, B)$. It is easy to see that $\underline{\lambda}_B^{-1} \circ \tilde{\tilde{\phi}}_\pm \circ e_A = \underline{\lambda}_B^{-1} \circ \underline{e}_{\mathcal{K} \otimes B} \circ \phi_\pm$. But then $\Lambda\Phi[\phi_+, \phi_-] = [\phi_+, \phi_-]$ by Lemma 4.3.16.

Now let $(\phi_+, \phi_-) \in \mathbb{G}(A, B)$. Let $\sigma_1, \sigma_2 : \mathcal{K} \otimes A \to \mathcal{K} \otimes \mathcal{K} \otimes A$ be the two *-homomorphisms given on simple tensors by $\sigma_1(k \otimes a) = k \otimes e \otimes a$ and $\sigma_2(k \otimes a) = e \otimes k \otimes a$, $k \in \mathcal{K}$, $a \in A$, respectively. (Note that $\sigma_2 = e_{\mathcal{K} \otimes A}$). They are both quasi-unital so we can consider their unique strictly continuous extensions $\underline{\sigma}_1, \underline{\sigma}_2 : M(\mathcal{K} \otimes A) \to M(\mathcal{K} \otimes \mathcal{K} \otimes A)$. It is easy to see that $\widetilde{(\phi_\pm \circ e_A)} = (id_{\mathcal{K}} \otimes \phi_\pm) \circ \sigma_1$ on $\mathcal{K} \otimes A$. Note that $id_{\mathcal{K}} \otimes (\phi_\pm|_{\mathcal{K} \otimes A})$ satisfies the assumption of Corollary 1.1.15 with $p = 1 \otimes \phi_\pm(1) \in M(\mathcal{K} \otimes \mathcal{K} \otimes B)$.

Let $\psi_\pm$ denote the strictly continuous extensions of $id_\mathcal{K} \otimes (\phi_\pm|_{\mathcal{K} \otimes A})$. By strict continuity and the density of $\mathcal{K} \otimes A$ in $\mathcal{M}(\mathcal{K} \otimes A)$ we conclude that $\widetilde{(\phi_\pm \circ e_A)} = \psi_\pm \circ \underline{\sigma}_1$. Since $\underline{\sigma}_1 = \operatorname{Ad} U \circ \underline{\sigma}_2$ for some unitary $U \in \mathcal{M}(\mathcal{K} \overline{\otimes \mathcal{K} \otimes} A)$ and the unitary group of $\mathcal{M}(\mathcal{K} \otimes \mathcal{K} \otimes A)$ is connected in the strict topology by Lemma 1.3.7, using that we get $\Phi\Lambda[\phi_+, \phi_-] = [\lambda_B^{-1} \circ \psi_+ \circ \underline{\sigma}_2, \lambda_B^{-1} \circ \psi_- \circ \underline{\sigma}_2]$. Since $(id_\mathcal{K} \otimes \phi_\pm) \circ \sigma_2 = \underline{e}_{\mathcal{K} \otimes B} \circ \phi_\pm$ on $\mathcal{K} \otimes A$ we conclude by strict continuity that $\psi_\pm \circ \underline{\sigma}_2 = \underline{e}_{\mathcal{K} \otimes B} \circ \phi_\pm$. But then $\Phi\Lambda[\phi_+, \phi_-] = [\phi_+, \phi_-]$ by Lemma 4.3.16. $\qquad\square$

Theorem 4.3.18. *Assume that A is a separable C^*-algebra. Then $\Lambda : kK_h(A, B) \to KK_h(A, B)$ is an isomorphism.*

Proof. By Lemma 4.3.17 be can assume that A is non-unital. By adjoining a unit we get a short exact sequence

$$0 \longrightarrow A \xrightarrow{\ j\ } A^+ \xrightarrow{\ p\ } \mathbb{C} \longrightarrow 0$$

Note that there is a unital section $s : \mathbb{C} \to A^+$ for p. By Lemma 4.3.15 this gives us a commutative diagram

$$
\begin{array}{ccccc}
kK_h(\mathbb{C}, B) & \xrightarrow{p^*} & kK_h(A^+, B) & \xrightarrow{j^*} & kK_h(A, B) \\
{\scriptstyle\Lambda}\downarrow & & {\scriptstyle\Lambda}\downarrow & & {\scriptstyle\Lambda}\downarrow \\
KK_h(\mathbb{C}, B) & \xrightarrow[p^*]{} & KK_h(A^+, B) & \xrightarrow[j^*]{} & KK_h(A, B).
\end{array}
$$

The two left vertical arrows are isomorphisms by Lemma 4.3.17. Since $KK_h(\cdot, B)$ is split exact by Proposition 4.2.6, we know from Proposition C.4 that the lower sequence is exact and that p^* is injective and j^* is surjective in this sequence. To conclude that the right vertical arrow is an isomorphism, it therefore suffices to show that $j^* : kK_h(A^+, B) \to kK_h(A, B)$ is surjective.

Let $x \in kK_h(A, B)$ and let $r_A : \mathcal{M}(\mathcal{K} \otimes A^+) \to \mathcal{M}(\mathcal{K} \otimes A)$ be the $*$-homomorphism obtained by restricting multipliers to $\mathcal{K} \otimes A$, cf. E 1.1.9 and Section 4.2. r_A is obviously strictly continuous. Then it follows easily that $(r_A, r_A \circ id_\mathcal{K} \otimes s \circ id_\mathcal{K} \otimes p) \in \mathbb{G}(A^+, A)$. The corresponding element of $kK_h(A^+, A)$ is denoted by ϵ. Then $\epsilon \cdot x \in kK_h(A^+, B)$ and $j^*(\epsilon \cdot x) = j^*(\epsilon) \cdot x$ by Theorem 4.3.14. So by the same theorem it suffices to show that $j^*(\epsilon) = 1_A$. To see this, note that Lemma 4.2.2 provides us with an isometry $v \in \mathcal{M}(\mathcal{K} \otimes A)$ such that $v^* r_A(\mathcal{K} \otimes A^+) = \mathcal{K} \otimes A$. It follows that $v(\mathcal{K} \otimes A) = vv^* r_A(\mathcal{K} \otimes A^+)$. Since r_A is injective (A is essential in A^+)

and $vv^* r_A(\mathcal{K} \otimes A^+) \subseteq r_A(\mathcal{K} \otimes A^+)$, there is a projection $p \in \mathcal{M}(\mathcal{K} \otimes A^+)$ such that $r_A(p) = vv^*$. It follows that $(id_\mathcal{K} \otimes j) \circ \operatorname{Ad} v : \mathcal{K} \otimes A \to \mathcal{K} \otimes A^+$ is quasi-unital with relative unit p. Thus

$$j^*(\epsilon) = [r_A \circ \underline{(id_\mathcal{K} \otimes j)} \circ \operatorname{Ad} v, r_A \circ \underline{(id_\mathcal{K} \otimes s)} \circ \underline{(id_\mathcal{K} \otimes p)} \circ \underline{(id_\mathcal{K} \otimes j)} \circ \operatorname{Ad} v].$$

By checking on $\mathcal{K} \otimes A$ one finds that $r_A \circ \underline{(id_\mathcal{K} \otimes j)} \circ \operatorname{Ad} v = \operatorname{Ad} v$ and $\underline{(id_\mathcal{K} \otimes p)} \circ \underline{(id_\mathcal{K} \otimes j)} \circ \operatorname{Ad} v = 0$, i.e. $j^*(\epsilon) = [\operatorname{Ad} v, 0]$. By connecting v to 1 through a strictly continuous path of isometries in $\mathcal{M}(\mathcal{K} \otimes A)$, cf. Lemma 1.3.7, we see that $j^*(\epsilon) = 1_A$. $\qquad\square$

Since the isomorphism $\Lambda : kK_h(A, B) \to KK_h(A, B)$ is natural we can transfer the "composition product" of Theorem 4.3.14 to $KK_h(A, B)$. The resulting bilinear pairing agrees with the Kasparov product by Theorem 4.2.9, at least when all C^*-algebras involved are separable.

4.3.19 *Notes and remarks.*

This section is taken from [31].

Exercise 4.3

E 4.3.1 (Difficult)

Let A, B be separable and C, D σ-unital C^*-algebras. Show directly (i.e. without using the results of Chapter 2 and Theorem 4.2.1) that the bilinear pairing from Theorem 4.3.14 is associative i.e. that $(x \cdot y) \cdot z = x \cdot (y \cdot z)$, $x \in kK_h(A, B)$, $y \in kK_h(B, C)$, $z \in kK_h(C, D)$.

CHAPTER 5

Cuntz's Picture of KK-Theory

5.1. qA

In this chapter we shall make use of free products of C^*-algebras. The necessary prerequisites on this subject are gathered in Appendix B.

Let A be an arbitrary C^*-algebra and consider $QA = A * A$, the free product of A with itself. We let $i : A \to QA$ and $\bar{i} : A \to QA$ denote the two canonical inclusions of A as a C^*-subalgebra of QA.

Definition 5.1.1. We define qA to be the closed two-sided ideal in QA generated by the set $\{i(x) - \bar{i}(x) : x \in A\}$.

Let $\phi, \psi : A \to B$ be two *-homomorphisms. By the universal property of QA there is a unique *-homomorphism $Q(\phi, \psi) : QA \to B$ such that $Q(\phi, \psi) \circ i = \phi$ and $Q(\phi, \psi) \circ \bar{i} = \psi$. We let $q(\phi, \psi)$ denote the restriction of $Q(\phi, \psi)$ to qA. Note that if $J \subseteq B$ is an ideal in B, then $Q(\phi, \psi)$ maps qA into J if and only if $\phi(x) - \psi(x) \in J$ for all $x \in A$. So in this case, $q(\phi, \psi) \in \operatorname{Hom}(qA, J)$. This way of obtaining *-homomorphisms defined on qA will be formalized below.

To simplify notation we set $qx = i(x) - \bar{i}(x) \in qA$, $x \in A$.

Lemma 5.1.2. *Elements of the form*

$$qx_1 qx_2 \cdots qx_n \quad \text{or} \quad i(x_0) qx_1 qx_2 \cdots qx_n,$$

$x_0, x_1, x_2, \ldots, x_n \in A$, $n \in \mathbb{N}$, *span a dense *-subalgebra $\mathcal{C}$ in qA. Furthermore, $qA = \ker Q(id_A, id_A)$.*

Proof. Clearly $\mathcal{C} \subseteq qA$. Since $Q(id_A, id_A)(qx) = x - x = 0$, $qA \subseteq \ker Q(id_A, id_A)$. Thus it suffices to show that $\mathcal{C}$ is dense in $\ker Q(id_A, id_A)$. The relation $qxy + qxqy = i(x)qy + qxi(y)$, $x, y \in A$, shows that the linear span of elements of the form $i(x_0)$, $i(x_0)qx_1qx_2\cdots qx_n$ and $qx_1qx_2\cdots qx_n$, $x_0, x_1, x_2, \ldots, x_n \in A$, $n \in \mathbb{N}$, form a *-subalgebra C_0 of QA. Since the ranges of i and $\bar{i}$ generate QA as C^*-algebra, and $\bar{i}(x) = i(x) - qx$, $x \in A$, we conclude that C_0 is dense in QA. So for any $x \in QA$, we can choose a sequence $\{x_n\} \in C_0$ such that $x_n \to x$. But then $x_n - i \circ Q(id_A, id_A)(x_n) \in C$ for all n, and if $x \in \ker Q(id_A, id_A)$, $x_n - i \circ Q(id_A, id_A)(x_n) \to x$. $\square$

An immediate but useful consequence of this lemma is that any *-homomorphism defined on qA is determined by its values on the elements in qA of the form $i(x_0)qx$ and qx, $x_o, x \in A$.

Let $\phi : A \to B$ be a *-homomorphism and let $\epsilon, \bar{\epsilon} : B \to QB$ be the canonical inclusions. The *-homomorphism $Q(\epsilon \circ \phi, \bar{\epsilon} \circ \phi) : QA \to QB$ will be denoted by $Q(\phi)$. Since $\epsilon \circ \phi(x) - \bar{\epsilon} \circ \phi(x) \in qB$ for all $x \in A$, $Q(\phi)$ restricts to a *-homomorphism $q(\phi) : qA \to qB$. We let $\gamma^A : qA \to A$ denote the restiction of $Q(id_A, 0) : QA \to A$ to qA. By using Lemma 5.1.2, we find that

$$(5.1.1) \qquad \gamma^B \circ q(\phi) = \phi \circ \gamma^A, \ \phi \in \mathrm{Hom}\,(A, B).$$

Definition 5.1.3.   A *prequasi-homomorphism* from A to B consists of a C^*-algebra E, two *-homomorphisms $\alpha, \bar{\alpha} : A \to E$, an ideal $J \subseteq E$ and a *-homomorphism $\mu : J \to B$, such that $\alpha(x) - \bar{\alpha}(x) \in J$, $x \in A$. In diagram form

$$A \ \underset{\bar{\alpha}}{\overset{\alpha}{\rightrightarrows}} \ E \triangleright J \ \overset{\mu}{\longrightarrow} \ B.$$

The reason for the importance of this notion is that a prequasi-homomorphism $(\alpha, \bar{\alpha}, J, E, \mu)$ from A to B gives rise to the *-homomorphism $\mu \circ q(\alpha, \bar{\alpha}) : qA \to B$, and that all elements of $\mathrm{Hom}\,(qA, B)$ arise this way.

Lemma 5.1.4. *Let $\psi \in \mathrm{Hom}\,(qA, B)$.   Then there is a prequasi-homomorphism from A to B, given by the diagram*

$$A \ \underset{\phi \circ \bar{i}}{\overset{\phi \circ i}{\rightrightarrows}} \ QA/\ker \psi \triangleright qA/\ker \psi \ \overset{\mu}{\longrightarrow} \ B$$

where $\phi : QA \to QA/\ker \psi$ is the quotient map and $\mu : qA/\ker \psi \to B$ is induced by ψ (i.e. $\mu(x + \ker \psi) = \psi(x)$, $x \in qA$). Furthermore, $\psi = \mu \circ q(\phi \circ i, \phi \circ \bar{i})$.

Proof. It is straightforward to check that the indicated diagram gives a prequasi-homomorphism. The equality $\psi = \mu \circ q(\phi \circ i, \phi \circ \bar{i})$ follows then from Lemma 5.1.2. $\square$

Definition 5.1.5.

$$KK_c(A, B) = [qA, \mathcal{K} \otimes B].$$

In other words we define $KK_c(A, B)$ to be the homotopy classes of *-homomorphisms from qA to $\mathcal{K} \otimes B$, cf. Section 1.3. By Lemma 1.3.12 (and E 1.3.3) we already know that $KK_c(A, B)$ is an abelian semigroup with 0 element represented by the 0 homomorphism. The addition, $+$, in $KK_c(A, B) = [qA, \mathcal{K} \otimes B]$ is given by choosing an inner *-isomorphism $\Theta_B : M_2(\mathcal{K} \otimes B) \to \mathcal{K} \otimes B$ and defining

$$(5.1.2) \quad [\phi] + [\psi] = \left[\Theta_B \circ \begin{bmatrix} \phi & 0 \\ 0 & \psi \end{bmatrix} \right], \qquad \phi, \psi \in \mathrm{Hom}\,(qA, \mathcal{K} \otimes B).$$

Because of the special nature of qA this actually defines an abelian group structure on $KK_c(A, B)$.

Theorem 5.1.6. $KK_c(A, B)$ is an abelian group.

Proof. By Lemma 1.3.12 we only have to prove the existence of inverses. So let $\psi \in \mathrm{Hom}\,(qA, \mathcal{K} \otimes B)$. By Lemma 5.1.4 there is a prequasi-homomorphism

$$A \xrightarrow[\bar{\alpha}]{\alpha} E \triangleright J \xrightarrow{\mu} \mathcal{K} \otimes B$$

such that $\psi = \mu \circ q(\alpha, \bar{\alpha})$. Set $\lambda = \mu \circ q(\bar{\alpha}, \alpha) \in \mathrm{Hom}\,(qA, B)$. Then

$$[\psi] + [\lambda] = \left[\Theta_B \circ \begin{bmatrix} \psi & 0 \\ 0 & \lambda \end{bmatrix} \right].$$

By using Lemma 5.1.2 one finds that

$$\Theta_B \circ \begin{bmatrix} \psi & 0 \\ 0 & \lambda \end{bmatrix} = \Theta_B \circ (\mu \otimes id_{M_2(\mathbb{C})}) \circ q\left(\begin{bmatrix} \alpha & 0 \\ 0 & \bar{\alpha} \end{bmatrix}, \begin{bmatrix} \bar{\alpha} & 0 \\ 0 & \alpha \end{bmatrix} \right).$$

Let R_t, $t \in [0, 1]$, be the "rotation unitaries" from (1.3.3). By considering these unitaries as multipliers of $M_2(E)$ we can define $\beta_t \in \mathrm{Hom}\,(A, M_2(E))$ by

$$\beta_t = \mathrm{Ad}\, R_t \circ \begin{bmatrix} \bar{\alpha} & 0 \\ 0 & \alpha \end{bmatrix}, \quad t \in [0, 1].$$

It is then straightforward to check that $\begin{bmatrix} \alpha & 0 \\ 0 & \bar{\alpha} \end{bmatrix}(a) - \beta_t(a) \in M_2(J)$ for

all $t \in [0,1]$. Thus $q\left(\begin{bmatrix} \alpha & 0 \\ 0 & \bar{\alpha} \end{bmatrix}, \beta_t \right)$, $t \in [0,1]$, is a homotopy connecting

$q\left(\begin{bmatrix} \alpha & 0 \\ 0 & \bar{\alpha} \end{bmatrix}, \begin{bmatrix} \bar{\alpha} & 0 \\ 0 & \alpha \end{bmatrix} \right)$ to $q\left(\begin{bmatrix} \alpha & 0 \\ 0 & \bar{\alpha} \end{bmatrix}, \begin{bmatrix} \alpha & 0 \\ 0 & \bar{\alpha} \end{bmatrix} \right) = 0$. By composing this

homotopy with $\Theta_B \circ (\mu \otimes id_{M_2(\mathbb{C})})$ we find that $\Theta_B \circ \begin{bmatrix} \psi & 0 \\ 0 & \lambda \end{bmatrix} \sim 0$. Thus

$[\psi] + [\lambda] = 0$. $\square$

Let $f : C \to A$ be a *-homomorphism of C^*-algebras. We can
then define $f^* : KK_c(A,B) \to KK_c(C,B)$ by $f^*[\phi] = [\phi \circ q(f)]$, for
$\phi \in \text{Hom}\,(qA, \mathcal{K} \otimes B)$. It is clear that f^* defines a group homomorphism and
in E 5.1.1 the reader is asked to check that this definition makes $KK_c(\cdot, B)$
into a contravariant homotopy invariant functor. f gives also rise to a map
$f_* : KK_c(B,C) \to KK_c(B,A)$ defined by $f_*[\phi] = [(id_\mathcal{K} \otimes f) \circ \phi]$, for
$\phi \in \text{Hom}\,(qB, \mathcal{K} \otimes C)$. By E 1.3.4 f_* is a group homomorphism and in
E 5.1.1 the reader is asked to check that $KK_c(B, \cdot)$ is a covariant homotopy
invariant functor.

We now want to "symmetrize" the definiton of $KK_c(A,B)$ by showing
that $KK_c(A,B) \simeq [\mathcal{K} \otimes qA, \mathcal{K} \otimes qB]$ (when A is separable). This requires
a considerable amount of preparations.

Set $q^2 A = q(qA)$, i.e. $q^2 A$ is the kernel of $Q(id_{qA}, id_{qA})$ in $Q(qA)$.

Lemma 5.1.7. Let $\chi, \bar{\chi} : QA \to Q(QA)$ be the canonical inclu-
sions. Then $Q(\chi|_{qA}, \bar{\chi}|_{qA}) : Q(qA) \to Q(QA)$ is injective. Furthermore
we have the identities $Q(id_{QA}, id_{QA}) \circ Q(\chi|_{qA}, \bar{\chi}|_{qA}) = Q(id_{qA}, id_{qA})$ and
$Q(id_{qA}, 0) = Q(id_{QA}, 0) \circ Q(\chi|_{qA}, \bar{\chi}|_{qA})$.

Proof. Let $\alpha, \bar{\alpha} : qA \to Q(qA)$ be the canonical inclusions and let ϕ be
a faithful representation of $Q(qA)$ on the Hilbert space $\mathcal{H}$. Then $\phi \circ \alpha$ and
$\phi \circ \bar{\alpha}$ are faithful represenations of qA. By E 5.1.2 there are representations
$\psi, \bar{\psi} : QA \to \mathcal{B}(\mathcal{H})$ such that

$$
\begin{array}{ccc}
QA & \xrightarrow{\psi} & \mathcal{B}(\mathcal{H}) \\
\cup & & \| \\
qA & \xrightarrow{\phi \circ \alpha} & \mathcal{B}(\mathcal{H})
\end{array}
\qquad \text{and} \qquad
\begin{array}{ccc}
QA & \xrightarrow{\bar{\psi}} & \mathcal{B}(\mathcal{H}) \\
\cup & & \| \\
qA & \xrightarrow{\phi \circ \bar{\alpha}} & \mathcal{B}(\mathcal{H})
\end{array}
$$

commute. Thus $Q(\psi, \bar{\psi}) : Q(QA) \to \mathcal{B}(\mathcal{H})$ is a representation such that
$Q(\psi, \bar{\psi}) \circ Q(\chi|_{qA}, \bar{\chi}|_{qA}) = \phi|_{qA}$. Since ϕ is injective, it follows that so is
$Q(\chi|_{qA}, \bar{\chi}|_{qA})$.

The identities of *-homomorphisms asserted in the lemma follow by checking on $\alpha(x)$ and $\bar{\alpha}(x)$, $x \in qA$. □

Theorem 5.1.8. *Assume that A is separable. Let $j_1 : qA \to M_2(qA)$ and $j_2 : q^2 A \to M_2(q^2 A)$ be the embeddings into the upper left-hand corners.*

*Then there exists a *-homomorphism $\phi_A : qA \to M_2(q^2 A)$ such that $(\gamma^{qA} \otimes id_{M_2(\mathbb{C})}) \circ \phi_A \sim j_1$ and $\phi_A \circ \gamma^{qA} \sim j_2$.*

Proof. Let $\chi, \bar{\chi} : QA \to Q(QA)$ denote the canonical inclusions of QA into $Q(QA)$. Set $D_1 = Q(\chi|_{qA}, \bar{\chi}|_{qA})(Q(qA))$ and $D_2 = Q(\chi|_{qA}, \bar{\chi}|_{qA})(q^2 A)$. D_2 is then an ideal in D_1, in fact $D_2 = \ker Q(id_{QA}, id_{QA}) \cap D_1$ by Lemma 5.1.2 and Lemma 5.1.7. Since $Q(\chi|_{qA}, \bar{\chi}|_{qA})$ is injective by Lemma 5.1.7, we can prove the theorem by producing a *-homomorphism $\psi_A : qA \to M_2(D_2)$ such that $\psi_A \circ \gamma^{qA} \sim (Q(\chi|_{qA}, \bar{\chi}|_{qA}) \otimes id_{M_2(\mathbb{C})}) \circ j_2$ and $(Q(id_{QA}, 0) \otimes id_{M_2(\mathbb{C})}) \circ \psi_A \sim j_1$. When such a ψ_A is constructed, $\phi_A = (Q(\chi|_{qA}, \bar{\chi}|_{qA}) \otimes id_{M_2(\mathbb{C})})^{-1} \circ \psi_A$ will have the right properties by Lemma 5.1.7. Set $R_1 = \chi(qA)$ and $R_2 = \bar{\chi}(qA)$ and let R be the C^*-subalgebra of $M_2(D_1)$ generated by

$$\begin{bmatrix} R_1 & R_1 R_2 \\ R_2 R_1 & R_2 \end{bmatrix}.$$

Let $V = \left\{ \begin{bmatrix} \chi \circ i(x) & 0 \\ 0 & \bar{\chi} \circ i(x) \end{bmatrix} : x \in A \right\}$ and let $\mathcal{D}$ denote the C^*-subalgebra of $M_2(Q(QA))$ generated by V and R. Since $VR \subseteq R$ and $RV \subseteq R$, we see that R is an ideal in $\mathcal{D}$. Thus we can define a *-homomorphism $\delta : \mathcal{D} \to \mathcal{M}(R)$ by $\delta(z)r = zr$, $z \in \mathcal{D}$, $r \in R$, cf. E1.1.9. Let ϕ denote the restriction of $Q(id_{QA}, id_{QA}) \otimes id_{M_2(\mathbb{C})} : M_2(Q(QA)) \to M_2(QA)$ to R. It is clear that ϕ maps R onto $M_2(qA)$. Let $\underline{\phi} : \mathcal{M}(R) \to \mathcal{M}(M_2(qA))$ be the strictly continuous extension of ϕ, cf. Corollary 1.1.15. It is straightforward to check that

$$\underline{\phi} \circ \delta \begin{bmatrix} \chi \circ i(x) & 0 \\ 0 & \bar{\chi} \circ i(x) \end{bmatrix}$$

is left multiplication by $\begin{bmatrix} i(x) & 0 \\ 0 & i(x) \end{bmatrix}$ on $M_2(qA)$, $x \in A$. Let R_t, $t \in [0,1]$, denote the rotation matrices from (1.3.3) and consider them as multipliers on $M_2(qA)$. Then for all $t \in [0,1]$, the R_t's lie in the relative commutant, $\underline{\phi}(\delta(V))' \cap \mathcal{M}(M_2(qA))$, of $\underline{\phi}(\delta(V))$ in $\mathcal{M}(M_2(qA))$. Note that

$R_t = e^{itA}$, $t \in [0,1]$, where $A = i2^{-1} \begin{bmatrix} 0 & -\pi \\ \pi & 0 \end{bmatrix} \in \underline{\phi}(\delta(V))' \cap \mathcal{M}(M_2(qA))$

is self-adjoint. Using Theorem 1.1.26 we can find a self-adjoint

$$A_0 \in \{m \in \mathcal{M}(R) : mf - fm \in \ker\phi, \ f \in \delta(V)\}$$

so that $\underline{\phi}(A_0) = A$. Setting $U_t = e^{itA_0}$, $t \in [0,1]$, we obtain a continuous path of unitaries in $\mathcal{M}(R)$ such that $U_t \delta(\mathcal{D})U_t^* = \delta(\mathcal{D})$, $U_0 = 1$ and $\underline{\phi}(U_t) = R_t$, $t \in [0,1]$. Let σ_t denote the *-automorphism of $\delta(\mathcal{D})$ obtained by conjugation with U_t.

Now define $\alpha, \bar{\alpha} : A \to \mathcal{D}$ by

$$\alpha(a) = \begin{bmatrix} \chi \circ i(a) & 0 \\ 0 & \bar{\chi} \circ \bar{i}(a) \end{bmatrix} = \begin{bmatrix} \chi \circ i(a) & 0 \\ 0 & \bar{\chi} \circ i(a) \end{bmatrix} - \begin{bmatrix} 0 & 0 \\ 0 & \bar{\chi}(i(a) - \bar{i}(a)) \end{bmatrix}$$

and

$$\bar{\alpha}(a) = \begin{bmatrix} \chi \circ \bar{i}(a) & 0 \\ 0 & \bar{\chi} \circ i(a) \end{bmatrix} = \begin{bmatrix} \chi \circ i(a) & 0 \\ 0 & \bar{\chi} \circ i(a) \end{bmatrix} - \begin{bmatrix} \chi(i(a) - \bar{i}(a)) & 0 \\ 0 & 0 \end{bmatrix},$$

for all $a \in A$.

We assert that $\delta \circ \alpha(a) - \sigma_t \circ \delta \circ \bar{\alpha}(a) \in R$ for all $a \in A$, $t \in [0,1]$. To see this note that

$$\sigma_t \circ \delta \begin{bmatrix} \chi \circ i(a) & 0 \\ 0 & \bar{\chi} \circ i(a) \end{bmatrix} = \delta \begin{bmatrix} \chi \circ i(a) & 0 \\ 0 & \bar{\chi} \circ i(a) \end{bmatrix} \bmod \ker\phi \subseteq R.$$

Thus

$$\delta \circ \alpha(a) - \sigma_t \circ \delta \circ \bar{\alpha}(a)$$
$$= -\delta \begin{bmatrix} 0 & 0 \\ 0 & \bar{\chi}(i(a) - \bar{i}(a)) \end{bmatrix} + \sigma_t \circ \delta \begin{bmatrix} \chi(i(a) - \bar{i}(a)) & 0 \\ 0 & 0 \end{bmatrix}$$
$$= 0 \bmod R,$$

proving the assertion.

It follows that we can define a continuous path α_t, $t \in [0,1]$, in $\mathrm{Hom}\,(qA, R)$ by $\alpha_t = q(\delta \circ \alpha, \sigma_t \circ \delta \circ \bar{\alpha})$. For the particular value $t = 1$, we have that $\delta \circ \alpha(a) - \sigma_1 \circ \delta \circ \bar{\alpha}(a) \in M_2(D_2)$. To see this note that $\ker(Q(id_{QA}, id_{QA}) \otimes id_{M_2(\mathbb{C})}) \cap R \subseteq M_2(D_2)$. Thus it suffices to show that $\phi(\delta \circ \alpha(a) - \sigma_1 \circ \delta \circ \bar{\alpha}(a)) = 0$. But

$$\phi(\delta \circ \alpha(a) - \sigma_1 \circ \delta \circ \bar{\alpha}(a)) = \underline{\phi}(\delta \circ \alpha(a)) - \mathrm{Ad}\,R_1 \circ \underline{\phi}(\delta \circ \bar{\alpha}(a))$$
$$= -\begin{bmatrix} 0 & 0 \\ 0 & i(a) - \bar{i}(a) \end{bmatrix} + \mathrm{Ad}\,R_1 \begin{bmatrix} i(a) - \bar{i}(a) & 0 \\ 0 & 0 \end{bmatrix}$$
$$= 0.$$

It follows that we can define $\psi_A : qA \to M_2(D_2)$ by $\psi_A = q(\delta \circ \alpha, \sigma_1 \circ \delta \circ \bar{\alpha})$, i.e. $\psi_A = \alpha_1$.

Set $\psi_t = (Q(id_{QA}, 0) \otimes id_{M_2(\mathbb{C})}) \circ \alpha_t \in \text{Hom}\,(qA, M_2(qA))$, $t \in [0, 1]$. Then $\psi_1 = (Q(id_{QA}, 0) \otimes id_{M_2(\mathbb{C})}) \circ \psi_A$. Note that

$$\psi_0(i(a) - \bar{i}(a)) = (Q(id_{QA}, 0) \otimes id_{M_2(\mathbb{C})})(\delta(\alpha(a) - \bar{\alpha}(a)))$$

$$= (Q(id_{QA}, 0) \otimes id_{M_2(\mathbb{C})})\left(\begin{bmatrix} \chi(i(a) - \bar{i}(a)) & 0 \\ 0 & 0 \end{bmatrix} - \begin{bmatrix} 0 & 0 \\ 0 & \bar{\chi}(i(a) - \bar{i}(a)) \end{bmatrix}\right)$$

$$= \begin{bmatrix} i(a) - \bar{i}(a) & 0 \\ 0 & 0 \end{bmatrix} = j_1(i(a) - \bar{i}(a)), \qquad a \in A.$$

Similarly we find that $\psi_0(i(a_0)(i(a) - \bar{i}(a)) = j_1(i(a_0)(i(a) - \bar{i}(a)))$, $a_0, a \in A$. Thus we conclude from Lemma 5.1.2 that $\psi_0 = j_1$. It follows that ψ_t, $t \in [0, 1]$, is a homotopy connecting j_1 to $(Q(id_{QA}, 0) \otimes id_{M_2(\mathbb{C})}) \circ \psi_A$.

Consider the following *-homomorphisms in $\text{Hom}\,(A, M_2(\bar{\chi}(QA)))$:

$$\begin{bmatrix} \bar{\chi} \circ i & 0 \\ 0 & \bar{\chi} \circ \bar{i} \end{bmatrix} \quad \text{and} \quad \text{Ad}\,R_t \circ \begin{bmatrix} \bar{\chi} \circ \bar{i} & 0 \\ 0 & \bar{\chi} \circ i \end{bmatrix}, \; t \in [0, 1],$$

where we consider the rotation matrices R_t from (1.3.3) as multipliers of $M_2(\bar{\chi}(QA))$. It is straightforward to check that

$$\begin{bmatrix} \bar{\chi} \circ i(a) & 0 \\ 0 & \bar{\chi} \circ \bar{i}(a) \end{bmatrix} - R_t \begin{bmatrix} \bar{\chi} \circ \bar{i}(a) & 0 \\ 0 & \bar{\chi} \circ i(a) \end{bmatrix} R_t^* \in M_2(\bar{\chi}(qA)) \subseteq M_2(D_1)$$

for all $a \in A$. Therefore

$$\beta_t = q\left(\begin{bmatrix} \bar{\chi} \circ i & 0 \\ 0 & \bar{\chi} \circ \bar{i} \end{bmatrix}, \; \text{Ad}\,R_t \circ \begin{bmatrix} \bar{\chi} \circ \bar{i} & 0 \\ 0 & \bar{\chi} \circ i \end{bmatrix}\right) \in \text{Hom}\,(qA, M_2(D_1)),$$

for $t \in [0, 1]$. Note that $\beta_1 = 0$.

As shown above α_t takes values in $R \subseteq M_2(D_1)$. We assert that $\alpha_t(x) - \beta_t(x) \in M_2(D_2)$, $x \in qA$, $t \in [0, 1]$. To see this it suffices to check that $(Q(id_{QA}, id_{QA}) \otimes id_{M_2(\mathbb{C})}) \circ \alpha_t = (Q(id_{QA}, id_{QA}) \otimes id_{M_2(\mathbb{C})}) \circ \beta_t$. Let $a_1, a_2 \in A$. Then

$$(Q(id_{QA}, id_{QA}) \otimes id_{M_2(\mathbb{C})}) \circ \alpha_t(i(a_1)qa_2)$$

$$= \phi \circ \alpha_t(i(a_1)qa_2)$$

$$= \underline{\phi} \circ \delta \circ \alpha(a_1)(\underline{\phi} \circ \delta \circ \alpha(a_2) - \text{Ad}\,R_t \circ \underline{\phi} \circ \delta \circ \bar{\alpha}(a_2))$$

$$= \begin{bmatrix} c_t^2 i(a_1)qa_2 & s_t c_t i(a_1)qa_2 \\ s_t c_t \bar{i}(a_1)qa_2 & -c_t^2 \bar{i}(a_1)qa_2 \end{bmatrix},$$

where $s_t = \sin \frac{\pi}{2} t$ and $c_t = \cos \frac{\pi}{2} t$, $t \in [0, 1]$. The same expression is easily obtained for $(Q(id_{QA}, id_{QA}) \otimes id_{M_2(\mathbb{C})}) \circ \beta_t(i(a_1)qa_2)$. Similarly we find that $(Q(id_{QA}, id_{QA}) \otimes id_{M_2(\mathbb{C})}) \circ \alpha_t$ and $(Q(id_{QA}, id_{QA}) \otimes id_{M_2(\mathbb{C})}) \circ \beta_t$ agree on qa_2. Thus the two *-homomorphisms agree by Lemma 5.1.2.

Set $\lambda_t = q(\alpha_t, \beta_t) \in \mathrm{Hom}\,(q^2 A, M_2(D_2))$, $t \in [0, 1]$. Let $\mu, \bar{\mu} : qA \to Q(qA)$ denote the canonical inclusions. Then

$$Q(\alpha_1, \beta_1)(\mu(x)) = \alpha_1(x)$$
$$= \psi_A(x)$$
$$= \psi_A \circ Q(id_{qA}, 0)(\mu(x))$$

and

$$Q(\alpha_1, \beta_1)(\bar{\mu}(x)) = \beta_1(x)$$
$$= 0 = \psi_A \circ Q(id_{qA}, 0)(\bar{\mu}(x)),$$

for all $x \in qA$. It follows that $\lambda_1 = \psi_A \circ \gamma^{qA}$. To calculate λ_0 we first use the explicit expressions for α and $\bar{\alpha}$ to conclude that

$$\alpha_0(z) = \begin{bmatrix} \chi(z) & 0 \\ 0 & (-1)^n \bar{\chi}(z) \end{bmatrix}$$

for $z \in qA$ of the form $z = qa_1 qa_2 \cdots qa_n$ or $z = i(a_0)qa_1 qa_2 \cdots qa_n$ where $a_0, a_1, \ldots, a_n \in A$.

Similarly we find that

$$\beta_0(z) = \begin{bmatrix} \bar{\chi}(z) & 0 \\ 0 & (-1)^n \bar{\chi}(z) \end{bmatrix}$$

for $z \in qA$ of the same form. It follows that

$$\alpha_0(z) - \beta_0(z) = \begin{bmatrix} \chi(z) - \bar{\chi}(z) & 0 \\ 0 & 0 \end{bmatrix} \quad \text{for all} \quad z \in qA$$

since the z's of the form above span a dense subspace of qA by Lemma 5.1.2. From the above we also obtain

$$\alpha_0(z_1)(\alpha_0(z_2) - \beta_0(z_2)) = \begin{bmatrix} \chi(z_1)(\chi(z_2) - \bar{\chi}(z_2)) & 0 \\ 0 & 0 \end{bmatrix},$$

for $z_1, z_2 \in qA$.

Now it follows from Lemma 5.1.2 that

$$\lambda_0 = (Q(\chi|_{qA}, \bar{\chi}|_{qA}) \otimes id_{M_2(\mathbb{C})}) \circ j_2.$$

Thus λ_t, $t \in [0,1]$, is a homotopy connecting $(Q(\chi|_{qA}, \bar{\chi}|_{qA}) \otimes id_{M_2(\mathbb{C})}) \circ j_2$ to $\psi_A \circ \gamma^{qA}$. The proof is complete. $\square$

Let $\rho_B = q(id_{\mathcal{K}} \otimes \epsilon, id_{\mathcal{K}} \otimes \bar{\epsilon}) : q(\mathcal{K} \otimes B) \to \mathcal{K} \otimes qB$ where $\epsilon, \bar{\epsilon} : B \to QB$ are the canonical inclusions, and let Θ_{qB} denote an inner *-isomorphism $\Theta_{qB} : M_2(\mathcal{K} \otimes qB) \to \mathcal{K} \otimes qB$. We shall need some lemmas relating γ^B, ρ_B, Θ_{qB} and Θ_B .

Lemma 5.1.9. $\Theta_B \circ (id_{\mathcal{K}} \otimes \gamma^B \otimes id_{M_2(\mathbb{C})}) \sim (id_{\mathcal{K}} \otimes \gamma^B) \circ \Theta_{qB}$.

Proof. Since γ^B is surjective, so is $id_{\mathcal{K}} \otimes \gamma^B$. Thus the strictly continuous extension $\underline{id_{\mathcal{K}} \otimes \gamma^B} : \mathcal{M}(\mathcal{K} \otimes qB) \to \mathcal{M}(\mathcal{K} \otimes B)$ is unital. It follows that, when V_1, V_2 are the isometries in $\mathcal{M}(\mathcal{K} \otimes qB)$ used to define Θ_{qB}, then $W_i = \underline{id_{\mathcal{K}} \otimes \gamma^B}(V_i)$, $i = 1, 2$, defines an inner *-isomorphism $\Phi : M_2(\mathcal{K} \otimes B) \to \mathcal{K} \otimes B$ satisfying $\Phi \circ (id_{\mathcal{K}} \otimes \gamma^B \otimes id_{M_2(\mathbb{C})}) = (id_{\mathcal{K}} \otimes \gamma^B) \circ \Theta_{qB}$. The lemma therefore follows from Lemma 1.3.9 and Lemma 1.3.7. $\square$

Lemma 5.1.10. $(id_{\mathcal{K}} \otimes \gamma^B) \circ \rho_B = \gamma^{\mathcal{K} \otimes B}$ in $\mathrm{Hom}\,(q(\mathcal{K} \otimes B), \mathcal{K} \otimes B)$.

Proof. Let $\alpha, \bar{\alpha} : \mathcal{K} \otimes B \to Q(\mathcal{K} \otimes B)$ denote the canonical inclusions. For $k \in \mathcal{K}$, $b \in B$ we find

$$
\begin{aligned}
(id_{\mathcal{K}} \otimes \gamma^B) \circ \rho_B(\alpha(k \otimes b) - \bar{\alpha}(k \otimes b)) &= (id_{\mathcal{K}} \otimes \gamma^B)(k \otimes \epsilon(b) - k \otimes \bar{\epsilon}(b)) \\
&= k \otimes b \\
&= \gamma^{\mathcal{K} \otimes B}(\alpha(k \otimes b) - \bar{\alpha}(k \otimes b)).
\end{aligned}
$$

In a similar way we find that the two *-homomorphisms agree on elements $\alpha(k_0 \otimes b_0)(\alpha(k \otimes b) - \bar{\alpha}(k \otimes b))$ when $k_0 \in \mathcal{K}$, $b_0 \in B$ is another pair. It is now easy to conclude from Lemma 5.1.2 that the *-homomorphisms agree. $\square$

Lemma 5.1.11. $\rho_B \circ q(id_{\mathcal{K}} \otimes \gamma^B) \sim \gamma^{\mathcal{K} \otimes qB}$ in $\mathrm{Hom}\,(q(\mathcal{K} \otimes qB), \mathcal{K} \otimes qB)$.

Proof. Define $\Phi, \Psi, \Psi_t \in \mathrm{Hom}\,(B, M_2(QB))$ by

$$
\Phi(x) = \begin{bmatrix} \epsilon(x) & 0 \\ 0 & 0 \end{bmatrix},
$$

$$
\Psi(x) = \begin{bmatrix} \bar{\epsilon}(x) & 0 \\ 0 & 0 \end{bmatrix} \quad \text{and}
$$

$$
\Psi_t(x) = R_t \Psi(x) R_t^*, \quad x \in B, \ t \in [0,1],
$$

where R_t, $t \in [0,1]$, are the rotation matrices from (1.3.3), considered as unitaries in $M_2(\mathcal{M}(QB))$. Set $\alpha_t = q(\Phi, \Psi_t)$ and $\beta_t = q(\Psi, \Psi_t)$, for $t \in [0,1]$. Then $\alpha_t, \beta_t \in \mathrm{Hom}\,(qB, M_2(QB))$, for $t \in [0,1]$. Since

$$(Q(id_B, id_B) \otimes id_{M_2(\mathbb{C})}) \circ \alpha_t = (Q(id_B, id_B) \otimes id_{M_2(\mathbb{C})}) \circ \beta_t,$$

we conclude from Lemma 5.1.2 that $id_{\mathcal{K}} \otimes \alpha_t - id_{\mathcal{K}} \otimes \beta_t$ maps $\mathcal{K} \otimes qB$ into $\mathcal{K} \otimes M_2(qB)$. Thus $s_t = q(id_{\mathcal{K}} \otimes \alpha_t, id_{\mathcal{K}} \otimes \beta_t)$ is a *-homomorphism from $q(\mathcal{K} \otimes qB)$ into $\mathcal{K} \otimes M_2(qB)$ for all $t \in [0,1]$.

To calculate s_0, we first observe that $\beta_0 = 0$ and that $\alpha_0 = j_1$, the embedding of qB into the upper left-hand corner of $M_2(qB)$. Let $\mu, \bar{\mu}$: $\mathcal{K} \otimes qB \to Q(\mathcal{K} \otimes qB)$ be the canonical inclusions. For $k \in \mathcal{K}$, $x \in qB$, we find

$$
\begin{aligned}
s_0(\mu(k \otimes x) - \bar{\mu}(k \otimes x)) &= k \otimes \alpha_0(x) - k \otimes \beta_0(x) \\
&= k \otimes \alpha_0(x) = k \otimes j_1(x) \\
&= (id_{\mathcal{K}} \otimes j_1) \circ \gamma^{\mathcal{K} \otimes qB}(\mu(k \otimes x) - \bar{\mu}(k \otimes x)).
\end{aligned}
$$

It follows that $s_0(\mu(z) - \bar{\mu}(z)) = (id_{\mathcal{K}} \otimes j_1) \circ \gamma^{\mathcal{K} \otimes qB}(\mu(z) - \bar{\mu}(z))$, $z \in \mathcal{K} \otimes qB$. Since $\beta_0 = 0$, we find that

$$
\begin{aligned}
s_0(\mu(z_0)(\mu(z) - \bar{\mu}(z)) &= s_0(\mu(z_0 z) - \bar{\mu}(z_0 z)) \\
&= (id_{\mathcal{K}} \otimes j_1) \circ \gamma^{\mathcal{K} \otimes qB}(\mu(z_0 z) - \bar{\mu}(z_0 z)) \\
&= (id_{\mathcal{K}} \otimes j_1) \circ \gamma^{\mathcal{K} \otimes qB}(\mu(z_0)(\mu(z) - \bar{\mu}(z)),
\end{aligned}
$$

for $z_0, z \in \mathcal{K} \otimes qB$. It follows that $s_0 = (id_{\mathcal{K}} \otimes j_1) \circ \gamma^{\mathcal{K} \otimes qB}$.

We assert that

$$
\begin{aligned}
(5.1.3) \qquad s_1 &= (id_{\mathcal{K}} \otimes j_1) \circ \rho_B \circ q(id_{\mathcal{K}} \otimes \gamma^B) \\
&= (id_{\mathcal{K}} \otimes j_1) \circ q(id_{\mathcal{K}} \otimes \epsilon \circ \gamma^B, id_{\mathcal{K}} \otimes \bar{\epsilon} \circ \gamma^B).
\end{aligned}
$$

The last equality follows from the definitions and is only included in order to check the first. Let $k_1, k_2 \in \mathcal{K}$, $x_1, x_2 \in qB$. Then

$$
\begin{aligned}
s_1(\mu(k_1 \otimes x_1)(\mu(k_2 \otimes x_2) - \bar{\mu}(k_2 \otimes x_2))) &= \\
k_1 \otimes \alpha_1(x_1)(k_2 \otimes (\alpha_1(x_2) - \beta_1(x_2))) &= \\
k_1 k_2 \otimes \alpha_1(x_1)(\alpha_1(x_2) - \beta_1(x_2)). &
\end{aligned}
$$

On the other hand

$$
\begin{aligned}
(id_{\mathcal{K}} \otimes j_1) \circ q(id_{\mathcal{K}} \otimes \epsilon \circ \gamma^B, id_{\mathcal{K}} \otimes \bar{\epsilon} \circ \gamma^B)(\mu(k_1 \otimes x_1)(\mu(k_2 \otimes x_2) - \bar{\mu}(k_2 \otimes x_2))) &= \\
k_1 \otimes j_1 \circ \epsilon \circ \gamma^B(x_1)(k_2 \otimes (j_1 \circ \epsilon \circ \gamma^B(x_2) - j_1 \circ \bar{\epsilon} \circ \gamma^B(x_2))) &= \\
k_1 k_2 \otimes (j_1 \circ \epsilon \circ \gamma^B(x_1)(j_1 \circ \epsilon \circ \gamma^B(x_2) - j_1 \circ \bar{\epsilon} \circ \gamma^B(x_2)). &
\end{aligned}
$$

So to conclude that the two sides of (5.1.3) agree on elements of the form

(5.1.4) $$\mu(z_1)(\mu(z_2) - \bar{\mu}(z_2)), \ z_1, z_2 \in K \otimes qB,$$

it suffices to check

(5.1.5) $\alpha_1(x_1)(\alpha_1(x_2) - \beta_1(x_2)) = j_1(\epsilon \circ \gamma^B(x_1)(\epsilon \circ \gamma^B(x_2) - \bar{\epsilon} \circ \gamma^B(x_2)),$

for $x_1, x_2 \in qB$. To prove (5.1.5) it suffices by Lemma 5.1.2 to check the following 4 cases (where $y_1, y_2, \ldots, y_n, u_1, u_2, \ldots, u_m \in B$):

(i) $x_1 = qy_1qy_2 \cdots qy_n, \ x_2 = qu_1qu_2 \cdots qu_m,$

(ii) $x_1 = \epsilon(y_1)qy_2qy_3 \cdots qy_n, \ x_2 = qu_1qu_2 \cdots qu_m,$

(iii) $x_1 = qy_1qy_2 \cdots qy_n, \ x_2 = \epsilon(u_1)qu_2qu_3 \cdots qu_m$

(iv) $x_1 = \epsilon(y_1)qy_2qy_3 \cdots qy_n, \ x_2 = \epsilon(u_1)qu_2qu_3 \cdots qu_m.$

To check (ii) we calculate

$$\alpha_1(\epsilon(y_1)qy_2qy_3 \cdots qy_n) = \begin{bmatrix} \epsilon(y_1y_2 \cdots y_n) & 0 \\ 0 & 0 \end{bmatrix},$$

and

$$\alpha_1(qu_1qu_2 \cdots qu_m) - \beta_1(qu_1qu_2 \cdots qu_m) =$$
$$\begin{bmatrix} \epsilon(u_1u_2 \cdots u_m) & 0 \\ 0 & (-1)^m \bar{\epsilon}(u_1u_2 \cdots u_m) \end{bmatrix} - \begin{bmatrix} \bar{\epsilon}(u_1u_2 \cdots u_m) & 0 \\ 0 & (-1)^m \bar{\epsilon}(u_1u_2 \cdots u_m) \end{bmatrix} =$$
$$\begin{bmatrix} q(u_1u_2 \cdots u_m) & 0 \\ 0 & 0 \end{bmatrix}.$$

It follows that the left-hand side of (5.1.5) for this particular choice of x_1 and x_2 equals $j_1(\epsilon(y_1y_2 \cdots y_n)qu_1u_2 \cdots u_m)$. We then calculate the righthand side of (5.1.5) in the same case:

$$\epsilon \circ \gamma^B(\epsilon(y_1)qy_2 \cdots qy_n) = \epsilon(y_1y_2 \cdots y_n),$$

and

$$\epsilon \circ \gamma^B(qu_1qu_2 \cdots qu_m) - \bar{\epsilon} \circ \gamma^B(qu_1qu_2 \cdots qu_m) =$$
$$\epsilon(u_1u_2 \cdots u_m) - \bar{\epsilon}(u_1u_2 \cdots u_m) = q(u_1u_2 \cdots u_m),$$

so we find $j_1(\epsilon(y_1y_2 \cdots y_n)q(u_1u_2 \cdots u_m))$ for the right hand side also. We leave the reader to check the other 3 cases. By doing so we conclude that the two *-homomorphisms of (5.1.3) agree on elements of the form (5.1.4). It is slightly simpler to check that they also agree on elements of the form $\mu(z) - \bar{\mu}(z)$, $z \in qB$, and we leave this to the reader. Now (5.1.3) follows from Lemma 5.1.2.

All in all we have shown that s_t, $t \in [0,1]$, is a homotopy connecting the homomorphisms $(id_{\mathcal{K}} \otimes j_1) \circ \gamma^{\mathcal{K} \otimes qB}$ and $(id_{\mathcal{K}} \otimes j_1) \circ \rho_B \circ q(id_{\mathcal{K}} \otimes \gamma^B)$ in $\mathrm{Hom}\,(q(\mathcal{K} \otimes qB), \mathcal{K} \otimes M_2(qB))$. By identifying $\mathcal{K} \otimes M_2(qB) = M_2(\mathcal{K} \otimes qB)$ in the usual way, $(id_{\mathcal{K}} \otimes j_1)$ becomes the embedding $\mathcal{K} \otimes qB \to M_2(\mathcal{K} \otimes qB)$ into the upper left-hand corner. Thus the proof is completed by an application of Lemma 1.3.11. $\square$

Theorem 5.1.12. *Assume that A is separable and $\phi_A : qA \to M_2(q^2 A)$ is the *-homomorphism given by Theorem 5.1.8.*

Define

$$\Phi_{A,B} : KK_c(A,B) \to KK_c(A, qB) = [qA, \mathcal{K} \otimes qB]$$

by

$$\Phi_{A,B}[\alpha] = [\Theta_{qB} \circ ((\rho_B \circ q(\alpha)) \otimes id_{M_2(\mathbb{C})}) \circ \phi_A], \quad \alpha \in \mathrm{Hom}\,(qA, \mathcal{K} \otimes B).$$

Then $\Phi_{A,B}$ is a group isomorphism with inverse

$$\Phi_{A,B}^{-1}[\beta] = [(id_{\mathcal{K}} \otimes \gamma^B) \circ \beta], \quad \beta \in \mathrm{Hom}\,(qA, \mathcal{K} \otimes qB).$$

In short $KK_c(A,B) \simeq [qA, \mathcal{K} \otimes qB]$.

Proof. Set $\lambda = \Phi_{A,B}$ and $\mu[\beta] = [(id_{\mathcal{K}} \otimes \gamma^B) \circ \beta]$, for any $\beta \in \mathrm{Hom}\,(qA, \mathcal{K} \otimes qB)$. We will prove the theorem by checking the following 3 things:

(i) μ is a homomorphism,

(ii) $\mu \circ \lambda = id$,

(iii) $\lambda \circ \mu = id$.

(i) : Let $\beta_1, \beta_2 \in \mathrm{Hom}\,(qA, \mathcal{K} \otimes qB)$. Then

$$\mu([\beta_1] + [\beta_2]) = \mu\left[\Theta_{qB} \circ \begin{bmatrix} \beta_1 & 0 \\ 0 & \beta_2 \end{bmatrix}\right]$$

$$= \left[(id_\mathcal{K} \otimes \gamma^B) \circ \Theta_{qB} \circ \begin{bmatrix} \beta_1 & 0 \\ 0 & \beta_2 \end{bmatrix}\right]$$

$$= \left[\Theta_B \circ (id_\mathcal{K} \otimes \gamma^B \otimes id_{M_2(\mathbb{C})}) \circ \begin{bmatrix} \beta_1 & 0 \\ 0 & \beta_2 \end{bmatrix}\right]$$

(by Lemma 5.1.9)

$$= \left[\Theta_B \circ \begin{bmatrix} (id_\mathcal{K} \otimes \gamma^B) \circ \beta_1 & 0 \\ 0 & (id_\mathcal{K} \otimes \gamma^B) \circ \beta_2 \end{bmatrix}\right]$$

$$= \mu[\beta_1] + \mu[\beta_2].$$

(ii) : Let $\alpha \in \mathrm{Hom}\,(qA, \mathcal{K} \otimes B)$. Then

$$\mu \circ \lambda[\alpha] = [(id_\mathcal{K} \otimes \gamma^B) \circ \Theta_{qB} \circ (\rho_B \circ q(\alpha) \otimes id_{M_2(\mathbb{C})}) \circ \phi_A]$$

$$= [\Theta_B \circ (id_\mathcal{K} \otimes \gamma^B \otimes id_{M_2(\mathbb{C})}) \circ (\rho_B \circ q(\alpha) \otimes id_{M_2(\mathbb{C})}) \circ \phi_A]$$

(by Lemma 5.1.9)

$$= [\Theta_B \circ (((id_\mathcal{K} \otimes \gamma^B) \circ \rho_B \circ q(\alpha)) \otimes id_{M_2(\mathbb{C})}) \circ \phi_A]$$

$$= [\Theta_B \circ ((\gamma^{\mathcal{K} \otimes B} \circ q(\alpha)) \otimes id_{M_2(\mathbb{C})}) \circ \phi_A] \quad \text{(by Lemma 5.1.10)}$$

$$= [\Theta_B \circ ((\alpha \circ \gamma^{qA}) \otimes id_{M_2(\mathbb{C})}) \circ \phi_A]$$

$$\text{(since } \gamma^{\mathcal{K} \otimes B} \circ q(\alpha) = \alpha \circ \gamma^{qA} \quad \text{by (5.1.1))}$$

$$= [\Theta_B \circ (\alpha \otimes id_{M_2(\mathbb{C})}) \circ j_1] \quad \text{(by Theorem 5.1.8)}$$

$$= [\alpha] \quad \text{(by Lemma 1.3.11).}$$

(iii) : Let $\beta \in \mathrm{Hom}\,(qA, \mathcal{K} \otimes qB)$. Then

$$\rho_B \circ q((id_\mathcal{K} \otimes \gamma^B) \circ \beta) = \rho_B \circ q(id_\mathcal{K} \otimes \gamma^B) \circ q(\beta) \sim \gamma^{\mathcal{K} \otimes qB} \circ q(\beta)$$

by Lemma 5.1.11. Since $\gamma^{\mathcal{K} \otimes qB} \circ q(\beta) = \beta \circ \gamma^{qA}$ by (5.1.1), we find that

$$\lambda \circ \mu[\beta] = [\Theta_{qB} \circ ((\beta \circ \gamma^{qA}) \otimes id_{M_2(\mathbb{C})}) \circ \phi_A]$$

$$= [\Theta_{qB} \circ (\beta \otimes id_{M_2(\mathbb{C})}) \circ j_1] \quad \text{(by Theorem 5.1.8)}$$

$$= [\beta] \quad \text{(by Lemma 1.3.11).} \qquad \square$$

From E4.1.4 and Theorem 5.1.6 we know that $[\mathcal{K} \otimes qA, \mathcal{K} \otimes qB]$ is an abelian group isomorphic to $KK_c(A, qB) = [qA, \mathcal{K} \otimes qB]$. The isomorphism E_{qA} is given by

$$E_{qA}[\alpha] = [\alpha \circ e_{qA}], \quad \alpha \in \mathrm{Hom}\,(\mathcal{K} \otimes qA, \mathcal{K} \otimes qB),$$

where $e_{qA} : qA \to \mathcal{K} \otimes qA$ is given by choosing a minimal projection e in $\mathcal{K}$; e_{qA} is defined by $e_{qA}(x) = e \otimes x$, $x \in qA$. Hence we have the following corollary to Theorem 5.1.12 :

Corollary 5.1.13. *Assume that A is separable. Then*

$$\Psi_{A,B} = E_{qA}^{-1} \circ \Phi_{A,B} : KK_c(A,B) \to [\mathcal{K} \otimes qA, \mathcal{K} \otimes qB]$$

is an isomorphism of abelian groups. $\square$

Now let $f : B \to C$ and $g : D \to A$ be homomorphisms. We can then define $f_* : [\mathcal{K} \otimes qA, \mathcal{K} \otimes qB] \to [\mathcal{K} \otimes qA, \mathcal{K} \otimes qC]$ by

$$f_*[\phi] = [(id_{\mathcal{K}} \otimes q(f)) \circ \phi], \quad \phi \in \mathrm{Hom}\,(\mathcal{K} \otimes qA, \mathcal{K} \otimes qB)$$

and $g^* : [\mathcal{K} \otimes qA, \mathcal{K} \otimes qB] \to [\mathcal{K} \otimes qD, \mathcal{K} \otimes qB]$ by

$$g^*[\phi] = [\phi \circ (id_{\mathcal{K}} \otimes q(f))], \quad \phi \in \mathrm{Hom}\,(\mathcal{K} \otimes qA, \mathcal{K} \otimes qB).$$

These definitions are compatible with the functoriality of $KK_c(A,B)$ under the isomorphism $\Psi_{A,B}$:

Lemma 5.1.14. *Assume that A and D are separable. Then*

$$\Psi_{D,B} \circ g^* = g^* \circ \Psi_{A,B} \quad and \quad \Psi_{A,C} \circ f_* = f_* \circ \Psi_{A,B}.$$

Proof. Let $\phi \in \mathrm{Hom}\,(\mathcal{K} \otimes qA, \mathcal{K} \otimes qB)$. Then

$$\Psi_{D,B}^{-1} \circ g^*[\phi] = \Phi_{D,B}^{-1} \circ E_{qD}[\phi \circ (id_{\mathcal{K}} \otimes q(g))]$$

$$= \Phi_{D,B}^{-1} \circ [\phi \circ (id_{\mathcal{K}} \otimes q(g)) \circ e_{qD}]$$

$$= [(id_{\mathcal{K}} \otimes \gamma^B) \circ \phi \circ (id_{\mathcal{K}} \otimes q(g)) \circ e_{qD}].$$

On the other hand

$$g^* \circ \Psi_{A,B}^{-1}[\phi] = g^* \circ \Phi_{A,B}^{-1} \circ E_{qA}[\phi]$$

$$= g^*[(id_{\mathcal{K}} \otimes \gamma^B) \circ \phi \circ e_{qA}]$$

$$= [(id_{\mathcal{K}} \otimes \gamma^B) \circ \phi \circ e_{qA} \circ q(g)].$$

Since $e_{qA} \circ q(g) = (id_{\mathcal{K}} \otimes q(g)) \circ e_{qD}$, this proves the first identity. The second follows in a similar way by using that $f \circ \gamma^B = \gamma^C \circ q(f)$, cf. (5.1.1). $\square$

For any C^*-algebra A, we set $1_A = [e_A \circ \gamma^A] \in [qA, \mathcal{K} \otimes A] = KK_c(A,A)$.

Theorem 5.1.15. *Let* A, B, C, D *be* C^*-*algebras. When* A *and* B *are separable, there is a bilinear pairing*

$$\cdot : KK_c(A, B) \times KK_c(B, C) \to KK_c(A, C)$$

satisfying the following relations

(i) *If* $f : D \to A$ *is a* *-*homomorphism and* D *is separable, then*

$$f^*(x \cdot y) = f^*(x) \cdot y, \quad x \in KK_c(A, B), \ y \in KK_c(B, C),$$

(ii) *If* $g : C \to D$ *is a* *-*homomorphism, then*

$$g_*(x \cdot y) = x \cdot g_*(y), \quad x \in KK_c(A, B), \ y \in KK_c(B, C),$$

(iii) *If* $h : B \to D$ *is a* *-*homomorphism and* D *is separable, then*

$$h_*(x) \cdot y = x \cdot h^*(y), \quad x \in KK_c(A, B), \ y \in KK_c(D, C),$$

(iv) $1_A \cdot x = x = x \cdot 1_B, \ x \in KK_c(A, B),$

(v) *If* C *is separable,*

$$x \cdot (y \cdot z) = (x \cdot y) \cdot z, \quad x \in KK_c(A, B), \ y \in KK_c(B, C), \ z \in KK_c(C, D).$$

Proof. Combine Corollary 5.1.13, Lemma 5.1.14 with E 1.3.5.   □

5.1.16   *Notes and remarks.*

The all important Theorem 5.1.8 was established in [5]. The idea of using Theorem 1.1.26 to lift the automorphism group $\mathrm{Ad}\, R_t$ in the proof seems to be new. Cuntz's proof used a derivation lifting result of Pedersen, which in turn was based on the Borchers-Arveson spectral theory for one-parameter groups of automorphisms. Corollary 5.1.13 was pointed out by Cuntz in [6].

Exercise 5.1

E 5.1.1

Show that $KK_c(A, B)$ is a two-variable functor from the category of C^*-algebras to the category of abelian groups, contravariant in the first variable, covariant in the second and homotopy invariant in both.

E 5.1.2

Let $\phi : A \to \mathcal{B}(\mathcal{H})$ be a representation of the C^*-algebra A on the Hilbert space $\mathcal{H}$.

(a) Let $m \in \mathcal{M}(A)$. Show that there is an element z in the weak closure $\phi(A)''$ of $\phi(A)$ such that $\phi(ma) = z\phi(a)$, $a \in A$.

[Hint : Let z be a weak condensation point for $\{\phi(u_i m)\}$ where $\{u_i\}$ is an approximate unit for A.]

(b) Show that there is a representation $\underline{\phi} : \mathcal{M}(A) \to \mathcal{B}(\mathcal{H})$ extending ϕ.

(c) Assume now that A is an ideal in B. Show that there is a representation $\psi : B \to \mathcal{B}(\mathcal{H})$ such that

$$
\begin{array}{ccc}
B & \xrightarrow{\psi} & \mathcal{B}(\mathcal{H}) \\
\cup & & \| \\
A & \xrightarrow[\phi]{} & \mathcal{B}(\mathcal{H})
\end{array}
$$

commutes.

E 5.1.3

In the next section we shall use the fact that any $*$-homomorphism $\phi : A \to B$ gives rise to the element $\{\phi\} \in KK_c(A,B)$ given by $\{\phi\} = [e_B \circ \gamma^B \circ q(\phi)] = [e_B \circ \phi \circ \gamma^A]$.

 (i) Show that $\Psi_{A,B}\{\phi\} = [id_\mathcal{K} \otimes q(\phi)]$.

(ii) Assume that $\psi : B \to C$ is a $*$-homomorphism and that A and B are separable. Prove that $\{\phi\} \cdot \{\psi\} = \{\psi \circ \phi\}$.

(iii) Prove that $x \cdot \{\psi\} = \psi_*(x)$, for all $x \in KK_c(A,B)$, and that $\{\phi\} \cdot y = \phi^*(y)$, $y \in KK_c(B,C)$.

5.2. $KK_c(A,B) \simeq KK^0(A,B)$

The strategy for proving that the bifunctor KK_c is another version of KK-theory as developed in the previous chapters will be to show that (for A separable) $KK_c(A, \cdot)$ is the universal covariant functor which is stable, homotopy invariant and split exact. It is trivial that $KK_c(A, \cdot)$ is homotopy invariant (for all C^*-algebras A). We establish the other properties below.

Lemma 5.2.1. *For any C^*-algebra A, the covariant functor $KK_c(A, \cdot)$ is stable.*

Proof. We must show that $e_B : B \to \mathcal{K} \otimes B$ induce a group isomorphism $e_{B*} : [qA, \mathcal{K} \otimes B] \to [qA, \mathcal{K} \otimes \mathcal{K} \otimes B]$. Let $\lambda_B : \mathcal{K} \otimes B \to \mathcal{K} \otimes \mathcal{K} \otimes B$ be the *-isomorphism from Lemma 4.1.13. Then we can define $\lambda : [qA, \mathcal{K} \otimes \mathcal{K} \otimes B] \to [qA, \mathcal{K} \otimes B]$ by

$$\lambda[\phi] = [\lambda_B^{-1} \circ \phi], \quad \phi \in \mathrm{Hom}\,(qA, \mathcal{K} \otimes \mathcal{K} \otimes B).$$

Since $id_{\mathcal{K}} \otimes e_B \sim e_{\mathcal{K} \otimes B}$ by E 5.2.4 and $\lambda_B^{-1} \circ e_{\mathcal{K} \otimes B} \sim id_{\mathcal{K} \otimes B}$ by Lemma 4.1.13, we find that $\lambda \circ e_{B*} = id$ on $[qA, \mathcal{K} \otimes B]$. Since $(id_{\mathcal{K}} \otimes e_B) \circ \lambda_B^{-1} \sim e_{\mathcal{K} \otimes B} \circ \lambda_B^{-1} \sim id_{\mathcal{K} \otimes \mathcal{K} \otimes B}$, we find that $e_{B*} \circ \lambda = id$ on $[qA, \mathcal{K} \otimes \mathcal{K} \otimes B]$. $\square$

The following lemma will not be used explicitly in the following, but it is included because it shows how the crucial split-exactness can be established in the present setting.

Lemma 5.2.2. *For separable A, the functor $KK_c(A, \cdot)$ is split exact.*

Proof. Let $0 \to J \xrightarrow{j} E \xrightarrow{p} B \longrightarrow 0$ be a short split exact sequence. Let $s : B \to E$ be a splitting map, i.e. s is a *-homomorphism such that $p \circ s = id_B$. By E 5.2.1 it suffices to consider the case where $s : B \to E$ is unital.

Note that there is a commutative diagram

$$
\begin{array}{ccccccccc}
0 & \longrightarrow & J & \xrightarrow{j} & E & \xrightarrow{p} & B & \longrightarrow & 0 \\
& & e_J \downarrow & & e_E \downarrow & & e_B \downarrow & & \\
0 & \longrightarrow & \mathcal{K} \otimes J & \xrightarrow[id_{\mathcal{K}} \otimes j]{} & \mathcal{K} \otimes E & \xrightarrow[id_{\mathcal{K}} \otimes p]{} & \mathcal{K} \otimes B & \longrightarrow & 0
\end{array}
$$

Note also that the splitting map, $id_\mathcal{K} \otimes s$, for the lower extension is quasi-unital with unital extension $\underline{id_\mathcal{K} \otimes s} : \mathcal{M}(\mathcal{K} \otimes B) \to \mathcal{M}(\mathcal{K} \otimes E)$. Since we established stability above we can use this diagram to reduce to the case where J, E and B are stable and s is quasi-unital with $\underline{s} : \mathcal{M}(B) \to \mathcal{M}(E)$ unital. Therefore we assume that this is the case.

Consider $(p, id_E), (0, s \circ p) \in \mathrm{Hom}\,(E, B \oplus E)$. These *-homomorphisms are given by $(p, id_E)(e) = (p(e), e)$ and $(0, s \circ p)(e) = (0, s \circ p(e))$, respectively. Then

$$(p, id_E)(e) - (0, s \circ p)(e) = (p(e), e - s(p(e))) \in B \oplus j(J), \ e \in E.$$

Thus we can define $\phi : qE \to \mathcal{K} \otimes (B \oplus J)$ by

$$\phi = e_{B \oplus J} \circ (id_B \oplus j^{-1}) \circ q((p, id_E), (0, s \circ p)).$$

Now define $\lambda \in \mathrm{Hom}\,(B \oplus J, E)$ by $\lambda(b, x) = \Theta_E \begin{bmatrix} s(b) & 0 \\ 0 & j(x) \end{bmatrix}$, for $b \in B, x \in J$, where $\Theta_E : M_2(E) \to E$ is an inner *-isomorphism. We assert that the following hold

(5.2.1) $[\phi] \cdot \{\lambda\} = 1_E$ and

(5.2.2) $\{\lambda\} \cdot [\phi] = 1_{B \oplus J}.$

Here $\{\lambda\} \in KK_c(B \oplus J, E)$ is the element represented by

$$e_E \circ \gamma^E \circ q(\lambda) = e_E \circ \lambda \circ \gamma^{B \oplus J} \in \mathrm{Hom}\,(q(B \oplus J), \mathcal{K} \otimes E),$$

cf. E 5.1.3.

To prove (5.2.1) we first establish the identity

(5.2.3) $\lambda \circ (id_B \oplus j^{-1}) \circ q((p, id_E), (0, s \circ p)) =$

$$\Theta_E \circ q \left(\begin{bmatrix} s \circ p & 0 \\ 0 & id_E \end{bmatrix}, \begin{bmatrix} 0 & 0 \\ 0 & s \circ p \end{bmatrix} \right).$$

To check (5.2.3) it suffices by Lemma 5.1.2 to check on elements of the form $\epsilon(e_1)(\epsilon(e_2) - \bar\epsilon(e_2)), \epsilon(e_2) - \bar\epsilon(e_2)$, where $\epsilon, \bar\epsilon : E \to qE$ are the canonical embeddings. We check the first case here and leave the second to the reader.

$\lambda \circ (id_B \oplus j^{-1}) \circ q((p, id_E), (0, s \circ p))(\epsilon(e_1)(\epsilon(e_2) - \bar\epsilon(e_2))) =$

$\lambda \circ (id_B \oplus j^{-1})(p(e_1 e_2), e_1 e_2 - e_1 s \circ p(e_2)) =$

$\Theta_E \begin{bmatrix} s \circ p(e_1 e_2) & 0 \\ 0 & e_1(e_2 - s \circ p(e_2)) \end{bmatrix} =$

$\Theta_E \circ q \left(\begin{bmatrix} s \circ p & 0 \\ 0 & id_E \end{bmatrix}, \begin{bmatrix} 0 & 0 \\ 0 & s \circ p \end{bmatrix} \right) (\epsilon(e_1)(\epsilon(e_2) - \bar\epsilon(e_2))).$

Note that an obvious rotation argument shows that

$$
q\left(\begin{bmatrix} s \circ p & 0 \\ 0 & id_E \end{bmatrix}, \begin{bmatrix} 0 & 0 \\ 0 & s \circ p \end{bmatrix}\right) \sim q\left(\begin{bmatrix} id_E & 0 \\ 0 & s \circ p \end{bmatrix}, \begin{bmatrix} 0 & 0 \\ 0 & s \circ p \end{bmatrix}\right)
$$

$$
= q\left(\begin{bmatrix} id_E & 0 \\ 0 & 0 \end{bmatrix}, 0\right) = j_1 \circ \gamma^E,
$$

where $j_1 : E \to M_2(E)$ is the embedding into the upper left-hand corner. Thus

$$
\Theta_E \circ q\left(\begin{bmatrix} s \circ p & 0 \\ 0 & id_E \end{bmatrix}, \begin{bmatrix} 0 & 0 \\ 0 & s \circ p \end{bmatrix}\right) \sim \gamma^E
$$

in $\mathrm{Hom}\,(qE, E)$ by Lemma 1.3.11.

With the aid of (5.2.3) and E 5.1.3 we can now easily obtain (5.2.1) :

$$
[\phi] \cdot \{\lambda\} = \lambda_*[\phi] = [(id_{\mathcal{K}} \otimes \lambda) \circ e_{B \oplus J} \circ (id_B \oplus j^{-1}) \circ q((p, id_E), (0, s \circ p))]
$$

$$
= [e_E \circ \lambda \circ (id_B \oplus j^{-1}) \circ q((p, id_E), (0, s \circ p))]
$$

$$
= \left[e_E \circ \Theta_E \circ q\left(\begin{bmatrix} s \circ p & 0 \\ 0 & id_E \end{bmatrix}, \begin{bmatrix} 0 & 0 \\ 0 & s \circ p \end{bmatrix}\right)\right]
$$

$$
= [e_E \circ \gamma^E]
$$

$$
= 1_E.
$$

To prove (5.2.2) we observe first that because $\underline{s} : \mathcal{M}(B) \to \mathcal{M}(E)$ is unital we can assume that the isometries $V_1, V_2 \in \mathcal{M}(E)$ used to define Θ_E have the form $V_i = \underline{s}(W_i)$ for some isometries $W_1, W_2 \in \mathcal{M}(B)$. With such a choice we see that $\underline{s} \circ \underline{p}(V_i) = V_i$, $i = 1, 2$, because $\underline{p} \circ \underline{s} = id_{\mathcal{M}(B)}$. This will be handy in the following calculations. Let $b \in B$, $x \in J$. Then

$$
(p, id_E) \circ \lambda(b, x) = (p, id_E)(\mathrm{Ad}\, V_1 \circ s(b) + \mathrm{Ad}\, V_2 \circ j(x))
$$

$$
= (\mathrm{Ad}\,\underline{p}(V_1)(b), \mathrm{Ad}\, V_1 \circ s(b) + \mathrm{Ad}\, V_2 \circ j(x))
$$

and

$$
(0, s \circ p) \circ \lambda(b, x) = (0, s \circ p)(\mathrm{Ad}\, V_1 \circ s(b) + \mathrm{Ad}\, V_2 \circ j(x))
$$

$$
= (0, \mathrm{Ad}\,\underline{s} \circ \underline{p}(V_1) \circ s(b))
$$

$$
= (0, \mathrm{Ad}\, V_1 \circ s(b)).
$$

By using Lemma 5.1.2 and that $V_1^* V_2 = 0$ it follows now straightforwardly that $q((p, id_E) \circ \lambda, (0, s \circ p) \circ \lambda) = q(\mathrm{Ad}\,\underline{p}(V_1) \oplus \mathrm{Ad}\, V_2 \circ j, 0)$. Since $\underline{p}(V_1)$ and V_2 are isometries, we conclude from Lemma 1.3.7 that $\mathrm{Ad}\,\underline{p}(V_1) \sim id_B$ and $\mathrm{Ad}\, V_2 \circ j \sim j$. It follows that $q(\mathrm{Ad}\,\underline{p}(V_1) \oplus \mathrm{Ad}\, V_2 \circ j, 0) \sim q(id_B \oplus j, 0) = (id_B \oplus j) \circ \gamma^{B \oplus J}$.

We can now easily check (5.2.2) by using E 5.1.3 :

$$\{\lambda\} \cdot [\phi] = \lambda^*[\phi]$$
$$= [e_{B \oplus J} \circ (id_B \oplus j^{-1}) \circ q((p, id_E), (0, s \circ p)) \circ q(\lambda)]$$
$$= [e_{B \oplus J} \circ (id_B \oplus j^{-1}) \circ q((p, id_E) \circ \lambda, (0, s \circ p) \circ \lambda)]$$
$$= [e_{B \oplus J} \circ (id_B \oplus j^{-1}) \circ (id_B \oplus j) \circ \gamma^{B \oplus J}]$$
$$= [e_{B \oplus J} \circ \gamma^{B \oplus J}]$$
$$= 1_{B \oplus J}.$$

Theorem 5.1.15, (5.2.1) and (5.2.2) show that "multiplication" with $\{\lambda\}$ gives a group isomorphism $KK_c(A, B \oplus J) \to KK_c(A, E)$. Thus we see that $\lambda_* : KK_c(A, B \oplus J) \to KK_c(A, E)$ is an isomorphism.

Let $s_B : B \to B \oplus J$ and $i_J : J \to B \oplus J$ be the natural inclusions. Then

$$\lambda \circ i_J = \Theta_E \circ \begin{bmatrix} 0 & 0 \\ 0 & j \end{bmatrix} = \operatorname{Ad} V \circ j$$

for some isometry $V \in \mathcal{M}(E)$. Thus $\lambda \circ i_J \sim j$ by Lemma 1.3.7. Hence $\lambda_* \circ i_{J*} = j_*$. In the same way $\lambda \circ s_B \sim s$ and $\lambda_* \circ s_{B*} = s_*$. To show that $s_* \oplus j_* : KK_c(A, B) \oplus KK_c(A, J) \to KK_c(A, E)$ is an isomorphism is now the same as proving that $s_{B*} \oplus i_{J*} : KK_c(A, B) \oplus KK_c(A, J) \to KK_c(A, B \oplus J)$ is an isomorphism. This is left to the reader, cf. E 5.2.2. $\square$

Theorem 5.2.3. *Let F be a covariant functor from the category of σ-unital C^*-algebras to the category of abelian groups. Assume that F is stable, homotopy invariant and split exact. Let A be a separable C^*-algebra and $x \in F(A)$ an arbitrary element. Then there is a unique natural transformation $T : KK_c(A, \cdot) \to F(\cdot)$ such that $T_A(1_A) = x$.*

Proof. To prove uniqueness observe that we can define a map $S_B : KK_h(A, B) \to KK_c(A, B)$ by $S_B[\phi_+, \phi_-] = [q(\phi_+, \phi_-)]$, for $(\phi_+, \phi_-) \in \mathbb{F}(A, B)$. By E 5.2.3, $S : KK_h(A, \cdot) \to KK_c(A, \cdot)$ is a natural transformation and $S_A(1_A) = 1_A$. We assert that if T^1, T^2 are two natural transformations with $T_A^i(1_A) = x$, then $T^1 S = T^2 S$. To see this let $\Phi = (\phi_+, \phi_-) \in \mathbb{F}(A, B)$ where B is σ-unital. Then

$$T_B^i \circ S_B([\phi_+, \phi_-]) = T_B^i \circ S_B(\Phi_{KK_h(A, \cdot)}(1_A))$$
$$= \Phi_F \circ T_A^i \circ S_A(1_A)$$
$$= \Phi_F(x), \quad i = 1, 2,$$

by Proposition 4.2.7 and E 4.2.1 (v). Hence $T^1 \circ S = T^2 \circ S$ as asserted. To prove uniquenes it is therefore enough to prove that each $S_B : KK_h(A,B) \rightarrow KK_c(A,B)$ is surjective. Let $\phi \in \text{Hom}\,(qA, \mathcal{K} \otimes B)$. To show that $[\phi] \in im\, S_B$ we assume, by Lemma 1.3.19, that ϕ is quasi-unital. Thus we can extend ϕ to a map between multiplier algebras, and since qA is an ideal in QA this gives us a commutative diagram

$$
\begin{array}{ccc}
QA & \xrightarrow{\psi} & \mathcal{M}(\mathcal{K} \otimes B) \\
\cup & & \cup \\
qA & \xrightarrow[\phi]{} & \mathcal{K} \otimes B
\end{array}
$$

Let $i, \bar{i} : A \rightarrow QA$ be the canonical inclusions. It is then a simple matter to check that $(\psi \circ i, \psi \circ \bar{i}) \in \mathbb{F}(A,B)$ and that $S_B[\psi \circ i, \psi \circ \bar{i}] = [\phi]$. Hence S_B is surjective, completing the proof of the uniquenes of T.

To prove the existence we consider the extension

$$
0 \longrightarrow qA \xrightarrow{\ j\ } QA \xrightarrow{Q(id_A, id_A)} A \longrightarrow 0,
$$

where j denotes the inclusion of qA as an ideal in QA, cf. Lemma 5.1.2. The extension is split by any of the two canonical inclusions $i, \bar{i} : A \rightarrow QA$. When $\phi : B \rightarrow C$ is a *-homomorphism we let $F(\phi) : F(B) \rightarrow F(C)$ denote the homomorphism determined by the given functor F. Since F is split exact and $F(Q(id_A, id_A)) \circ (F(i) - F(\bar{i})) = F(id_A) - F(id_A) = 0$ we see that $F(i) - F(\bar{i})$ maps $F(A)$ into the image of $F(j)$. As $F(j)$ is injective because F is split exact, we can define $F(j)^{-1} \circ (F(i) - F(\bar{i}))$. Let $\phi \in \text{Hom}\,(qA, \mathcal{K} \otimes B)$. Since F is stable, we get a homomorphism $\mathcal{F}(\phi) : F(A) \rightarrow F(B)$ defined by $\mathcal{F}(\phi) = F(e_B)^{-1} \circ F(\phi) \circ F(j)^{-1} \circ (F(i) - F(\bar{i}))$. Since F is homotopy invariant we get a map $T_B : KK_c(A,B) \rightarrow F(B)$ by setting $T_B[\phi] = \mathcal{F}(\phi)(x)$, $\phi \in \text{Hom}\,(qA, \mathcal{K} \otimes B)$. T_B is a homomorphism by E 5.2.5. To see that T is a natural transformation, let $f : B \rightarrow C$ be a *-homomorphism. Then $f_*[\phi] = [(id_\mathcal{K} \otimes f) \circ \phi]$ so

$$
\begin{aligned}
T_C \circ f_*[\phi] &= F(e_C)^{-1} \circ F(id_\mathcal{K} \otimes f) \circ F(\phi) \circ F(j) \circ (F(i) - F(\bar{i}))(x) \\
&= F(f) \circ F(e_B)^{-1} \circ F(\phi) \circ F(j) \circ (F(i) - F(\bar{i}))(x) \\
&= F(f) \circ T_B[\phi], \quad \phi \in \text{Hom}\,(qA, \mathcal{K} \otimes B),
\end{aligned}
$$

since $(id_\mathcal{K} \otimes f) \circ e_B = e_C \circ f$.

Note finally that

$$
\begin{aligned}
T_A(1_A) &= F(e_A)^{-1} \circ F(e_A \circ \gamma^A) \circ F(j)^{-1} \circ (F(i) - F(\bar{i}))(x) \\
&= F(\gamma^A) \circ F(j)^{-1} \circ (F(i) - F(\bar{i}))(x) \\
&= F(Q(id_A, 0)) \circ (F(i) - F(\bar{i}))(x) \\
&= F(id_A)(x) \\
&= x,
\end{aligned}
$$

where we have used that $\gamma^A = Q(id_A, 0) \circ j$. $\square$

We can now prove that $KK_c(A, B)$ agrees with $KK_h(A, B)$ when A is separable and B is σ-unital.

Theorem 5.2.4. *Let A be a separable C^*-algebra and B a σ-unital C^*-algebra. Then the map $S_B : KK_h(A, B) \to KK_c(A, B)$ given by $S_B[\phi_+, \phi_-] = [q(\phi_+, \phi_-)]$, $(\phi_+, \phi_-) \in \mathbb{F}(A, B)$, is an isomorphism.*

Proof. In the proof of Theorem 5.2.3 we proved the surjectivity of S_B. Consider the natural transformation $S : KK_h(A, \cdot) \to KK_c(A, \cdot)$ of functors. Since $KK_h(A, \cdot)$ is split-exact, homotopy invariant and stable by the results of Chapter 4, Theorem 5.2.3 shows that there is a natural transformation $T : KK_c(A, \cdot) \to KK_h(A, \cdot)$ such that $T_A(1_A) = 1_A$. Then $TS : KK_h(A, \cdot) \to KK_h(A, \cdot)$ is a natural transformation fixing 1_A. As in the proof of Theorem 5.2.3, we can now use Proposition 4.2.7 to conclude that TS is the identity transformation. In particular, $T_B \circ S_B = id$ on $KK_h(A, B)$, proving that S_B is injective. $\square$

The natural isomorphism $KK_c(A, B) \simeq KK_h(A, B)$ makes it possible to transfer the bilinear pairing described in Theorem 5.1.15 to KK_h. By Theorem 4.2.9 the bilinear pairing obtained in this way agrees with the Kasparov product (for separable C^*-algebras).

5.25   *Notes and remarks.*

This section is a mixture of Section 2 of [5] and Section 3 of [13].

Exercise 5.2

E 5.2.1

Let F be a covariant functor from the category of C^*-algebras to the category of abelian groups. Assume that F is split exact with respect to split exact sequences

$$0 \longrightarrow J \longrightarrow E \longrightarrow B \longrightarrow 0$$

with E and B unital and with a unital splitting *-homomorphism $s : B \rightarrow E$. Show that F is then split exact (with respect to arbitrary split exact extensions of C^*-algebras).

[Hint : Consider the commutative diagram

$$
\begin{array}{ccccccccc}
0 & \longrightarrow & J & \longrightarrow & E & \longrightarrow & B & \longrightarrow & 0 \\
& & \| & & \downarrow & & \downarrow & & \\
0 & \longrightarrow & J & \longrightarrow & E^+ & \longrightarrow & B^+ & \longrightarrow & 0 \\
& & & & \downarrow & & \downarrow & & \\
& & & & \mathbb{C} & = & \mathbb{C} & & \\
& & & & \downarrow & & \downarrow & & \\
& & & & 0 & & 0 & &
\end{array}
$$

where E^+ and B^+ denote the C^*-algebras obtained by a adjoining a unit.]

E 5.2.2

Let $E = A_1 \oplus A_2$ and let $i_1 : A_1 \rightarrow E$, $i_2 : A_2 \rightarrow E$ be the natural inclusions. Show that $i_{1*} \oplus i_{2*} : KK_c(B, A_1) \oplus KK_c(B, A_2) \rightarrow KK_c(B, E)$ is an isomorphism for all C^*-algebras B.

E 5.2.3

Fix a σ-unital C^*-algebra A. Prove that there is natural transformation $S : KK_h(A, \cdot) \rightarrow KK_c(A, \cdot)$ of functors defined on the category of σ-unital C^*-algebras given by

$$S_B[\phi_+, \phi_-] = [q(\phi_+, \phi_-)], \quad (\phi_+, \phi_-) \in \mathbb{F}(A, B).$$

Show that $S_A(1_A) = 1_A$.

E 5.2.4

Let B be an arbitrary C^*-algebra. Show that $id_{\mathcal{K}} \otimes e_B \sim e_{\mathcal{K} \otimes B}$.

E 5.2.5

Let F be a covariant functor from the category of σ-unital C^*-algebras to abelian groups. Assume that F is additive in the following sense. When A_1, A_2 are σ-unital C^*-algebras and $i_1 : A_1 \to A_1 \oplus A_2$, $i_2 : A_2 \to A_1 \oplus A_2$ the natural inclusions, then $i_{1*} \oplus i_{2*} : F(A_1) \oplus F(A_2) \to F(A_1 \oplus A_2)$ is an isomorphism.

Let $\phi, \psi : A \to B$ be *-homomorphisms between σ-unital C^*-algebras satisfying $\phi(a_1)\psi(a_2) = 0$, $a_1, a_2 \in A$. Show that $(\phi + \psi)_* = \phi_* + \psi_*$ as maps from $F(A)$ to $F(B)$.

Assume in addition that F is homotopy invariant and let $f_1, f_2 : A \to \mathcal{K} \otimes B$ be arbitrary *-homomorphisms and $\Theta_B : M_2(\mathcal{K} \otimes B) \to \mathcal{K} \otimes B$ an inner *-isomorphism. Show that $\left(\Theta_B \circ \begin{bmatrix} f_1 & 0 \\ 0 & f_2 \end{bmatrix} \right)_* = f_{1*} + f_{2*}$.

APPENDIX A

In this appendix we prove a result on Fredholm operators used in the proof of Kasparov's homotopy invariance theorem, Theorem 2.2.17, and which is not included in [10]. First we need a lemma.

Lemma A.1. *Let* $\pi : A \to B$ *be a surjective *-homomorphism between unital C^*-algebras. If u is a unitary in B which can be connected to the identity of B through a continuous path of unitaries, then there is a unitary $w \in A$ such that $\pi(w) = u$.*

Proof. Let u_t, $t \in [0,1]$, be a continuous path of unitaries such that $u_0 = 1$, $u_1 = u$. By norm continuity there are points $t_0, t_1, t_2, \ldots t_n \in [0,1]$, such that $\|u_{t_i} - u_{t_{i-1}}\| < 1$, $i = 1, 2, 3, \ldots, n$, $t_0 = 0$ and $t_n = 1$. Since $\|1 - u_{t_i} u_{t_{i-1}}^*\| < 1$, there is a gap in the spectrum of the unitary $u_{t_i} u_{t_{i-1}}^*$ so that we can define $\log u_{t_i} u_{t_{i-1}}^*$, $i = 1, 2, \ldots, n$. Thus $d_j = -i \log u_{t_j} u_{t_{j-1}}^*$ is self-adjoint and satisfies $\exp i d_j = u_{t_j} u_{t_{j-1}}^*$, $j = 1, 2, 3, \ldots, n$. Thus $u = \exp i d_1 \exp i d_2 \cdots \exp i d_n$. Choose self-adjoint elements $c_1, c_2, \ldots, c_n \in A$ such that $\pi(c_j) = d_j$, $j = 1, 2, 3, \ldots, n$. Then $w = \exp i c_1 \exp i c_2 \cdots \exp i c_n$ is the desired unitary. $\qquad\square$

Proposition A.2 *Let $\mathcal{H}$ be a separable infinite dimensional Hilbert space. For each $n \in \mathbb{Z}$ the set of Fredholm operators of index n which are unitary mod $\mathcal{K}(\mathcal{H})$ is an arc-wise connected subset of $\mathcal{B}(\mathcal{H})$.*

Proof. We can assume that $n \geq 0$. If $U \in \mathcal{B}(\mathcal{H})$ is unitary mod $\mathcal{K}(\mathcal{H})$ and has index n, then $(U_+)^n U$ is unitary mod $\mathcal{K}(\mathcal{H})$ and has index 0. (Here U_+ denotes the unilateral shift.) Since $U = (U_+^*)^n (U_+)^n U$ we can assume that U has index 0. We can conclude the proof by showing that U in this case can be connected to the identity 1 of $\mathcal{B}(\mathcal{H})$ through a path of operators that are unitary mod $\mathcal{K}(\mathcal{H})$ and have index 0. By [10], Theorem 5.36, the image of U in the Calkin algebra $\mathcal{B}(\mathcal{H})/\mathcal{K}(\mathcal{H})$ can be connected to the identity of $\mathcal{B}(\mathcal{H})/\mathcal{K}(\mathcal{H})$ through a path of invertible elements. Since the image of U is unitary it follows by taking the unitary part of such a path that the image of U can be connected to the identity of $\mathcal{B}(\mathcal{H})/\mathcal{K}(\mathcal{H})$ through a path of unitaries. Hence the preceding lemma give a unitary $W \in \mathcal{B}(\mathcal{H})$ and a compact operaror $K \in \mathcal{K}(\mathcal{H})$ such that $U = W + K$. Then $U_t = W + tK$, $t \in [0,1]$, gives a path of operators that are unitary mod $\mathcal{K}(\mathcal{H})$ and connects U to W. By [10], Lemma 5.20, the Fredholm operators in this path all have index 0. Finally, we note that W can be connected by a path of unitaries to the identity of $\mathcal{B}(\mathcal{H})$ by e.g. [10], Proposition 5.29. $\square$

APPENDIX B

Free Products of C^*-Algebras

Let A and B be two C^*-algebras. A C^*-algebra C is called *the free product of A and B* if there are *-homomorphisms $i_A : A \to C$ and $i_B : B \to C$ with the following (universal) property: Given *-homomorpisms $\phi_A : A \to D$ and $\phi_B : B \to D$ mapping A and B into the same C^*-algebra D, there is a unique *-homomorphism $\phi : C \to D$ such that $\phi \circ i_A = \phi_A$ and $\phi \circ i_B = \phi_B$. The *-homomorphisms i_A and i_B are referred to as *the canonical inclusions*.

Theorem B.1. *For any pair of C^*-algebras A and B the free product C of A and B exists. The free product of A and B is unique in the following sense:*

*Let C' be another free product of A and B and let $i_A : A \to C$, $i_B : B \to C$, $i'_A : A \to C'$ and $i'_B : B \to C'$ be the corresponding canonical inclusions. Then there is a *-isomorphism $\psi : C \to C'$ such that $\psi \circ i_A = i'_A$ and $\psi \circ i_B = i'_B$.*

Proof. Let us prove uniquenes first. By the universal property of C there is a *-homomorphism $\psi : C \to C'$ such that $\psi \circ i_A = i'_A$ and $\psi \circ i_B = i'_B$. Similarly the universal property of C' gives a *-homomorphism $\phi : C' \to C$ such that $\phi \circ i'_A = i_A$ and $\phi \circ i'_B = i_B$. We assert that $\psi \circ \phi = id_{C'}$ and that $\phi \circ \psi = id_C$. But since $\psi \circ \phi \circ i'_A = \psi \circ i_A = i'_A = id_{C'} \circ i'_A$ and $\psi \circ \phi \circ i'_B = id_{C'} \circ i'_B$, the uniquenes condition appearing in the universal property implies that $\psi \circ \phi = id_{C'}$. In the same way $\phi \circ \psi = id_C$. Thus ψ is a *-isomorphism (with ϕ as inverse).

The proof of the existence of the free product of A and B is much longer since it involves some constructions.

For each $n = 1, 2, 3, \ldots$ we consider the following algebraic tensor products:

$A_n = A \otimes_{\mathbb{C}} B \otimes_{\mathbb{C}} A \otimes_{\mathbb{C}} B \otimes_{\mathbb{C}} A \cdots \otimes_{\mathbb{C}} B$ (n tensor factors), if n is even,

$A_n = A \otimes_{\mathbb{C}} B \otimes_{\mathbb{C}} A \otimes_{\mathbb{C}} B \otimes_{\mathbb{C}} A \cdots \otimes_{\mathbb{C}} A$($n$ tensor factors), if n is odd,

$B_n = B \otimes_{\mathbb{C}} A \otimes_{\mathbb{C}} B \otimes_{\mathbb{C}} A \otimes_{\mathbb{C}} B \cdots \otimes_{\mathbb{C}} A$($n$ tensor factors), if n is even,

$B_n = B \otimes_{\mathbb{C}} A \otimes_{\mathbb{C}} B \otimes_{\mathbb{C}} A \otimes_{\mathbb{C}} B \cdots \otimes_{\mathbb{C}} B$($n$ tensor factors), if n is odd.

Consider the vector space direct sum

$$E_0 = \sum_n \oplus (A_n \oplus B_n).$$

For n and m even there is a bilinear map $\cdot : A_n \times B_m \to A_{n+m-1}$ given on simple tensors by

$$a_1 \otimes_{\mathbb{C}} b_2 \otimes_{\mathbb{C}} a_3 \otimes_{\mathbb{C}} \cdots \otimes_{\mathbb{C}} b_n \cdot b_1 \otimes_{\mathbb{C}} a_2 \otimes_{\mathbb{C}} b_3 \otimes_{\mathbb{C}} \cdots \otimes_{\mathbb{C}} a_m =$$

$$a_1 \otimes_{\mathbb{C}} b_2 \otimes_{\mathbb{C}} a_3 \otimes_{\mathbb{C}} \cdots \otimes_{\mathbb{C}} a_{n-1} \otimes_{\mathbb{C}} b_n b_1 \otimes_{\mathbb{C}} a_2 \otimes_{\mathbb{C}} b_3 \cdots \otimes_{\mathbb{C}} a_m, \ a_i \in A, \ b_j \in B.$$

For n odd and m even $\cdot : A_n \times B_m \to A_{n+m}$ is given by

$$a_1 \otimes_{\mathbb{C}} b_2 \otimes_{\mathbb{C}} a_3 \otimes_{\mathbb{C}} \cdots \otimes_{\mathbb{C}} a_n \cdot b_1 \otimes_{\mathbb{C}} a_2 \otimes_{\mathbb{C}} b_3 \otimes_{\mathbb{C}} \cdots \otimes_{\mathbb{C}} a_m =$$

$$a_1 \otimes_{\mathbb{C}} b_2 \otimes_{\mathbb{C}} a_3 \otimes_{\mathbb{C}} \cdots \otimes_{\mathbb{C}} b_{n-1} \otimes_{\mathbb{C}} a_n \otimes_{\mathbb{C}} b_1 \otimes_{\mathbb{C}} a_2 \cdots \otimes_{\mathbb{C}} a_m.$$

Similarly for n and m even there is a bilinear map $\cdot : B_n \times A_m \to B_{n+m-1}$ given by

$$b_1 \otimes_{\mathbb{C}} a_2 \otimes_{\mathbb{C}} b_3 \otimes_{\mathbb{C}} \cdots \otimes_{\mathbb{C}} a_n \cdot a_1 \otimes_{\mathbb{C}} b_2 \otimes_{\mathbb{C}} a_3 \cdots \otimes_{\mathbb{C}} b_m =$$

$$b_1 \otimes_{\mathbb{C}} a_2 \otimes_{\mathbb{C}} b_3 \otimes_{\mathbb{C}} \cdots \otimes_{\mathbb{C}} b_{n-1} \otimes_{\mathbb{C}} a_n a_1 \otimes_{\mathbb{C}} b_2 \otimes_{\mathbb{C}} \cdots \otimes_{\mathbb{C}} b_m.$$

For n odd and m even $\cdot : B_n \times A_m \to B_{n+m}$ is given by

$$b_1 \otimes_{\mathbb{C}} a_2 \otimes_{\mathbb{C}} b_3 \otimes_{\mathbb{C}} \cdots \otimes_{\mathbb{C}} b_n \cdot a_1 \otimes_{\mathbb{C}} b_2 \otimes_{\mathbb{C}} \cdots \otimes_{\mathbb{C}} b_m =$$

$$b_1 \otimes_{\mathbb{C}} a_2 \otimes_{\mathbb{C}} b_3 \otimes_{\mathbb{C}} \cdots \otimes_{\mathbb{C}} b_n \otimes_{\mathbb{C}} a_1 \otimes_{\mathbb{C}} b_2 \otimes_{\mathbb{C}} \cdots \otimes_{\mathbb{C}} b_m.$$

In an analogous way we get bilinear maps $\cdot : A_n \times B_m \to A_{n+m-1}$ for n even and m odd, $\cdot : A_n \times B_m \to A_{n+m}$ for n and m odd, $\cdot : B_n \times A_m \to B_{n+m-1}$ for n even and m odd, $\cdot : B_n \times A_m \to B_{n+m}$ for n and m odd, $\cdot : A_m \times A_m \to A_{n+m-1}$ for n odd, $\cdot : A_n \times A_m \to A_{n+m}$ for n even, $\cdot : B_n \times B_m \to B_{n+m-1}$ for n odd, and $\cdot : B_n \times B_m \to B_{n+m}$ for n even.

We can then define a bilinear map $\cdot : E_0 \times E_0 \to E_0$ by

$(x_1, y_1, x_2, y_2, \dots) \cdot (x_1', y_1', x_2', y_2', \dots) = \sum_{i,j} (x_i \cdot x_j' + y_i \cdot x_j' + x_i \cdot y_j' + y_i \cdot y_j')$
$x_i, x_i' \in A_i, \ y_i, y_i' \in B_i.$

It is then straightforward (but rather tedious) to check that E_0 is an algebra with the product $\cdot$

To define an involution on E_0 we start by letting $* : A_1 \to A_1$ and $* : B_1 \to B_1$ be the involution on $A_1 = A$ and $B_1 = B$, respectively. For n even we define $* : A_n \to B_n$ as the conjugate linear map satisfying

$$(a_1 \otimes_\mathbb{C} b_1 \otimes_\mathbb{C} a_2 \cdots \otimes_\mathbb{C} a_{\frac{n}{2}} \otimes_\mathbb{C} b_{\frac{n}{2}})^* = b_{\frac{n}{2}}^* \otimes_\mathbb{C} a_{\frac{n}{2}}^* \otimes_\mathbb{C} \cdots \otimes_\mathbb{C} b_1^* \otimes_\mathbb{C} a_1^*,$$

$a_1, a_2, \ldots, a_{\frac{n}{2}} \in A$, $b_1, b_2, \ldots, b_{\frac{n}{2}} \in B$. $* : B_n \to A_n$ is then defined as the inverse of $* : A_n \to B_n$.

For n odd, we define $* : A_n \to A_n$ by

$$(a_1 \otimes_\mathbb{C} b_1 \otimes_\mathbb{C} a_2 \cdots \otimes_\mathbb{C} a_{\frac{n-1}{2}} \otimes_\mathbb{C} b_{\frac{n-1}{2}} \otimes_\mathbb{C} a_0)^* =$$

$$a_0^* \otimes_\mathbb{C} b_{\frac{n-1}{2}}^* \otimes_\mathbb{C} a_{\frac{n-1}{2}}^* \otimes_\mathbb{C} \cdots \otimes_\mathbb{C} a_2^* \otimes_\mathbb{C} b_1^* \otimes_\mathbb{C} a_1^*,$$

$a_0, a_1, a_2, \ldots, a_{\frac{n-1}{2}} \in A$, $b_1, b_2, \ldots, b_{\frac{n-1}{2}} \in B$. $* : B_n \to B_n$ is defined similarly.

We can then define $* : E_0 \to E_0$ by setting

$$(x_1, y_1, x_2, y_2, x_3, y_3, \ldots)^* = (x_1^*, y_1^*, y_2^*, x_2^*, x_3^*, y_3^*, y_4^*, x_4^*, \ldots),$$

$x_i \in A_i$, $y_i \in B_i$. Then we have turned E_0 into a *-algebra.

Suppose $\phi_A : A \to D$ and $\phi_B : B \to D$ are two *-homomorphisms. Define for n even $\phi_n : A_n \to D$ as the linear map given on simple tensors by

$$\phi(a_1 \otimes b_1 \otimes a_2 \otimes b_2 \otimes \cdots a_{\frac{n}{2}} \otimes b_{\frac{n}{2}}) =$$
$$\phi_A(a_1)\phi_B(b_1)\phi_A(a_2)\phi_B(b_2) \cdots \phi_A(a_{\frac{n}{2}})\phi_B(b_{\frac{n}{2}}).$$

In a similar way we define $\phi_n : A_n \to D$ for odd n and $\phi_n : B_n \to D$ for all n. Then we can define a *-homomorphism $\tilde{\phi} : E_0 \to D$ by $\tilde{\phi}(x_1, y_1, x_2, y_2, \ldots) = \sum_n \phi_n(x_n) + \phi_n(y_n)$, $x_n \in A_n$, $y_n \in B_n$.

A *-representation π of E_0 is a *-homomorphism $\pi : E_0 \to \mathcal{B}(\mathcal{H})$ from E_0 into the bounded operators on the Hilbert space $\mathcal{H}$. For any two C^*-algebras A and B we can find a Hilbert space $\mathcal{H}$ and *-representations of both A and B on $\mathcal{H}$. From these we can construct a *-representation of E_0 on $\mathcal{H}$ as above, so there are *-representations of E_0. For each $x \in E_0$, we set

$$\|x\| = \sup \{\|\pi(x)\| : \pi \text{ is a *-representation of } E_0\}.$$

Note that $x \to \pi(x, 0, 0, \ldots)$, $x \in A$, and $y \to \pi(0, y, 0, 0, \ldots)$, $y \in B$, define *-representations of A and B, respectively. Thus $\|\pi(x, 0, 0, \ldots)\| \leq \|x\|$, $x \in A$, and $\|\pi(0, y, 0, 0, \ldots)\| \leq \|y\|$, $y \in B$. Then since every element of E_0 is a finite sum of products of elements from $\{(x, 0, 0, \ldots), (0, y, 0, 0, \ldots) : x \in A, y \in B\}$, it follows that $\|z\|$ is finite for all $z \in E_0$. Note that $N = \{x \in E_0 : \|x\| = 0\}$ is a two-sided *-ideal in E_0 so that E_0/N is a *-algebra in the natural way and a norm is given on E_0/N by $\|x + N\| = \|x\|$, $x \in E_0$. Let C denote the completion of E_0/N in this norm. Then C is a C^*-algebra.

Define $i_A : A \to C$ and $i_B : B \to C$ by $i_A(x) = (x, 0, 0, 0, \ldots) + N$ and $i_B(y) = (0, y, 0, 0, 0, \ldots) + N$, respectively. We now only need to check that C, i_A and i_B have the universal property. So let $\phi_A : A \to D$ and $\phi_B : B \to D$ be two *-homomorphisms, then we can define $\tilde{\phi} : E_0 \to D$ as above.

Since D has a faithful representation on a Hilbert space it is clear that $\|\tilde{\phi}(x)\| \leq \|x\|$, $x \in E_0$. Thus $\tilde{\phi}$ induces a *-homomorphism $\phi : C \to D$ in the obvious way. Note that $\phi \circ i_A = \phi_A$ and $\phi \circ i_B = \phi_B$. So to finish the proof it suffices to check that ϕ is unique with these properties. To see this it suffices to prove that the ranges of i_A and i_B in C generate a dense *-subalgebra of C. But this is clear from the construction of E_0. $\square$

The free product of A and B will be denoted by $A * B$.

The material in this appendix is taken from [4].

APPENDIX C

Homotopy Invarians, Stability and Split-Exactness

Let F be a covariant functor from a full subcategory of C^*-algebras to the category of abelian groups. Let G be a contravariant functor between the same categories.

The only subcategories of the category of C^*-algebras we are interested in here are the category itself or the full subcategories consisting of either the σ-unital C^*-algebras or the separable C^*-algebras. Recall that "full" means that there are no restrictions on the *-homomorphisms. Let $\mathcal{C}$ denote the subcategory we consider. For the purposes in this appendix the only restriction we will require is that $\mathcal{K} \otimes A$ is an object in the subcategory $\mathcal{C}$ when A is.

Definition C.1. F is called *homotopy invariant* when $f_* = g_* :$ $F(A) \to F(B)$ whenever $f, g \in \mathrm{Hom}\,(A, B)$ are homotopic.

G is called *homotopy invariant* when $f^* = g^* : G(B) \to G(A)$ whenever $f, g \in \mathrm{Hom}\,(A, B)$ are homotopic.

Let $e \in \mathcal{K}$ be a minimal projection and define a *-homomorphism $e_A : A \to \mathcal{K} \otimes A$ by $e_A(a) = e \otimes a$, $a \in A$, $A \in \mathcal{C}$. It follows from some of the results presented in Section 1.3 that any other choice of a minimal projection in $\mathcal{K}$ gives rise to a *-homomorphism homotopic to e_A, cf. the remarks preceding Lemma 4.1.13.

Definition C.2. Assume that F and G are homotopy invariant. F is called *stable* when $e_{A*} : F(A) \to F(\mathcal{K} \otimes A)$ is an isomorphism for all $A \in \mathcal{C}$. Similarly, G is called *stable* when $e_A^* : G(\mathcal{K} \otimes A) \to G(A)$ is an isomorphism for all $A \in \mathcal{C}$.

Consider now three C^*-algebras $A, B, C \in \mathcal{C}$ and let $f : A \to B$ and $g : C \to B$ be *-homomorphisms. We can then define a homomorphism

$f_* \oplus g_* : F(A) \oplus F(C) \to F(B)$ by

$$f_* \oplus g_*(x,y) = f_*(x) + g_*(y), \quad x \in F(A), \ y \in F(C),$$

and a homomorphism $f^* \oplus g^* : G(B) \to G(A) \oplus G(C)$ by

$$f^* \oplus g^*(x) = (f^*(x), g^*(x)), \quad x \in G(B).$$

Definition C.3. F is called *split-exact* when the following condition is satisfied: For every split-exact sequence

(C 1) $$0 \longrightarrow A \xrightarrow{f} B \xrightarrow{p} C \longrightarrow 0$$

with splitting *-homomorphism $g : C \to B$ and $A, B, C \in \mathcal{C}$, the map $f_* \oplus g_* : F(A) \oplus F(C) \to F(B)$ is an isomorphism.

G is called *split-exact* when $f^* \oplus g^* : G(B) \to G(A) \oplus G(C)$ is an isomorphism for all such split-exact sequences.

Proposition C.4. *Consider the split exact sequence (C 1). Then the following are equivalent :*

(i) $f_* \oplus g_* : F(A) \oplus F(C) \to F(B)$ *is an isomorphism.*

(ii) *The sequence*

$$0 \longrightarrow F(A) \xrightarrow{f_*} F(B) \xrightarrow{p_*} F(C) \longrightarrow 0$$

is exact and splits with $g_* : F(C) \to F(B)$ as splitting map.

Proof.

(i) $\Rightarrow$ (ii) We must check that (i) implies that

(a) f_* is injective,

(b) $\ker p_* = im f_*$, and

(c) $p_* \circ g_* = id_{F(C)}$.

(c) is trivial since $p \circ g = id_C$. To prove (a), assume $x \in F(A)$ and $f_*(x) = 0$. Then $f_* \oplus g_*(x, 0) = f_*(x) = 0$. Thus $x = 0$ by assumption. To prove (b) it suffices to prove that $\ker p_* \subseteq im f_*$ since the other inclusion is obvious. So let $x \in F(B)$ and assume that $p_*(x) = 0$. By assumption $x = f_*(y) + g_*(z)$ for some $(y, z) \in F(A) \oplus F(C)$. Thus $0 = p_*(x) = p_* \circ g_*(z) = z$. Thus $x = f_*(y)$, proving (b).

(ii) $\Rightarrow$ (i) Let $x \in F(B)$. Then $x - g_* \circ p_*(x) \in im\, f_*$ by (b) above since $p_*(x - g_* \circ f_*(x)) = 0$. Thus $f_* \oplus g_*$ is surjective. Assume that

$(y, z) \in F(A) \oplus F(C)$ and that $f_* \oplus g_*(y, z) = f_*(y) + g_*(z) = 0$. Then $0 = p_*(f_*(y) + g_*(z)) = z$. Thus $f_*(y) = 0$ and it follows that $y = 0$ since f_* is injective. Hence $(y, z) = 0$.   □

We leave the reader to prove the following contravariant version.

Proposition C.5. *Consider the split exact sequence (C 1). Then the following are equivalent:*

(i) $f^* \oplus g^* : G(B) \to G(A) \oplus G(C)$ *is an isomorphism,*

(ii) *the sequence*

$$0 \longrightarrow G(C) \xrightarrow{p^*} G(B) \xrightarrow{f^*} G(A) \longrightarrow 0$$

is exact and splits with $g^* : G(B) \to G(C)$ *as splitting map.*   □

REFERENCES

The reader should notice that the following list only contains references to work that has been used to write this book or is not contained in the references to Blackadars book, [2] below.

[1] W. Arveson, *Notes on extensions of C*-algebras*, Duke Math. J. **44** (1977), 329–355.

[2] B. Blackadar, *K-theory for Operator Algebras*, MSRI Publications, Springer Verlag, New York, 1986.

[3] R. Busby, *Double centralizers and extensions of C*-algebras*, Trans. Amer. Math. Soc. **132** (1968).

[4] E. Christensen, E. Effros, and A. Sinclair, *Completely bounded multilinear maps and C*-algebraic cohomology*, Invent. Math. **90** (1987), 279–296.

[5] J. Cuntz, *A new look at KK-theory*, K-theory 1, (1987), 31–51.

[6] J. Cuntz, *A survey of some aspects of KK-theory*, preprint, Marseille, 1988.

[7] J. Cuntz and N. Higson, "Kuiper's theorem for Hilbert modules", in *Operator Algebras and Mathematical Physics*, Contemporary Mathematics, **62**, Amer. Math. Soc., Providence, 1987.

[8] J. Cuntz and G. Skandalis, *Mapping cones and exact sequences in KK-theory*, J. Operator Theory, **15** (1986), 163–180.

[9] J. Dixmier and A. Douady, *Champs continue d'espaces Hilbertiens et de C*-algèbres*, Bull. Soc. Math. France, **91** (1963), 227–284.

[10] R. Douglas, *Banach Algebra Techniques in Operator Theory*, Academic Press, New York/London, 1972.

[11] M.J. Dupré and P.A. Fillmore, *Triviality theorems for Hilbert modules*, Topics in Modern Operator Theory, 5th International Conference on Operator Theory, Birkhäuser Verlag, Basel, Boston, Stuttgart, 1981.

[12] N. Higson, *On a technical theorem of Kasparov*, J. Func. Anal. **73**, (1987), 107–112.

[13] N. Higson, *A characterization of KK-theory*, Pacific J. Math. **126** (1987) 253–276.

[14] N. Higson, *Algebraic K-theory of stable C*-algebras*, Advances in Math., **67** (1988) 1–140.

[15] N. Higson, *A primer on KK-theory*, Notes of lectures given at the University of Warwick, 1986.

[16] N. Higson, *A primer on KK-theory*, preprint, Philadelphia.

[17] R.V. Kadison and J.R. Ringrose, *Fundamentals of the theory of operator algebras I, II*, Academic Press, 1986.

[18] G.G. Kasparov, *Hilbert C*-modules: theorems of Stinespring and Voiculescu*, J. Operator Theory **4** (1980), 133–150.

[19] G.G. Kasparov, *The operator K-functor and extensions of C*-algebras*, Math. USSR Izvestija **16** (1981), 513–572.

[20] G.G. Kasparov, *Equivariant KK-theory and the Novikov conjecture*, Invent. Math. **91** (1988), 147–201.

[21] J. Mingo and W. Philips, *Equivariant triviality theorems for Hilbert C*-modules*, Proc. Amer. Math. Soc. **91** (1984), 225–230.

[22] C. Olsen and G.K. Pedersen, *Corona C*-algebras and their applications to lifting problems*, Math. Scand. **64** (1989), 63–86.

[23] W. Paschke, *Inner product modules over B*-algebras*, Trans. Amer. Math. Soc., **182** (1973), 443–468.

[24] G.K. Pedersen, *C*-algebras and their Automorphism Groups*, Academic Press, London/New York/San Francisco, 1979.

[25] J. Rosenberg, *K-theory of group C*-algebras, foliation C*-algebras, and crossed products*, Contemporary Math. **70** (1988), 251–301.

[26] J. Rosenberg, *K and KK: Topology and Operator Algebras*, preprint, Maryland, Jan. 1989.

[27] J. Rosenberg and C. Schochet, *The Künneth theorem and the universal coefficient theorem for Kasparov's generalized K-functor*, Duke J. Math. **55** (1987), 337–347.

[28] G. Skandalis, *Some Remarks on Kasparov Theory*, J. Functional Analysis **56** (1984), 337–347.

[29] G. Skandalis, *Une notion de nucléarité en K-théorie*, K-theory **1**(1988), 549–573.

[30] K. Thomsen, *Hilbert C*-modules, KK-theory and C*-extensions*, Various Publication Series No. 38, Aarhus Universitet, Oktober 1988.

[31] K. Thomsen, *Homotopy classes of *-homomorphisms between stable C*-algebras and their multiplier algebras*, Duke J. Math. **61**(1990), 67–104.

[32] R. Zekri, *A new description of Kasparov's theory of C*-extensions*, J. Func. Anal. **84** (1989), 441–471.

[33] R. Zekri, *Abstract Bott periodicity in KK-theory*, K-Theory **3**(1990), 543–559.

Index

Index of Symbols